STAGE AND STRUCTURE

Reopening the Debate

edited by

IRIS LEVIN

Tel Aviv University, Israel

Human Development

The Tel Aviv Annual Workshop in Human Development

Sidney Strauss, Series Editor

ABLEX PUBLISHING CORPORATION
NORWOOD, NEW JERSEY

Printed in the United States of America.

Library of Congress Cataloging-in-Publication Data

Main entry under title:

Stage and structure.

(Human Development)
Bibliography: p.
Includes indexes.
1. Cognition. 2. Developmental psychology.
I. Levin, Iris.
BF311.S6772 1985 155.4′13 85-15624
ISBN 0-89391-224-7

Ablex Publishing Corporation
355 Chestnut Street
Norwood, New Jersey 07648

Contents

To Shaul and Max

Human Development: Preface to the Series

Sidney Strauss
Series Editor

This book series is the product of annual workshops held by the Unit for Human Development and Education in the School of Education at Tel-Aviv University. Whenever a new publication comes out, the person responsible for it feels compelled to answer the question: Why a new X (journal, book, book series)? I do not feel exonerated from asking and answering that question. The purpose of the series is to address topics of theory, as well as conceptual and methodological issues that reflect the diversity and complexity of human development. It aims to address four needs in our field, as they relate to edited books.

There are few attempts to deal with important developmental phenomena in a concentrated manner in most of the book series in our field. They generally contain a collection of chapters on unrelated topics. There is a need to have leading experts present their views about relatively circumscribed issues. Our books will be organized around a topic where contributors write chapters that fit into a larger picture the editor will be illuminating.

A second need in our field is to have more interdisciplinary work. We generally read books written by developmental psychologists for developmental psychologists. Restricting inquiry to the psychological realm promotes a rather narrow view of human development. The view of man as a biological entity and cultural producer and consumer as well as a psychological being has been deeply neglected, and an element that will characterize our books is the inclusion of representatives of intellectual disciplines to discuss topics we believe to be of importance for our field.

Most work in developmental psychology is published in English and written by Americans. Much of the Western and Eastern European work is unknown to the vast majority of developmental psychologists, and the same can be said with more emphasis for work being done in Latin America, Asia, and Africa. Whenever possible, we will bring leading intellectual figures from these continents to contribute to our deliberations.

The fourth need is to attend to the applied parts of developmental psychology. Some of the later books will revolve around issues relating developmental psychology and education. We chose this out of the conviction that educational issues force one to face the depth of human complexity and its development in ways that traditional academic developmental psychology often turns away from.

The topics for the books are chosen with an eye towards directions our field is taking or we think should be taking. In some instances we will have a book about a topic that has been discussed in the literature and is ripe for a summary or stock-taking. In other instances the book will be around a topic that has not been discussed at all or has only begun to be written about. In still other cases the book will be about a topic that was once a subject of inquiry and then dropped. We will publish a book around that topic if we believe it should be resurrected because of developments in the field.

The support given by Ablex Publishing Corporation, the Tel-Aviv University Committee for Conferences, and different heads of the School of Education (Rina Shapira and Shimon Reshef) have made the workshops and this book series possible, and their support is gratefully acknowledged. This is especially so in light of the financial hardships in the publishing field and in Israel. Continuing support of this series is a welcome sign that despite economic adversity, there is a commitment to the intellectual pursuits that are of interest to us all.

It is the hope of the members of the Unit for Human Development and Education that this series will reflect the lively and stimulating workshops that have been held at Tel-Aviv University, and that the community of developmental psychologists will find interest in the issues discussed in these pages.

Sidney Strauss
Ramat Hasharon
August 1985

Introduction

Iris Levin

Cognitive development could have been a very tidy discipline had our studies unanimously agreed with the notions that children pass through a series of well-defined stages and that being at a certain stage determined children's ways of coping with each and every problem. In such a simple world, you could examine children's performance on any task or skill you fancied, be it the classical number conservation task, open-ended story telling, or perhaps linguistic skills, and conclude what stage they are at, and moreover what their performance would be in any other domain. This of course is very far from what we seem to know about cognitive development.

The present volume presents various attempts to provide appropriate organizing frameworks for cognitive developmental psychology. They all agree with major features of the Piagetian structure-based stage model but liberate it from certain constraints so as to make it fit the current state of the art. Many common issues are found in the various chapters, but they are frequently dealt with in very different frameworks. One such issue concerns the question of what differentiates one stage or level of development from another. Claims made here include views that stages manifest cognitive organizations of elements or different structures, with increasing logical-mathematical power; that they are characterized by different modes of representations—from perceptual to formal; that they involve increase in attentional scope and integrational ability; and that they manifest increasing access to a priori competencies along with the adoption of new strategies.

Another issue concerns the question of the generality of stages. General stages are seen to emerge simultaneously across-the-board or, alternatively, to appear at different ages on different tasks. In contrast, domain specific stages reflecting the structure of knowledge in different domains are suggested, along with models which incorporate both general and domain specific stages.

The issue of behavioral growth patterns—the occurrence of "jumps" as well as regressions—is traced either as a critical feature of an adequate developmental model on the one hand, or as a factor to be ignored as an artifact of measurement on the other hand. Pluralism still reigns!

The subject matter discussed in this volume covers a very wide range of areas. You may find here examinations of visuo-motor coordinations, social physical and mathematical concepts (balance scale appears no less than three times!), notational systems, map drawing, linguistic skills in English, French, and Hebrew, role playing, story telling, and free drawing. In the rest of this Introduction major themes of each chapter will be highlighted.

The neo-Piagetian model of development portrayed by CASE, MARINI, MCKEOUGH, DENNIS, and GOLDBERG (Chapter 1) is in major respects the closest in our book to the classical stage model of Piaget. Case and his colleagues argue that children's performance exhibits age-related horizontal structures. They support their argument by providing evidence that tasks of a similar format and structure, though differing in content (e.g., social vs. non-social) produce synchronized qualitative shifts across ages, so that the mean performance at each particular age is actually the same across tasks.

While their claim for horizontal structures is in line with the Piagetian stage model, Case et al. interpret the phenomenon differently than do orthodox structuralists. The present authors do not explain the emergence of age-related structures as an outcome of a "structure of the whole" which determines the child's approach to each and every task. Rather, they claim that although children's problem solving evolves relatively independently in different domains, their performances yield to age-related constraints such as working memory capacity. Hence, the resulting responses on various tasks exhibit the same control structures tailored by the same cognitive limitations.

The structures by which Case et al. qualify different age groups substantially resemble those of Piaget. Children's thought is claimed to progress from predimensional, to unidimensional, to unintegrated bidimensional, to bidimensional thought integrated by compensation. The stages qualified by these structures are illustrated in Chapter 1, not only by examining responses to Piagetian tasks (such as the balance beam), but more boldly by analyzing the structure of children's open-ended productions such as their stories and drawings.

In line with Case et al., MOUNOUD (Chapter 2) describes development as a fixed series of overarching stages, common to all areas of knowledge. There are two crucial differences between Mounoud's model and the structure-based stage model of Piaget. The first is that the general stages are viewed by Mounoud as emerging at more or less fixed ages, since they originate from the maturation of the nervous system, the effect of environmental variations being non-specific and limited only to the modulation of the natural unfolding of stages. The second difference relates to what differentiates between stages. In contrast to Piaget, who attributes progress from one stage to the next to the reorganization of thought into new structures, Mounoud attributes this evolution to the construction of new representations of knowledge. These systems of representations go from perceptual through conceptual to semiotic-formal internal organizations. Furthermore, the phases in which a system of representation is constructed out of the previous one repeat themselves for all stages. They start with syncretical representations, which are at first dissociated into elementary partial and local representations, are then integrated into rigid non-segmentable representations, and finally become units which have established relationships with the whole.

The data presented in the chapter illustrate these phases in various realms such as word and sentence segmentation, construction of tools, and planning and controlling movements. The equivalence between phases is assessed by analyzing the fundamental process involved in children's behavior in each phase, and their synchrony is determined by the finding that the same qualitative shift occurs for each, around the same age (e.g., 6 years for the conceptual stage). Mounoud concludes that the decision to teach reading and writing at the age of 6 years exhibits a sound pedagogical insight about children's cognitive development. Similarities between stages are illustrated as well.

Both Case et al. and Mounoud insist on synchronized development within the framework of overarching stages. But they support their claim of synchrony in very different ways. Case et al. argue that synchrony in development can be shown if the structure and format of the tasks used are carefully controlled and matched. Moreover, they believe that consistency should be looked for on a group level rather than on an individual level. Each individual can have particular talents or undergo special experiences which will modulate to some extent his or her level of performance in some tasks but not in others. Thus, stage-dependent evolution can be most clearly detected in group results which neutralize these individual variations.

Mounoud examines synchrony in development on a more molar level. He finds tasks which assess competencies from distant content domains (e.g., linguistics, visuo-motor coordination, perception) which, in his analysis, provide an opportunity for the emergence of the same qualitative shifts in representational formats. He then points out that in the different studies the shifts took place at more or less the same age range. Mounoud's molar analysis emphasizes the hypothesized similarity between the processes involved in the various competencies, while Case et al. accentuate the importance of careful matching of the tasks so that they are equivalent in terms of format or structure after molar levels are determined. Interestingly, the molar approach is also recommended by Feldman (Chapter 12, this volume) as the appropriate level of analysis for finding significant stages.

Piaget's structuralist approach and that of information processing—exemplified by Siegler's rule assessment approach—are compared in Chapter 3, contributed by STRAUSS and EPHRON-WERTHEIM. Rather than taking the conventional stance of preferring one approach over the other, the authors attempt to reconcile the two without obfuscating the distinctions which seem to them irreconcilable. In a study which constitutes the empirical background for their discussion, the authors assess children's solutions on the same task—the balance scale—via two methodologies, which grew from the aforementioned theoretical models.

Following Piaget's methodology, children actively explore the scale in an attempt to balance its arms by taking into account different weights placed at different distances from the fulcrum. The combination of action and explanation is then categorized into the intuitive, concrete, and formal stages of cognition.

Following Siegler's rule assessment approach, children are presented with a carefully designed series of tasks each including a configuration of certain weights placed at certain distances. Children's pattern of judgments across tasks about which of the arms, if either, would go down, indicates which rule they could be using in producing their judgments. The emerging rules agreed with those previously established by Siegler, and involved taking into account either one dimension (i.e., weight), considering both dimensions without integrating them, or using the correct rule of multiplying weight by distance (see a similar, though not identical sequence of "rules" established by Case et al. in this volume).

The very same children exhibit stage-like behavior when assessed by the Piagetian tasks and use specific rules when faced with the Sieglerian series of tasks. Moreover, both stage-like behaviors and rules progress with age in the predicted fashion, and to a great extent match each other, i.e., higher stage children "using" higher rules. The main conclusion drawn by the authors is that the knowledge-structure, knowledge-state, or strategy exhibited by children's behavior is to a great extent determined or even produced by the methodology used.

While both methodologies provide an appropriate framework for describing changes with age in children's behavior, their explanatory constructs differ in their level of specificity. Piaget explains behavioral changes in terms of across-the-board formal structures, while Siegler explains them in terms of specific rules, not necessarily generalizable across domains or even contents. The authors claim that Siegler gains precision and reliability by remaining close to the surface behavior but seems to lose generality in comparison with the structuralist approach.

Strauss and Ephron-Wertheim conclude their presentation by discussing the question of whether the two frameworks are compatible. They believe that the frameworks can be viewed as either compatible or incompatible, or even as "maybe compatible," and they do not seem to prefer one of these conclusions over the others. It could be argued, on the one hand, that the frameworks are incompatible because Piaget's general overarching stages are non-existent or because Siegler's too-specific rules are insignificant epistemologically or educationally. On the other hand, Piaget's and Siegler's models could be seen as compatible, measuring the same phenomena at a different level of specificity, so that choosing between them becomes a matter of taste. The two frameworks could also be viewed as "maybe compatible," leaving the answer to future research. The possibility of general cognitive structures, probably not of the formalism proposed by Piaget, should not be ruled out yet. Since it is at present difficult to conceptualize and measure the overarching cognitive structures, the pragmatic way to come closer to them could be through accumulating particular rules, waiting to see if they do lead to general structures.

While the current literature in the Piagetian or in the information processing traditions is rich with illustrations of fixed developmental sequences within concepts (see Strauss & Wertheim-Ephron, Turiel & Davidson, Karmiloff-Smith, Berman, and Feldman in this volume), developmental synchronies or fixed sequences between concepts are rather rare, Case et al. and Mounoud's discussions being an exception. Yet, from an epistemological point of view, the discovery of sequences between concepts seems to be of special significance. A consistent developmental sequence among different concepts or operations may provide a clue to the functional relations between them, clarifying how one achievement leads to the next. In Chapter 4 LEVIN and SIMONS focus on the difficulties inherent in the attempt to determine between-concept sequences, via their analysis of the development of time, speed, and distance. Levin and Simons argue that sequence in performance is determined not only by the subject's structure of knowledge, but also by the interplay between that structure and the structure of the task. They show that while speed was previously found to be the easiest concept on tasks using linear motions, it emerges as the most difficult concept in their task involving rotational motions.

Within the framework of their task, subjects from a wide age range (third-graders to university students) and different trainings (humanities vs. physics–mathematics majors) reveal 'childish' misconceptions about speed, and even to some extent about distance. Many children and adults lack an objective scale for describing the speed of a moving object. So, after estimating how far and for how long an object is moving, they have no idea how to go about estimating its speed (though speed can be directly derived from distance and time: $V = s/t$). Moreover, when asked to compare the movements of two objects rotating around the same axis, many subjects use linear and rotational frames of reference interchangeably, often being quite unaware of their inconsistency.

Overall, Levin and Simons's study can be seen to cast some doubt on stage-like models which emphasize qualitative shifts in structures, processes, or representations that occur with development. Queries arise from the surprising similarities found between the types of responses given by children and many adults, particularly those without specialized training in physics-mathematics. One might claim however, that developmental stages do in fact emerge when optimal performance is assessed, rather than performance constrained by lack of experience, expertise, or environmental support (see Case et al., Fischer & Canfield, Globerson, and Feldman, this volume). If this should be the case, then the higher functioning of adults compared to children, particularly in complex tasks and in contexts of low support, can be viewed as dependent to a great extent on planned interventions such as schooling.

The developmental literature is rich in findings of development as asynchronic, inconsistent, and heterogenous across tasks. This trend is taken by Levin and Simons like many others to imply that a model of development as a sequence of structure-based stages should be questioned. TURIEL and DAVIDSON (Chapter 5) acknowledge these findings of asynchrony, inconsistency, and heterogeneity; nevertheless they adhere to the stage and structure model of development. They do so by proposing that structures do not characterize the individual's thought in a global fashion but rather that they qualify specific knowledge systems or domains. While homogeneity and consistency should characterize tasks which belong to the same domain, one should not expect to find these properties for tasks which represent distinct domains or which involve domain mixture or conflict.

Turiel and Davidson base their differentiation between domains on epistemological considerations along with distinctions utilized by their subjects. They argue for broad domains such as convention vs. morality in the social realm and physical vs. logical-mathematical knowledge in the non-social realm. They further claim that stages are exhibited via structural-qualitative reorganizations of thought within domains, and that these reorganizations are gradual and continuous, even though the emergent changes may themselves be abrupt. With development, the evolution of structures within domains coupled with coordination between domains may bring about higher order homogeneity of thought.

While Turiel and Davidson's hypothesis of domain specificity is diametrically opposed to the horizontal structure hypothesis pursued by Case et al., yet it should be noted that these hypotheses are supported by their presenters in very different ways. Case and his colleagues search for consistency across domains by looking for synchrony in development across tasks taken to represent different domains. Turiel and Davidson dismiss synchrony as an irrelevant criterion to assess consistency, instead analyzing how subjects distinguish between and function within domains. Their finding that children from a very young age differentiate between domains is taken by them as supporting the validity of domain distinction.

Since the assignment of tasks to domains is carried out by different means in these two frameworks, the same task can be categorized by each to a different domain. It is the case, I believe, that the tasks used by Case et al. to represent different content domains (social vs. non-social) can be reinterpreted from Turiel and Davidson's viewpoint as belonging to the same non-social domain. Thus, Case's data can be analyzed as supporting, or at least as not refuting, Turiel and Davidson's domain specificity approach. This is not to say that Case and his colleagues cannot come up with tasks attributed unequivocally to different domains, and still reveal the horizontal structure which they argue for.

According to KEIL's analysis (Chapter 6), the main debate between stage and non-stage models of development concerns the emergence of qualitative shifts in modes of representation or of processing. Stage theorists share the assumption that qualitative shifts occur throughout development. These theorists differ, however (as noted above), on whether they believe such shifts emerge synchronically across the board, across domains, or within local knowledge structures, and furthermore whether the shifts are of the same nature in different categories. Non-stage theorists, in contrast, assume that the same representational or computational systems characterize cognition at all ages, the difference in performances between age groups being attributed to increasing access to the same structure, gradual differentiation, and adoption of new strategies.

Keil's theory of development incorporates features of both stage and non-stage models, leading him to conclude that pervasive structural qualitative shifts do occur but that they are not dependent on general overarching structures, and that they are not age-dependent. His analysis focuses on constraints imposed on hypotheses generation and hence on knowledge acquisition. Keil assumes, in line with non-stage theorists, that there are universal constraints on knowledge that determine how knowledge is acquired and represented from a very young age. With development, however, these overarching constraints are progressively enriched with domain specific constraints that typify broad range domains of the kind dealt with by Turiel and Davidson, as well as by post-priori expertise constraints (such as illustrated in the work of Levin and Simons). Thus, the more advanced the representations of knowledge, the more pluralistic in nature they become, as they come to increasingly depend on the particular structure of the knowledge concerned.

Keil illustrates the shift from a single to plural representational formats in the domain of word meanings. While young children represent words by characteristic features, older children shift to various representations—some based on defining features, others on functions, and still others on causal theories—depending on the particular category of words concerned. These shifts are structure-dependent in the sense that they take the same form and emerge synchronically for terms of the same semantic field. But at the same time they are nonsynchronic and have varying formats across local knowledge structures within the same broad domain.

The approaches of Keil and of Turiel and Davidson argue for independent development of competencies related to different domains, but their hypothesis of domain specificity differs in the following respect. While Turiel and Davidson claim that differentiation between major domains exists from a very young age on (e.g., issues of morality and convention are dealt with differently even by preschoolers), Keil believes that differentiation between domains grows with age, or more specifically with the accumulation of knowledge. With development and learning, the properties of knowledge in different domains gradually diverge, and domain specific structures gain dominance over general a priori structures.

The model of development pursued by KARMILOFF-SMITH (Chapter 7) agrees with that of Keil in acknowledging the existence both of all-encompassing structural stages and of domain specific levels based on different local structures or systems. However, her main focus is on a process-oriented analysis of development suggesting a three-phase model. She claims that development, across-the-board, exhibits three phases of information processing. These phases characterize macrodevelopmental sequences in different domains. They emerge nonsynchronically across domains, but are more or less age-dependent within domains. The very same phases, however, also characterize microdevelopmental steps which emerge within a single learning session when individuals, be they children or adults, attempt to solve a problem or acquire a skill which is within their cognitive capacity but which requires them to work out a new solution or strategy.

The first "procedural" phase is predominantly a data-driven process progressively reaching adaptation of behavior to environmental information. It is limited by a deficient overall organization of behavioral units. The second phase is predominantly a top-down process in which the units are progressively integrated at the expense of their richness, elaboration, and degree of match with the environmental input. The third "conceptual" phase involves a coordination of the data-driven and top-down processes, leading to organized representational systems.

This model is illustrated in the present chapter in macrodevelopments of linguistic competencies (e.g., acquisition of the French determiner system) and cognitive competencies (e.g., the use of the torque principle in a balance task) as well as in a combination of micro and macrodevelopments (e.g., production of notational systems as mnemonic devices). Karmiloff-Smith argues that the successful application of this process-oriented model to both the linguistic and the cognitive realms indicates that their interrelationship can be fruitfully researched within her framework.

The process of integration of information or of representation, which characterizes Karmiloff-Smith's second phase, plays a central role in the neo-Piagetian cognitive theories of Case and of Mounoud and in Berman's analysis of linguistic development. Furthermore, Case, Mounoud, and Berman, like Karmiloff-Smith, all believe that the process of integration is recurrent in development. Yet Karmiloff-Smith further notes (as does Berman) the interesting feature that reorganization can lead to a U-shaped behavioral growth, i.e., to temporary decreases in levels of performance or in the match between mental organization and environmental input (see also Feldman).

BERMAN (Chapter 8) presents a five-step model of language acquisition which shares several major features with the three-phase processing model of Karmiloff-Smith. Berman based her model on a rich variety of data concerning the acquisition of morphology in Hebrew, a language which encodes within words much information that may be encoded across words in other languages. The model is illustrated by developmental permutations in how children construe words as documented in longitudinal and cross-sectional studies dealing with different linguistic subsystems such as consonantal roots, noun pluralization, and tense marking on verbs.

According to this model, during the first two ''pre-grammatical'' steps the child's knowledge is largely item and context bound, starting with unanalyzed amalgams later giving way to fairly restricted alterations of highly familiar items. The second two steps of ''grammar acquisition'' are characterized by the progressive application of productive rules, first non-normative and deficient, later conventional, though not entirely appropriately constrained. At the final end-state step, the learner exhibits adult norms and conventions supplemented by the development of individual style and register.

This model is claimed to be applicable across linguistic subsystems, across languages, and possibly even across other cognitive acquisitions. In this respect Berman's model is in line with the framework presented by Karmiloff-Smith, both of which suggest moreover a two-way interaction between cognition and language acquisition. The sequence of steps is presented as obligatory in the Piagetian sense that no real regression is conceivable. When regressions do emerge they are of a superficial nature and stem mainly from an advanced grammatization. While the sequence is unidirectional, not all steps are construed as necessary phases across individuals or subsystems. Steps may be skipped by particular individuals or by all learners in certain domains.

The steps are said to be age-dependent in only a very broad sense, with the first two steps occurring predominantly, though not exclusively, up to 2 years of age, and the preceding steps taking place mainly up to puberty, depending on the particular linguistic structure to be acquired. Moreover, the steps are viewed as continuous rather than discrete, so that even within the subsystem there is an overlap between steps. Berman's model of nonsynchronous qualitative development via gradual shifts, modulated in pace by domains, accords surprisingly in agreement in overall terms with the model proposed by Turiel and Davidson for cognitive development.

DROMI (Chapter 9) examines the issue of stages in development with reference to the linguistic domain. But while Dromi and Berman deal with the same general content domain, their assumptions as to what constitutes a stage differ in major respects. Dromi documented her child's acquisition of Hebrew as a mother tongue during the "one word" period by means of a detailed diary based on systematic observations of her child's naturalistic verbal behavior. Her data include the initial emergence of words, phonological and morphological changes in words, and changes in the context in which each word was produced across time. This rich database is analyzed by Dromi in terms of whether the initial period of "talking in single words" constitutes a discrete developmental stage.

According to Dromi, the stage concept is especially useful within the context of hierarchically organized complex systems such as language. She claims that for a period to constitute a developmental stage it should have a well defined starting and end point and it should cover a substantial span of time. The emergent behaviors of a new stage should be novel, have underlying homogeneity, sometimes despite superficial heterogeneity, and should constitute a necessary condition for the subsequent stages. These criteria are thus stricter and more limiting than those usually applied in the linguistic or the cognitive domains (see Karmiloff-Smith, Berman, Turiel & Davidson, and Feldman in this volume). Nevertheless, the data she presents on the "one word period" support the conclusion that this period is indeed a stage, even according to such strict criteria.

For Dromi, the one-word stage starts when the child purposefully produces a meaningful word, and ends with the sudden emergence of productive two- and multi-word combinations. The entire stage she observed lasted for about 8 months, a relatively longer period than the following one of producing mainly two-word combinations.

Dromi analyzes her diary data from several perspectives: rate of acquiring new words, phonological and morphological forms of words, their category of reference (e.g., object-word, action-word), and range of their semantic extension. Generally speaking, the analyses point to a stage with distinct boundaries consisting of two phases. The first and relatively long "preparatory" phase is characterized by a slow acquisition of new words, as though each is learned as a special case. The next shorter phase of "consolidation" starts with an enormous increase in the number of new words; it involves phonological improvements and morphological changes; the meaning of words comes to approximate regular adult extensions, along with the emergence of over-extensions. Thus, the entire stage is characterized by non-monotonic behavioral growths.

Non-monotonic changes in behavior, such as appear in many of Dromi's analyses, are presented by FISCHER and CANFIELD's skill theory (Chapter 10) as the most critical criteria for the transition between cognitive-developmental levels. They believe that people go through ten general levels between infancy and young adulthood. The transition between levels is marked by spurts that emerge in the pattern of behavioral growth. These spurts, however, can be traced only when the behavior assessed manifests the optimal performance of the child. To ensure optimal performance, the child should be well-motivated and practiced in the task s/he is coping with. The task itself should be supportive for high performance, and its level of complexity should not be far beyond the current cognitive level of the child. Continuous developmental growth patterns, which are so prevalent in the developmental literature, emerge when the elicited performance is below its optimal level.

The predictions of the optimal performance hypothesis are supported in Chapter 10 by a study tracing the developmental growth pattern between third grade and college students in arithmetic concepts. Performance showed abrupt improvement at particular age levels, but only under the optimal conditions of practiced subjects working in a supportive context. Moreover, the spurts emerged at the particular age levels seen as transitional within the general framework of skill theory. Under less optimal conditions, of either less practiced subjects and/or low supportive context, performance grew continuously across age levels.

As seen from the above, in some major respects Fischer's skill theory agrees with Piaget's stage theory, as well as with neo-Piagetian models. In other central respects, however, skill theory is quite different from the others. On the one hand, skill theory agrees with stage theories in describing development as a fixed series of levels, more or less age related. Moreover, the levels described differ, inter alia, in the number of variables integrated by the child, as is true for Piaget, and as is central for Case and for Globerson (this volume). On the other hand, in contrast with the aforementioned theories Fischer and Canfield argue for individual and group differences in the sequence in which the levels are passed. Organismic factors of maturation, motivation, and emotion, and the learning history of the child, are claimed to determine not only the current level of the child's performance and its possible zone of variation, but also the sequence of levels that s/he will pass through. Allowing individual differences in the sequence of levels fits nicely with data which are otherwise troublesome for stage theories (see Levin & Simons in this volume) by broadening the concepts of stage and structure to a degree which is beyond their accepted meaning for many stage theorists.

Behavioral spurts play a central role also for GLOBERSON (Chapter 11) as a criterion for structural changes. Globerson claims that spurts emerge in behavior particularly if optimal performance is measured in short intervals on a continuous scale. These conditions seem to be particularly well satisfied in the study of Dromi (this volume) which found spurts in natural word acquisition in data collected on a day-to-day basis.

Drawing from a wide range of resources, including scientific disciplines other than psychology such as physics and biology, Globerson attempts to clarify two issues basic to cognitive developmental theories: What constitutes structural change, most particularly a change which is perceived as significant by the theorist; and under what conditions can structural changes be detected. This analysis leads her to propose a research program for studying structural changes which utilizes the construct of mental capacity.

Globerson claims that significant structural changes are changes in the relationship between the elements that constitute the structure, rather than in the elements themselves. Though structural changes are inferred from behavior, they are not easily detectable. A change in a structure can occur without a corresponding change in behavior, and a behavioral change is not necessarily the result of a structural change (see Berman and Karmiloff-Smith in this volume for similar arguments). However, Globerson offers a way out of this blind alley by concluding that a behavioral change is necessarily accompanied by a structural change when the change in behavior is abrupt, i.e., when there is discontinuity in the behavioral function or in its first derivative.

To study structural changes Globerson suggests using the measurement of mental capacity, i.e., the limited attentional capacity for processing information, a construct which plays a major role in the developmental model of Case as well (see Case et al. in this volume). Mental capacity is measured on a continuous scale under optimal conditions, has been claimed to have a strong maturational factor, and has recently emerged as having a growth function of alternating plateaus and inclines (i.e., abrupt changes). Globerson recommends that an initial starting point for analyzing structural changes could be in the study of behavioral changes in measures of mental capacity itself as well as in the performance of other tasks, which are predicted by mental capacity.

In line with Piaget's equilibration model, and with Globerson's scheme of behavioral growth which indicates structural changes, FELDMAN closes this volume (Chapter 12) by portraying development as an alternating sequence between periods of relative internal instability and periods of stability, the stable periods becoming more so from one level to the next. However, on the basis of microanalyses of a series of maps drawn by children in a longitudinal study, Feldman introduces new refinements to the equilibration model. First, at any point in time the system seems to be undergoing transformations and includes elements from different levels (see also Fischer & Canfield). Second, at the peak of change, when many elements progress, quite a number of elements regress as well. Third, at that point or thereafter, breakthroughs are made, in the sense that few elements, not necessarily of the highest current level, reach a new level altogether. Feldman's new insights into children's new insights are claimed to have the potential to contribute to our understanding of creativity, both at the level of the child who invents what is already well known in the community and of the truly original scientist or artist.

Feldman suggests a novel distinction between domains, which is not based on their content (such as that proposed by Turiel & Davidson, Keil, Karmiloff-Smith, and Berman) but rather on the criterion of their universality. Research in developmental psychology deals deliberately, almost exclusively, with competencies achieved by all normal adults in all cultures. This is true for most of the work described in the present book, as well. Feldman, in his attempt to analyze the process of transformation from one stage or level to the next, rather than documenting age or stage differences, comes to the conclusion that studying only universal achievements can be limiting, and thus recommends the study of non-universal domains, i.e., those competencies which are neither universally achieved nor spontaneously acquired. Studying non-universal rather than universal domains, under the assumption that the processes of transformation are substantially common for both, is argued to illuminate both individual differences and the effect of experience on cognitive progress. The call for shifting the focus of research to non-universal domains is in line with Mounoud's recommendation to study non-fragmented achievements acquired through schooling, as well as with the new trend to study the novice–expert distinction (dealt with in the present volume by Globerson, Keil, and Levin & Simons). Progressing in a non-universal domain and becoming an expert involves exposure to special intervention, including interventions taking place in school.

The short presentations of the major themes of the series of chapters composing this volume, as well as the tentative attempts to draw some comparisons between them, indicate clearly that the issues surrounding stage and structure are neither settled nor dead. Some of the issues raised here have already and, in the future, will lead to interesting discoveries about children's cognitive development. Other issues will probably have the fate of many thought-provoking problems in the past, and either lead to a blind alley or simply become too boring to the community at large. However, the basic questions of how and what develops, which is concealed behind the concepts of stage and structure, are still the crux of our domain. So, the debate is reopened. . . .

1

Horizontal Structure in Middle Childhood: Cross-Domain Parallels in the Course of Cognitive Growth

Robbie Case
Zopito Marini
Anne McKeough
Sonja Dennis
Jill Goldberg
The Ontario Institute for Studies in Education
University of Toronto

In a recent Minnesota symposium, John Flavell made a distinction between vertical and horizontal structure in children's development (Flavell, 1980). According to his distinction, two tasks or domains were said to share a common *vertical* structure when children went through the same sequence of general stages on each, and they were said to share a common *horizontal* structure when children went through the same sequence during the same age range. Since the present chapter will be concerned primarily with the notion of horizontal structure, it is important to be clear from the outset what this notion implies. A group of tasks, A, B, and C, may be said to share a common horizontal structure, if and only if children function at one level (e.g., Level 1) on all three tasks at one age, at a higher level (e.g., Level 2) on all three tasks at a later age, at a still higher level (e.g., Level 3) on all three tasks at a still later age, and so on.

In evaluating the empirical evidence on vertical and horizontal structure, Flavell (1980) offered the following comment:

> I suspect that a surprisingly large number of developments may proceed quite independently of one another, and that the maturity of the child's thinking—however defined—may be quite uneven across these developments. . . . The structure in children's thought is of the vertical (sequential) rather than the horizontal variety.

In a recent presidential address to The Society for Research in Child Development, Flavell has softened his original evaluation somewhat (Flavell, 1982). Nevertheless, the fact remains that it is his original evaluation which represents the majority opinion in North American developmental psychology today. By and large, North Americans are convinced that horizontal "décalage" or "uneveness" is the rule in children's development and not horizontal structure.

In the present chapter, we would like to argue against this majority point of view. The central thesis we shall advance is that, although Piaget's particular view of horizontal structure may have been slightly inaccurate, there exists a very similar view—and one which preserves the central core of his thinking—which is entirely accurate. According to this view, there *are* general stages in children's thinking, and children at any particular age *do* tend to function at the same stage across a wide variety of tasks and content domains. In fact, providing that their thinking is assessed in an appropriate manner, it is horizontal structure and not décalage which is the rule in their development.

The present chapter will be divided into four broad sections. In the first, we shall briefly review the nature of Piaget's position on horizontal structure and the supposedly massive evidence which has been arrayed against it. Then we shall offer a slightly recast version of Piaget's formulation and show that the existing evidence, far from contradicting the revised position, is actually congruent with it. In the second section, we shall present our own "neo-Piagetian" theory of children's development. In the third section we shall present the data from three studies we have conducted in order to evaluate this theory and assess the evidence on horizontal structure in a fresh manner. Finally, in the last section, we shall consider the data from three further studies, and suggest that the horizontal structure in children's thinking may be even more extensive than either Piaget or we ourselves had previously realized.[1]

Piaget's Theory and the Problem of Décalages

According to Piaget's theory, children's reasoning at the age of 3 or 4 is qualitative and intuitive. By the age of 5 or 6, they become capable of a logic which he calls the logic of functions. That is, they become capable of isolating quantitative relations of the form $X = f(Y)$. By the age of 7 or 8, they become capable of decentering and focusing on a second quantitative relation as well. Their logic might therefore be expressed as follows $X = f\,(Y \text{ or } Z)$. Finally, at the age of 9 or 10, they become capable of establishing a quantitative compensation between two opposing variables. This form of logic might be represented in abbreviated form as $X = f\,(Y - Z)$, since the actual compensation is normally accomplished via subtraction or addition and not by multiplication.[2] As even the most casual reader of Piaget is aware, a further assumption of his theory is that this last form of logic is necessary for a "true" understanding of conservation.

Unfortunately, if the last ten years of research have demonstrated anything, it is that this latter conceptualization is incorrect. Depending on the task, conservation can be found at any one of the above ages, with Gelman's simplified number

[1] For readers of our recent volume (Case, 1985) the first two sections will be largely a review. The last two sections, however, will contain a good deal of new material.

[2] In Piaget's terminology, this is usually written $X = f\,(Y, Z)$ or $X = f\,(Y \times Z)$. The notation is introduced to clarify the nature of the compensation which is involved.

conservation task being passed at the age of 3 to 4 (Gelman, 1972), the standard conservation of number task being passed at the age of 5 or 6 (Piaget, 1952), the standard liquid conservation task being passed at 7 to 8, and certain other liquid conservation tasks which require estimating the new level of a transformed liquid, not being passed until around the age of 9 or 10 (Gelman & Weinberg, 1972). Similar research findings have been uncovered for decentration tasks, which may be passed as early as 2 or as late as 15, depending on the particular task situation (Flavell, Bolkin, Fry, Wright, & Jarvis, 1968).

What is to be made of these findings? One of the first conclusions which was suggested was that the notion of a general stage was not a viable one (Gelman, 1979; Brainerd, 1978). Increasingly, however, investigators are beginning to realize that the stage concept may simply have to be reconceptualized (cf. Case, 1978; Halford, 1980; Fischer, 1980). In fact, the present position is that tests which are simply assigned global labels such as "conservation" or "decentration" do not really test Piaget's core assertion. Rather, what they test (and falsify) is his auxiliary notion that certain types of insight or capability are an automatic consequence of certain particular levels of thinking. In this regard, the following points are important to note.

1. Gelman's simplified number conservation task only requires that children scan two arrays globally and realize that the one with three items (mice) is always called a winner.
2. The standard conservation of number task requires that children establish a functional relation between "amount" and "numerosity." As such, it requires thinking of the form $X = f(Y)$.
3. The standard conservation of liquid task requires that children establish a relationship between amount (X) and variation along two dimensions: namely, height (Y) and width (Z) (see Case, 1977).
4. Finally, the "level estimation" task requires that children actively manipulate height and width and form a tradeoff between them, i.e., it requires a form of thought which might be symbolized as $X = f(Y-Z)$.

In fact, then, the existing results are reasonably congruent with Piaget's core assertions concerning the structure of children's reasoning at different ages, as long as one does not accept his auxiliary assumption that conservation (or for that matter any concept or ability) is an automatic consequence of one and only one of these forms of reasoning. Further evidence that Piaget's core assertion may have been correct has come from Siegler's (1976) work on the balance-beam task, and the follow-up work which we have done on this task in our own research group (e.g. Liu, 1981; Marini, 1984). What we have shown is (a) that 3- to 4-year-olds can make a global determination of weight on a balance beam; (b) that 5- to 6-year-olds can establish a single quantitative relationship between balance tilt and the quantity of weights on each side, $X = f(Y)$; (c) that 7- to 8-year-olds can relate tilt either to the quantity of weights or to the distance from the fulcrum, $Y = f(Y \text{ or } Z)$; and (d) that 9- to 10-year-olds can form an approximate compensation between the two variables via some sort of formal procedure such as addition or subtraction $X = f(Y-Z)$[3].

What we believe these data indicate, then, is that the issue of horizontal structure needs to be reexamined with a more careful analysis being made of the demands which are placed on children's reasoning by different variations of the same task. It could well be the case that, when this is done, children will indeed be found to function at the same level across a wide variety of different content domains and that the notion of general stages will have to be revived, albeit in an altered and somewhat more conservative form.

A Neo-Piagetian Theory of Development

The theory on which our own studies have been based is neo-Piagetian in the sense that it is based on many of Piaget's original conceptions, but that it also aims to go beyond them (Case, 1985). In its "qualitative" description of children's thinking from 4 to 10, the theory stays quite close to the formulations advanced by Piaget.

Substage 0: Predimensional Thought (3½–5 years)

At this substage, it is presumed that children consolidate control structures for scanning the basic events or perceptual situations to which they are exposed and placing them in global categories. Very often these categories are polar dimensions, such as heavy–light, big–little, good–bad, more–less, etc. However, they need not necessarily be. Categories such as "three" versus "*two*," "red versus green," and "dinner time" versus "play time" also fall under the same general rubric and can be appropriately handled at this stage.

Substage 1: Unidimensional Thought (5–7 years)

At the next substage, children become capable of treating the elementary structures which they consolidated at the previous substage as units, and coordinating them into higher order structures, very often ones with some sort of means–ends format. It is presumed that the functional logic described by Piaget emerges from one such coordination, namely, that between number and polar categorization. When children can use quantification (via counting or precise visual estimation) as a means to determining whether one object is heavier, longer, or "has more" than another, then they can establish the sort of functional relationship which Piaget described in his formula $X = f(Y)$. In short, they can think in terms of a single "quantitative dimension."

---

[3] If the procedure is informal, or if the variables themselves need not be precisely quantified, then some sort of compensation can actually be achieved a good deal earlier (Wilkening & Anderson, 1980). Like the conservation tasks which can be passed at young ages, however, what these successes indicate is the presence of a capability which is formally (i.e., mathematically) similar, but which is actually being generated by a structurally simpler procedure.

Substage 2: Bidimensional Thought (7–9 years)

At the next substage, children again become capable of coordinating two of the sorts of units they considered at the previous substage. Thus, in quantitative tasks, they begin to focus on two quantitative dimensions rather than just one and to use the sort of logic we described above as $X = f(Y \text{ or } Z)$. Once again, however, the transition which takes place is not limited to quantitative tasks but also takes place in tasks where the primary units are more qualitative (e.g., categories, roles).

Substage 3: Elaborated Bidimensional Thought (9–11 years)

At the final substage, children introduce some additional elaboration into the control structures of the previous substage, often one in which the two elements which were dealt with at the previous stage are better integrated. These integrations include those described by Piaget, such as quantitative compensation via subtraction. However, once again, they are not necessarily limited to them. Once consolidated, these structures function as the elementary units of the structures during the fourth major stage of development, i.e., the stage which Piaget referred to as "formal operations."

The foregoing qualitative descriptions capture many of the important changes in children's thinking that take place during middle childhood. The present theory also has a quantitative aspect to it, however, which is derived from a hypothesis originally advanced by Pascual-Leone (1969).

The quantitative aspect of the theory depends on a somewhat more detailed analysis of the underlying structures which are responsible for governing children's problem solving. Table 1 presents an illustration of the sorts of items which children pass on the balance-beam task at different stages and the underlying executive structures which we presume control them. Note that each control structure has three components to it: a representation of some feature of the task, a representation of some goal which this feature occasions, and a representation of a strategy or set of operations for achieving this objective. Note further that each successive control structure is an elaboration of the previous one and requires the establishment of one additional goal or subgoal. The number of goals which must be kept in mind at each substage are 1, 2, 3, and 4, respectively, and the present position is that children's working memory for categorical or dimensional operations increases at this rate. In fact, the present position is that this increase *permits* children to assemble control structures of the complexity indicated, given appropriate experience and the chance to reflect on it. Thus, the reason that children's thinking has a common form in many different domains at the same age, according to our view, is not that children are developing a common underlying "structure-of-the-whole." Rather, it is that their thinking is subject to a common quantitative constraint, namely, a limitation in working memory.

Although we agree with Flavell that development in different domains proceeds independently, then, we also believe that it ends up at the same point in different domains at the same point in time, given sufficient exposure to a set of task situations within a common external structure. Note the implication—which Piaget

Table 1. Hypothesized Control Structures for Balance-Beam Task at Different Ages.

4 Years
Task Passed:

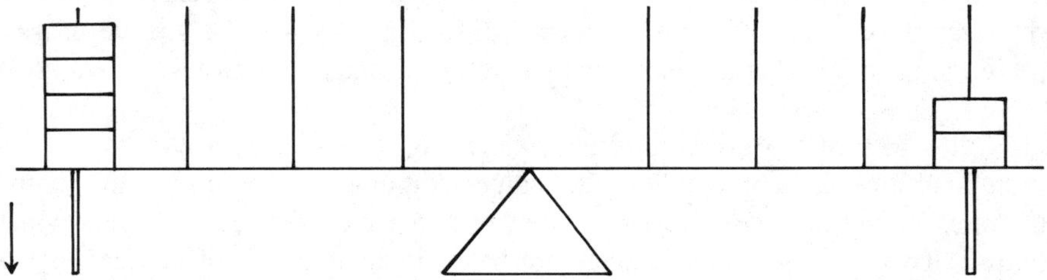

Control Structure:

Problem Situation
- Balance beam with an object on each arm

→ Objective
- Determine which side will go down

↙ Strategy
1. Look at each side. Predict that the one which looks heaviest will go down, the lightest up.

6 Years
Task Passed:

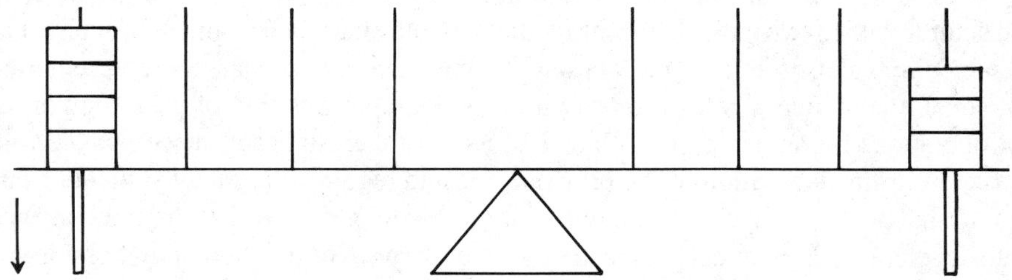

Control Structure:

Problem Situation
- Balance with stack of objects on each arm
- Each stack composed of a number of identical units

→ Objectives
- Predict which side will go down
- Determine which side has larger number of units

↙ Strategy
1. Count each set of units; note which side has the bigger number
2. Pick side with bigger number as the one which will weigh more (and therefore go down)

Table 1. (*Continued*)

8 Years
Task Passed:

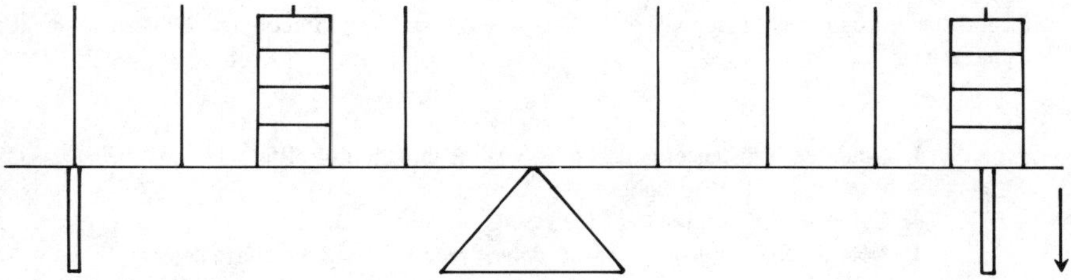

Control Structure:

Problem Situation
- Balance beam with stack of objects on each side
- Each object stack composed of a number of identical units
- Each object at a specifiable distance from fulcrum

Objectives
- Predict side which will go down
  - Determine side with greater number of objects
  - Determine side with weight at greater distance

Strategy
1. Count each set of weights: note which side has greater number
2. Repeat 1 for distance pegs
3. If the weights are about equal, predict that the greater distance will go down. Otherwise predict that the greater weight will go down

10 Years
Task Passed:

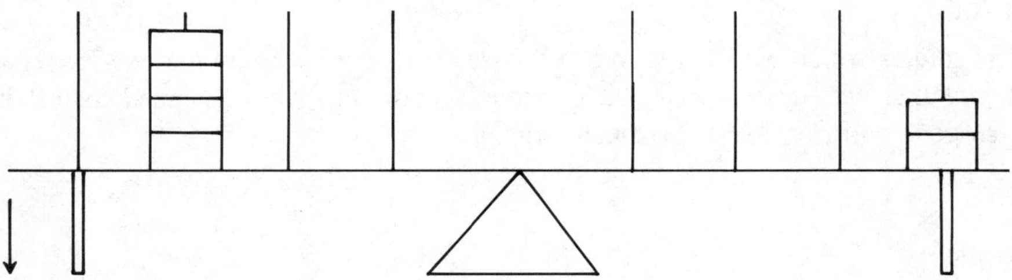

Control Structure:

Problem Situation
- Balance beam with stack of weights at various distances

Objectives
- Predict which side will go down

Table 1. (*Continued*)

• Action of weight and distance in opposite directions[a]	• Determine whether weight or distance has a greater effect
• Each weight stack composed of equal units	• Determine relative number of weights on each side
• Each distance composed of number of equal units	• Determine relative distance on each side

Strategy

1. Count each distance; note size as well as direction of difference
2. Repeat step 1 for weight.
3. Compare the magnitude of the results in steps 1 and 2.
4. Note dimension of greatest difference; pick side with greater quantity along this dimension.

[a]Note that this "feature" could only be "encoded" by applying the strategy of the previous substages; that is, by counting both weight and distance.

applied in practice, but rarely mentioned explicitly—that one must let children familiarize themselves with a situation and give them feedback on their performance, if one is to detect the horizontal structure which characterizes their thinking in two different domains. Either that, or one must use a situation that is relatively familiar to them to begin with. If one does not take either of these precautions, then the horizontal structure in children's problem-solving processes may very well be masked by surface differences, which arise due to variability in the knowledge or "content" on which these processes must operate.

New Data on the Horizontal Structure Question

The studies which we shall report in this section were based on the neo-Piagetian theory that was described in the previous section and were designed to test its assertions with regard to horizontal structure.

Study 1

The first study was conducted by Zopito Marini, as part of his master's thesis (Marini, 1981). Its object was to provide a preliminary test of the previously discussed notion concerning horizontal structure during the age range from 3½ to 11. More specifically, the goal was to determine whether children's level of performance on a newly designed task in one domain could be predicted on the basis of what was already known about their performance on a task in another domain.

Subjects. The subjects were drawn from a public school in a middle-class neighborhood in Toronto. Four different age groups were tested. The mean ages of the groups were 4 years, 10 months ($n = 19$), 6 years, 10 months ($n = 28$), 8 years, 10 months ($n = 27$) and 10 years, 10 months ($n = 30$).

Tasks. Two different tasks were used, one involving content that was social and one involving content that was nonsocial. The underlying structure of the operations in each task, however, was identical.

The test whose content was nonsocial was Noelting's Juice-Mixing Task (Noelting, 1982). This is a task where children are shown two empty pitchers, together with two sets of beakers filled with juice and water. They are then told that the first set of juice and water beakers will be poured into one pitcher, while the second set will be poured into the other pitcher. Finally, they are asked which mixture they think will taste juicier when the pouring is complete. A number of trials are then presented, in which the number of juice and water beakers in each set is varied from trial to trial.

During the period under consideration, children's strategies go through four distinct substages. At Level 0, the strategy is one of Global Juice Evaluation. Children look at each array of juice and water beakers and choose the set which has a large number of juice cups. If the two arrays have approximately the same number of juice cups, they respond that they will both taste the juiciest, or else give an irrelevant answer. At Level 1, children's strategy is one of Counting the Juice. Children carefully count the number of juice cups in each set and say that the one with the greater number will taste juicier, $X = f(Y)$. At Level 2, children's strategy is one of Counting the Juice or Counting the Water, $X = f(Y \text{ or } Z)$. Thus, when the number of juice cups is equal, they predict that the set with less water will taste juicier. Finally, at Level 3, children's strategy is one of Juice Versus Water Compensation, $X = f(Y-Z)$. When the side which has more juice also has more water, children pick the side with the greater excess of juice over water, rather than guessing or falling back on the answer which may be obtained by counting only the juice.

In order to investigate the question of horizontal structure, a new task was created whose content was quite different. In this task children were shown a picture of two boys, each of whom was smiling. They were told that the two boys lived in different cities but that each was having a birthday party and that each wanted a set of polished stones for a present.[4] They were then shown how many stones each boy wanted and how many he actually received. The question they were asked was, which child would be happier? The actual items which were presented and the scoring criteria which were used on the Birthday Party task were directly parallel to those presented in the Juice-Mixing Task. The order of presentation was counterbalanced.

---

[4] Polished stones were chosen as the gift, because they were a highly prized commodity in the school where the testing was conducted.

Predictions. In keeping with the horizontal structure hypothesis, it was predicted that children would go through four parallel stages on the birthday party task, namely, (0) Global Gift Evaluation, (1) Counting of Gifts Received, (2) Counting of Gifts-Received or Gifts-Wanted, and (3) Gifts-Received Versus Gifts-Wanted Compensation. It was also predicted that they would go through the four substages at the same rate and during the same age range.

Results. The mean scores on each task at each grade level are shown in Fig. 1. As may be seen, the mean scores of the 10-year-old group were somewhat lower than expected, perhaps due to the presence of a ceiling effect. However, the results with regard to horizontal structure were exactly as expected. An analysis of variance showed a highly significant effect due to age, but no significant effect due to task.

The absence of any task effect is of course quite unusual in developmental psychology. However, it is also unusual for two tasks to be designed so that they vary only in their content while maintaining a common underlying structure. What we believe the group results show, then, is that a substantial degree of horizontal structure exists across tasks in different content domains, and that the existence of such structure can be used as a basis for making predictions about novel tasks.

Of course, to say that two structurally equivalent tasks generate identical mean scores at different ages, or even that they generate identical distributions, is not to

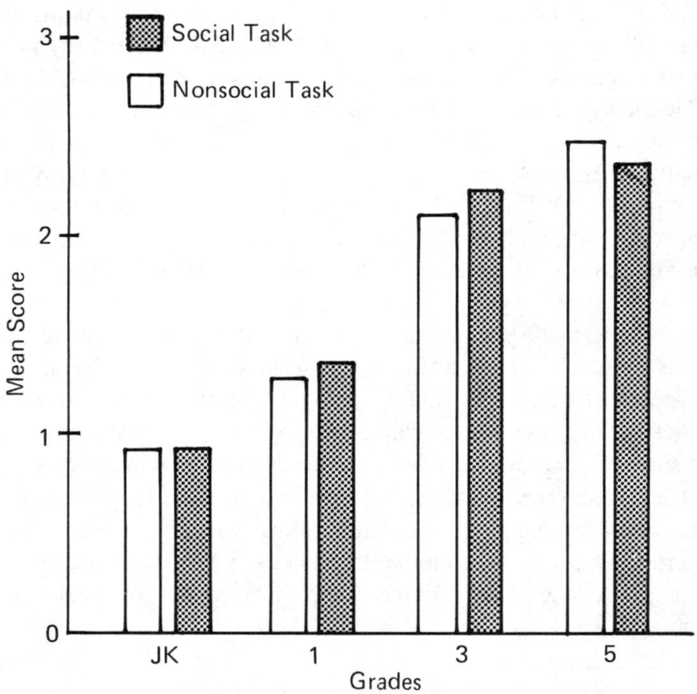

Figure 1. Mean scores on two tasks with social and nonsocial content (Study 1).

say that this commonality should be present at the individual level. As the reader may remember, the present claim is not that children acquire some general "structure of the whole" and then apply this structure to each new problem-solving situation they encounter, thus achieving an identical level of performance on each. Rather, the present claim is that children approach each problem-solving task relatively independently of each other task but tend to reach approximately the same level of performance on each, because they are subject to a common limitation in their working memory. If an individual child has some experiential deficit or advantage which is task-specific, then one would not expect him to function at an identical level on two structurally equivalent tasks. The same would be true if he had some task-specific talent or disability of a more endogenous nature. Thus, while the present theory predicts that the overall distribution of responses should be identical across structurally equivalent tasks, it does not predict that each individual child should occupy exactly the same point in that distribution. Table 2 presents the data which are of relevance to the question of individual performance. As may be seen, approximately half the children at the youngest two age levels did function at the same level across the two tasks, in spite of whatever endogenous or experiential variability may have been present. At the upper two age levels, however, this proportion was reduced to about a third, with the majority of the remaining subjects showing a split in functioning of one developmental level.[5] Reflecting this general trend, the correlation between the two tasks was highly significant (r = .53), but dropped substantially after age was partialled out (r = .11).

Discussion. Putting the group and the individual data together, the following three conclusions may be suggested:

1. Groups of children at different ages tend to function at the same general developmental level, across structurally equivalent tasks with different content matter. Thus, there is clear evidence of horizontal structure at the group level.
2. Individual children also frequently function at the same level across structurally

Table 2. Percentage of Subjects Showing Different Degrees of Developmental Unevenness at Different Age Levels (Study 1).

Age Level	Perfect Synchrony (i.e. same level on both tasks) (%)	Split of One Substage (%)	Split of Two Substages (%)	Split of Three Substages (%)
JK	47	42	11	5
GR I	57	39	4	0
GR III	30	67	3	0
GR V	37	43	20	0

[25] The term *split* rather than *décalage* is used because, in the sense in which Piaget used the term, décalage refers to a lawful difference in task difficulty, i.e., a difference for the average or "epistemic" subject, not a difference for an individual subject.

equivalent tasks. However, this is not necessarily the case. At certain ages or stages, asymmetries in their performance may appear.

3. Finally, the presence of such intraindividual variability notwithstanding, the absolute *amount* of such variation appears to be quite limited. Even at the end of the stage, splits in functioning of more than one substage appear to be quite rare, at least across structurally equivalent tasks in which the basic content is familiar, and this content must be put to some novel (i.e., problem-solving) use.

Study 2

The second study was also conducted by Zopito Marini as a part of his doctoral dissertation. Its primary goal was to replicate the results of the first study across a broader range of tasks. A secondary goal was to see if the mean scores of 10-year-olds would increase, when the artificial ceiling on their performance was eliminated. Finally, a third goal was to see if the number of ''splits'' at Levels 3 and 4 would be reduced, when a more carefully designed set of test items was developed. (For a description of the problems with the first set of items, see Marini, 1981.)

Subjects. The subjects were again drawn from a public school in a middle-class neighborhood in metropolitan Toronto. Twenty children were tested at each of four ages: 4, 6, 8, and 10.

Tasks. Four different tasks were used, namely the Balance-Beam Task, the Birthday Party Task, the Projection-of-Shadows Task, and a Reward-Sharing Task. The first two tasks have already been described. The third task was the adaptation of Inhelder & Piaget's (1958) Projection-of-Shadows Task which has been described by Siegler (1981). Four levels of response are possible on this task, namely Level 0: Global Evaluation of Size of Objects Casting Shadow; Level 1: Precise Quantification of Size of Objects Casting Shadow; Level 2: Quantification of Object Size or Distance from Wall; and Level 3: Object Size Versus Distance Compensation.

Finally, the last task was invented for the present study on the basis of earlier tasks used by Damon (1973) and DeMerssman (1976). The story that was told to each child was as follows:

> A teacher asked some of the kids in her class to come in after school, to make some cards for children who were in hospital. John came in on these days. (At this point the experimenter laid down several pieces of paper, each with a day of the week printed on it.[6] And here are the cards he made. (At this point the experimenter laid down a number of Christmas cards, each identical in appearance.) Mary came in on these days (as described), and here are the cards she made (as described). Now here are some candies (at this point the experimenter pulled out a bag of candies) which the teacher

[6] In order to succeed on the task, children did not actually have to read the days of the week. Rather they simply had to understand that each piece of paper stood for ''one day.'' This was explained to the youngest subjects orally, until it was clear that they understood, before the test continued.

> decided to give to John and Mary as a reward. Tell me, who do you think should get more candies? Do you think John or Mary should get more candies? Or should they get the same?

As will no doubt be apparent, this task has the same structure as the first three and can therefore be scored in the same fashion. The basic scoring categories which were used were as follows: Level 0: Global Output Evaluation (i.e., cards produced); Level 1: Quantification of Each Child's Output; Level 2: Quantification of Output or Input (i.e., number of days worked); and Level 3: Input Versus Output Compensation.

In addition to the four problem types, one additional problem type was added to each of the tasks in order to eliminate the possibility of ceiling effects. On this sort of problem, a correct answer could not be reached via subtraction or addition. Rather, a strategy which involved simple ratios (e.g., 1:2, 1:3) was necessary.

Scoring. Subjects were not given credit for an individual item unless they could give the correct answer to that item *and* justify their response in a fashion which indicated that they had used an appropriate strategy. In addition, if a subject got only one of the two items at his highest level correct, then this problem was discarded, and he was awarded credit for the next highest level instead. The intention behind this procedure was to obtain an estimate of subjects' developmental level which would reflect a thoroughly consolidated rather than a transitional form of functioning. All protocols proved classifiable by this procedure and, although there were occasional disagreements among raters, the interrater reliabilities for all four tasks were in the 90s.

Results. The mean scores on each task at each age level are shown in Fig. 2. As may be seen, all scores were very close to the predicted level. In addition, the mean scores at the first two age levels were virtually identical, with the greatest difference between any two tasks being two tenths of a point. At the upper two grade levels, the range in mean scores widened to four tenths of a point, with the children doing slightly better on the two tasks with social content than on the two tasks with content that was nonsocial. However, it was still the case that the majority of children performed at the predicted level on each task. And, once again, there was a strong age effect, but no task effect.

When the data on individual subjects were examined, it was found that they were similar to those in the first study in certain respects but different in others. What was similar was (a) the fact that few subjects showed a split of two substages, at any age level, and (b) the fact that the percentage of subjects showing perfect synchrony was somewhat lower at the 8- and 10-year-old levels than at the 4- and 6-year-old levels. What was different was the absolute percentages that were involved. Presumably as a result of the improved item design, a majority or near majority of subjects showed no asynchrony from one task to the next at any age level.

The data in Table 3 indicate the degree of synchrony across pairs of tasks and are

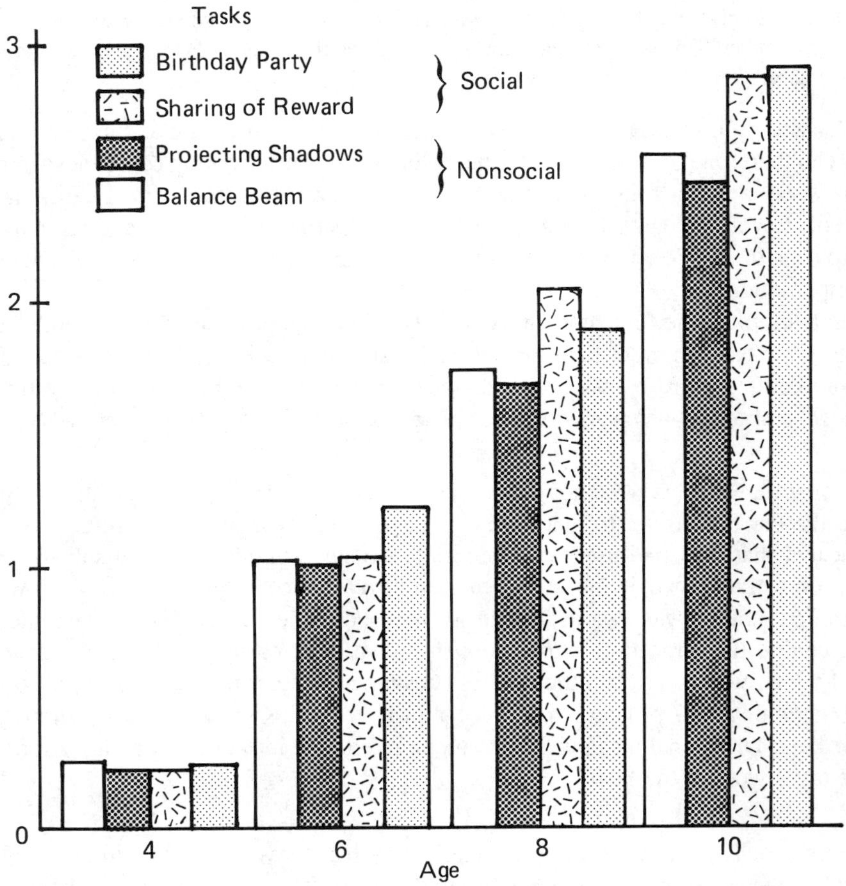

Figure 2. Mean scores on four tasks with social and nonsocial content (Study 2).

Table 3. Percentage of Subjects Showing Various Degrees of Developmental Unevenness on Two of the Four Tasks, Averaged Across All Six Possible Pairs of Tasks (Study 2).

Age	Perfect Synchrony (same level on both tasks) (%)	Split of One Substage (%)	Split of Two Substages (%)
4	60	39	1
6	86	14	0
8	46	46	8
10	53	42	5

thus directly comparable to those of Study 1. It is also possible to inquire, however, concerning the pattern of performance across all four tasks. When the data were examined in this fashion, it was found that—at all ages—the majority of subjects performed at the same level across the majority of tasks. However, only a minority (approximately 25%) actually performed at the same level across all four tasks. The most common pattern of performance was one in which children performed at the same level on three of the four tasks (e.g., Level 3), and at a level which was one removed from this (e.g., Level 2) on the fourth task. As might be expected given this pattern, the average intercorrelations among tasks was extremely high (.82), and the average age partialled correlation lower but still significant (.42).

Discussion. Since the data from the second study were so similar to those from the first, it would seem reasonable to draw the same three conclusions, namely (a) groups of children tend to function at the same general developmental level across structurally equivalent tasks with different content matter; (b) individual children will not *necessarily* function at the same level across such tasks; and (c) while intraindividual variability is reasonably common, the splits which are present nevertheless tend to be quite restricted. For the majority of subjects, there is a clear ''modal'' level of functioning, with only an occasional deviation of one substage from it.

Of course, these three conclusions follow only if one accepts the pattern of data in the table as representative. Thus, before proceeding, it is worthwhile to consider the data that have been obtained in other laboratories. As was mentioned in the introduction, the pattern of data that has been obtained when traditional Piagetian tests have been used has been quite different from that which was reported here. In virtually all the studies, the pattern has been of décalages and not horizontal structure. The reason that such décalages have been the rule, however, was also suggested, namely, that the tasks which have been employed have tapped problem solving at a variety of different developmental levels, as a function of variation in their stimulus conditions and question formats.

What about studies in which problem formats have been carefully varied and children's strategies and/or control structures carefully assessed? Here the results are quite different. In a study by Chandler, Mangione, & Moss (1977), subjects were given a test of proportional reasoning, and a test of their understanding of the concept of ''intelligence'', which required focusing on the two dimensions of ''quality of output'' and ''age.'' In this study, the results were essentially identical to those we have reported in the present chapter, both at the group and the individual level. Similar results were also obtained by DeMerssman (1976), in a study comparing children's moral judgments and behaviors in four different content areas. Finally, similar results have also been reported by Siegler (1981), for four out of 6 tests modelled on Piagetian items. The fifth task in Siegler's study was a conservation of number test, which behaved in the fashion one would expect if it was a unidimensional rather than a bidimensional task. Since, according to the present analysis, number conservation task *is* a undimensional task, these results can not be considered truly ''anomalous''. Finally, the sixth task (probability) did generate data that were anomalous. While the data deviated considerably from the data on the other tasks at the level of individual performance, however, they were still quite similar at the group level. Moreover, such deviations as were recorded—at least at the younger levels—seem explicable in terms of the novelty of the problem question, and might well disappear were a period of task familiarization to be included, or on-line feedback provided.[7]

The general conclusion which we feel to be warranted, then, on the basis not only of our own results but the majority of the results which have been obtained from other laboratories, is that a substantial amount of horizontal structure does exist across tasks during this age range. However, this structure can only manifest itself under conditions where tasks have been appropriately designed and where children have access to all the information that is necessary to solve them.

Study 3

The tasks which were used in our first two studies all involved a common underlying operation. They also involved a common problem-solving format and a single (convergent) answer.

The primary objective of the third study was to examine the pattern of results on tasks where the operations and task formats were variable and where the answers that were required were divergent. In order to accomplish this objective, two different tasks were chosen, one a story-telling task and one a picture-drawing task. In fact, the third study actually began as two separate experiments, each using one of the two tasks in question. It was only after we began to look at the parallel in the results that we realized the relevance of the two experiments to the horizontal structure question.

Subjects. The subjects were selected from a public school which served a middle-class neighborhood in metropolitan Toronto. They were prescreened to ensure that their academic performance and IQs were in the average range for such a population (98–121) and that their ages fell close to one of the modal ages for the dimensional period, namely 4, 6, 8, or 10 years of age. The numbers of subjects in each age group were 16, 22, 19, and 18, respectively.

Tasks. The first task was presented by Anne McKeough as a part of her master's thesis (McKeough, 1982, 1984). Children were administered several tests of general cognitive ability. Immediately afterward, they were asked to tell the experimenter three stories. The instructions for the story-telling task were as follows: ''Now, I want you to tell me a story about (e.g.,) a cute little lamb, kind of like you'd tell it to a boy or girl littler than you are.''

Several days later, the second task was presented to the same group of subjects. This time the children were presented with a set of colored pencils and paper and

[7] For the importance of such feedback see Case (1978), Siegler (1978).

given the following instructions: "Draw me a picture which shows a little girl (or boy) your age, doing something that makes her (or him) happy." The data from the second task were analyzed by Sonja Dennis as part of the work on her doctoral dissertation (Dennis, 1984).

Results. On the story-telling task, children's productions fell into one of four general categories.

Level 0: At the preliminary level, children told a story which contained a single global episode. No clear plot or theme was evident. Instead, the production described an isolated occurrence of the sort which might take place in the child's own life (e.g., the mummy and daddy started eating, the little lamb didn't want any, so he left the table). By adult standards, then, the children's responses sounded more like "scripts" than like stories.

Level 1: At the next level, children's productions underwent a clear qualitative shift, which had the following characteristics: (a) two episodes or events were mentioned, rather than just one; (b) these two episodes had a very clear relationship to each other; the first episode told of a sequence of events which led up to some problem, while the second episode told of how the problem was resolved and the happy events which followed; and (c) as a result, a clear plot emerged. This plot tends to have a clear and conventional theme (e.g., villainy or friendship), which was related in a straightforward nature to the type of character involved (e.g., a fox vs. a horse).

Level 2: At Level 2 there was another shift, which was more quantitative than qualitative in nature. Now children embedded a minor plot within their major plot. Moreover, they tended to do so in such a fashion that the minor plot was juxtaposed against the major one, thus presenting the major character with some additional difficulty (e.g., authority intervening in an attempt to establish friendship, and preventing it.)

Level 3.: Finally, at the last level, some additional elaboration was introduced, often one which produced a more fully elaborated solution to both the plot and the subplot.

On the picture-drawing task, children's productions also fell into four clear levels.

Level 0: At this preliminary level, children drew a global outline of a human figure. The figure had a face, a trunk, and arms or legs. If an object was shown as well, it was generally as a circle or blob attached to the figure.

Level 1: At this level a clear qualitative shift took place, which was characterized by the addition of a "background" with several well-defined elements such as sky, grass, sun, and flowers. The overall construction thus for the first time took on the quality of a "scene" or a "picture."

Level 2: At the next level, another set of changes took place, which were of a more minor but nevertheless significant nature. Now children introduced some element of depth into their pictures. They did so by a variety of artistic techniques, such as bringing the sky down to the ground (and hence making it look "behind" the figure), occluding some part of the body with a limb, drawing feet or hands of the figure sideways, etc.

Level 3: Finally, at Level 3, the children introduced an additional set of elaborations, which integrated the depth dimension and the planar dimension in a more satisfactory fashion. One of these elaborations was to draw the size of the figures in proportion, with figures in the background becoming smaller than those in the foreground. Another was to draw the sun reflecting objects in the foreground, thus changing their color.

In writing up their experiments, both McKeough and Dennis conducted detailed structural analyses of each of their four levels and related them directly to the previously described substages. In addition, they developed a reliable coding system and tested a number of predictions concerning performance on each of their tasks and the more general cognitive measures which were administered. The inter-rater reliability on the story-telling task was .91; on the figure drawing task, it was .96. What is of interest for the moment, however, is not the details of each procedure but the relationship of the resulting levels to each other.

The mean scores on each measure are illustrated in Fig. 3. As may be seen, development in each domain took place at a common rate. An analysis of variance once again revealed a highly significant effect due to age, and an insignificant effect due to task. Figure 4 graphs the mean scores against those on two working memory

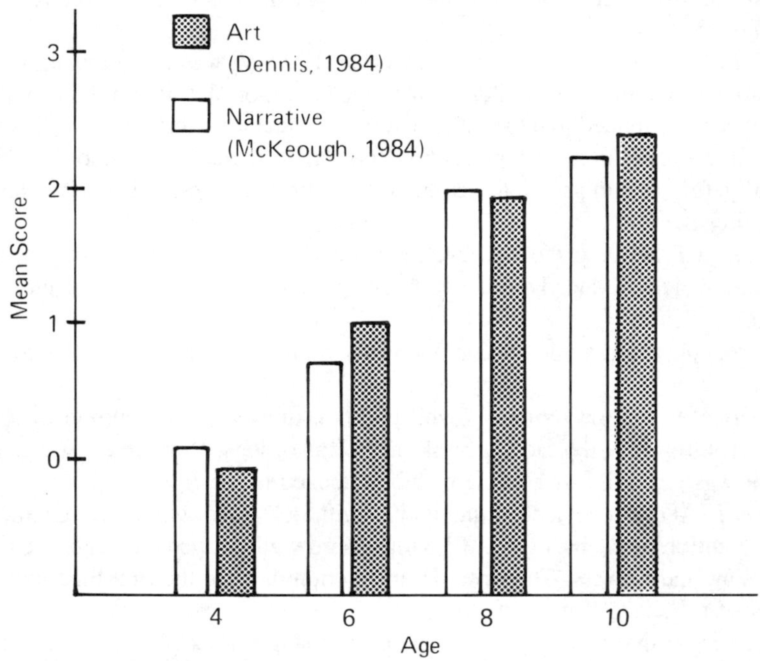

Figure 3. Mean Scores on story-telling and drawing tasks at different age levels (Study 3).

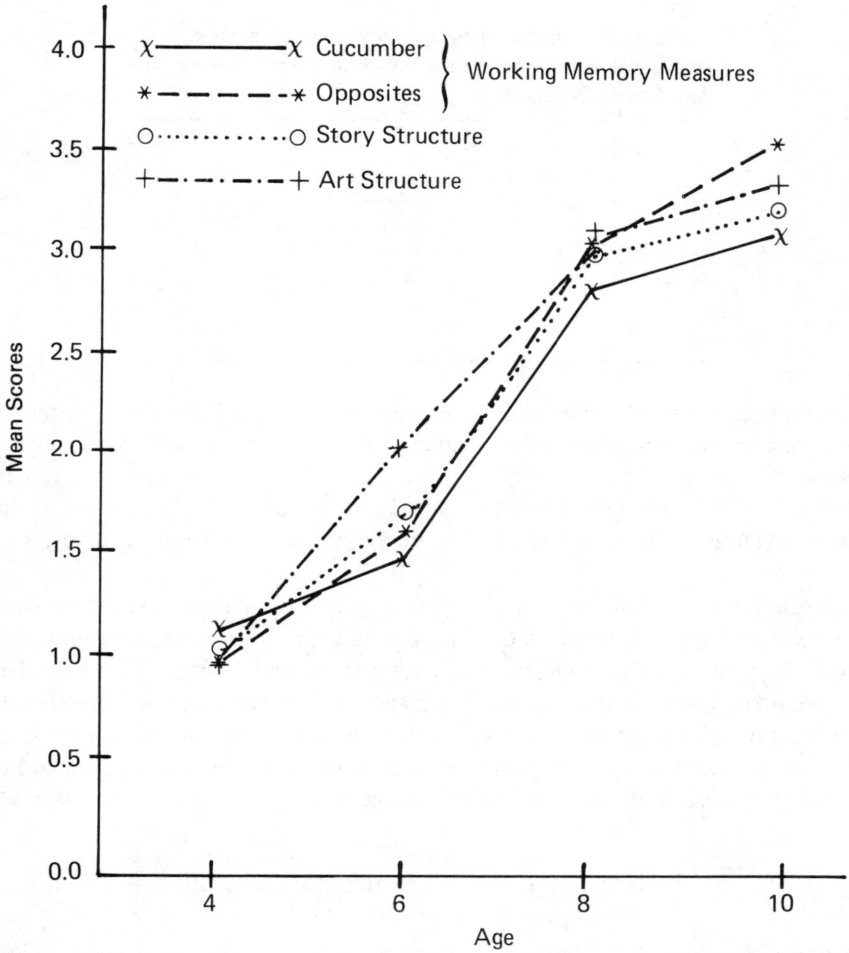

Figure 4. Mean scores on working memory tests and on the two tests of artistic production (Study 3).

measures.[8] As will be seen, the results once again indicated a very close parallel. In fact, from a statistical point of view, it was not even possible to distinguish the scores on the two working memory measures from those on the art or the story-telling tasks.

Finally, the pattern of intraindividual performance is presented in Table 4. As may be seen, this pattern was also similar to that which was observed on the previous set of tasks. At all ages, the majority of children performed at the same

[8] The two measures were "Mr. Cucumber," a test where children have to remember the position of spots on a "Cucumber Man," and the Opposites Test, a test where children are read a series of words and have to respond with their opposites.

Table 4. Percentage of Subjects Showing Various Degrees of Developmental Unevenness At Different Age Levels (Study 3)

Age	Perfect Synchrony (same level on both tasks) (%)	Split of One Substage (%)	Split of Two Substages (%)
4	75	25	
6	73	27	
8	58	42	
10	56	41	13

level on both the story telling and the picture drawing tasks. However, in the two upper age groups, the percentage of subjects who performed at the same level on both tasks appeared to be slightly less than at the lower two age groups. Finally, when subjects did not perform at the same level on both tasks, the typical pattern was a deviation of one substage, with splits larger than this being relatively rare.

Discussion. Since the results of the third study were quite close to those which were obtained in the first two studies, they would appear to warrant the same three conclusions; (a) at the group level, décalages is the exception and not the rule; (b) at the individual level, asymmetry in performance is also the exception, although it is an exception which appears to become somewhat more common at older ages; and (c) even at older ages, the magnitude of the asynchronies which are observed are relatively modest, with splits of more than one substage being relatively rare.

Probing the Limits on Horizontal Structure: Changes Within and Across Chunks

While the analyses that were reported in the previous section capture the global structure of children's thought as they move into and through the third major stage of their development, they leave out most of the details. Thus, they fail to capture much of the richness in the developmental changes that take place. This is particularly true for the sort of development that takes place in the areas of story telling and art. As even a cursory inspection of children's artistic productions will reveal, the type of change which takes place in these domains involves more than the simple coordination of elements. At the same time any two elements are coordinated, there is a change in the internal structure of each element as well. In Piagetian terms, one could say that the elements become more differentiated at the same time as they are coordinated.

Given that this is the case, a question which naturally arises is whether the differentiation that takes place is as lawful as the coordination. Moreover, at least within the present theoretical framework, this question gives rise to a second question as well: "Is the differentiation which takes place limited by the same quantitative parameter as the coordination?'' For if it should turn out that the same quantitative parameter does set a limit on both sorts of change, then it should be the case that considerable horizontal structure should be present *within* as well as *across* elements.

If one assumes that children aim for coherence in their stories and drawings, then it seems reasonable to suggest that horizontal structure should exist within elements as well as across them. For coherence among any set of elements—whether within or across chunks—can only be achieved by attending to those elements and setting up a relationship among them. And this in turn can only be done in working memory. Since the functional capacity of working memory grows with age, then a parallel change should take place in both within-element and across-element development.

In order to understand why this is so, it is worthwhile to consider in more detail the act of generating a coherent story. In describing the opening episode in their stories, children should be able to generate and interrelate a set of M-events. As they move on to the next epidsoe, they should be able to generate a second set of events of almost equivalent richness. We say ''almost,'' because if the second episode is also to be *related* to the first episode, then at some point this first episode will have to be taken into account, either in an ''on-line'' fashion or in review. And this in turn will require that some global feature of this episode will have to be retrieved and related to the events of the second episode. It is quite likely, then, that the second episode will still be reasonably rich, but not quite so rich as the first one, that is, if all individual elements are to be related both to each other and to the first episode.

Since a similar set of considerations applies to each successive episode, the result is that the total complexity of such a production may be given by the following formula, $M + (M-x) + (M-2x)$. . . where x is equal to the amount of working memory required for storing and retrieving a single pointer to the first episode.

How is one to estimate the value of x? The most obvious suggestion would be that a single pointer should require a single unit of storage space, i.e., that the value of x should be one. While this seems a reasonable approximation, however, it should also be possible to estimate this value more precisely by empirical means.

In our recent work, we have attempted to effect such an estimation, by creating a parallel situation with working memory measures to that experienced in story telling. In effect, we see the problem faced by a child in constructing a coherent set of elements—each of which is itself made up of a coherent set of subelements—as being analogous to that faced by a child who has just been given several working memory measures in a row, each with clearly distinctive internal content. While the subject should be able to retrieve elements from more than one set, there should be some diminution of the *number* of elements he can give back for each subsequent test, because the act of retrieving and outputting the elements from one test should interfere with his ability to preserve the elements of the previous test in a clearly organized and sequential fashion. In short, this ''multiple working memory function'' should also be $M + (M-x) + (M-2x)$ . . . etc.

In this final section, then, we shall report two further studies. The first one presented children with several working-memory measures in a row, in order to determine the shape of the "multiple span" function and to estimate the value of x. The second study looked at the internal structure of children's stories in more detail and compared this internal complexity with that found in the multiple-span test.

Study 4

The fourth study was conducted in collaboration with Jill Goldberg. Its object was to determine the amount by which children's total span would be decremented, in a situation where they were given several distinctive working-memory measures in a row and asked to give back the items from each.

Subjects. The subjects were 11 children with working memories of between 2 and 3, who were enrolled in grades 1 and 2 of a university laboratory school and who had served as the subjects in a series of earlier studies on working memory.

Stimulus Material. Two different sets of stimulus material were utilized. In the first set, what made the different types of items distinctive was that they were drawn from different semantic categories. All items were three-syllable nouns, with the different categories being (a) cars (e.g., Cadillac, Volkswagen); (b) fruit (e.g., pineapple, nectarine), and (c) countries (e.g., Italy, Australia). In the second set, what made the items distinctive was that they had different stimulus characteristics. The three different types of stimuli were (a) spatial positions (the test used was the Cucumber test described in footnote 8); (b) numbers (the test used was the Count-the-Spot test, which is described in Case, Kurland and Goldberg, 1982), and (c) visual gestures (e.g., "hands in prayer" or "thumbs up").

Experimental Procedure. Each set of stimulus materials was presented in a different experimental session, with the order of administration being counter-balanced across subjects. Subjects were first told about the purpose of the study. They were then given each of the three different span tests separately in the order previously mentioned. Each set of items was presented at a rate of one item per second, and subjects had to regenerate the items immediately following their presentation. Testing began with one-item sets and worked upward until a subject failed all three items at any one level. A subject's score was equal to the highest level at which he received a perfect score plus one third of a point for any item passed at higher levels.

Once the subject's span had been determined on each of these tests separately, he was then given two tests in immediate succession. First, the subject's span on the first type of item was "rounded down" to the nearest whole number. Then he was given this many (M) items, preceded by M−1 items from the second class of material. This second number was then adjusted up or down appropriately, until the maximum number of items which could be retained while still retaining M (novel) items from the other set had been determined. Then an item from the third set was introduced, again preceding the items from the two sets already considered.

On every trial, subjects had to repeat all the items from all of the sets. They were, however, free to give the actual items themselves in any order, providing they did not give back the last item first. For example, if presented with one country name, two fruit names, and three car names, they were allowed to first give back the three car names (in the correct order), then the two fruit names (in the correct order), followed by the one country name. Note that this procedure roughly parallels the constraints imposed in drawing a picture, where there is some natural grouping of elements and subelements, but where subjects are free to generate the elements and subelements in whatever order they may chose.[9]

Results. The results are presented in Fig. 5. As may be seen, there was a lawful decrement in performance in the multiple-span function in each condition of the sort which the theoretical analysis had suggested. As will also be seen, the magnitude of this decrement varied slightly from test to test but was just under one unit in each case.

Discussion. Since these results are only for one age group, it is impossible to say with any certainty what the corresponding numbers would be at different ages. However, if the theoretical analysis which was presented was correct, the values should be approximately as follows: 4 years: 1; 6 years: 2 + 1 (i.e., 3); 8 years: 3 + 2 + 1 (i.e., 6); 10–adult: 4 + 3 + 2 + 1 (i.e., 10).[10] Whether or not these extrapolated values are correct, of course, the 6-year-old results can be used as a baseline against which to compare the complexity of children's stories at that age. It is to the results of such an endeavor, then, that we now turn.

Study 5

Although it is reported here as a separate study, the fifth set of data were actually generated by a more detailed analysis of the stories and drawings that were described in Study 3. The first question of interest was whether there would be an increase in the complexity of children's productions within as well as across chunks. The answer to this question was a clear "yes," as may be seen from an inspection of Fig. 6a, 7a, 8, and 9 for children's stories (McKeough, 1982) and Fig. 6b and 7b for their art (Dennis, 1984).

The second question of interest was whether there might be any parallel in the changes taking place in the two domains of either a qualitative or a quantitative sort. As yet, we are not able to answer this in a very precise fashion, since we have just begun to tackle the problem of finding a common notational format in which to

---

[9] The reason the subjects were asked not to give the very last item back first, is that this item would still be in immediate memory and would never really have to be stored and retrieved.

[10] Jill Goldberg has tested adults and found that the result is, in fact, 4 + 3 + 2 + 1. As yet, however, the study has not been conducted with 4-, 8-, or 10-year-olds.

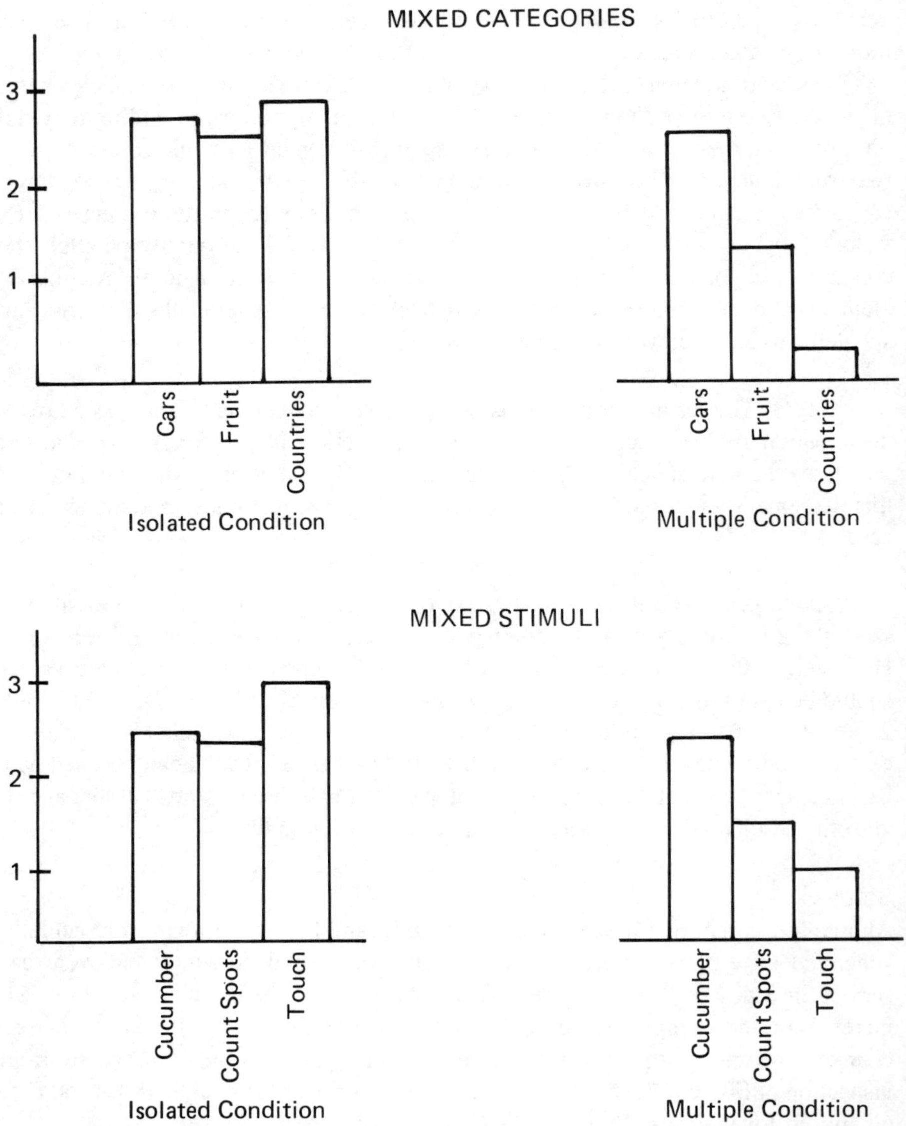

Figure 5. Mean scores of grade 1 and 2 students on three span measures given under isolated and mixed (multiple-span) conditions (Study 4).

graph the two sorts of structural changes. At least at a global level, however, and for the 4 to 6 transition, it would seem that the answer to this question is also "yes," as may be seen from an inspection of Fig. 6 and 7.

The final question was quantitative. Bearing in mind that 6-year-olds have a multiple span of approximately 2 + 1 units, one might expect that there would be a doubling of the complexity of their initial episodes in the story task and a "tacking

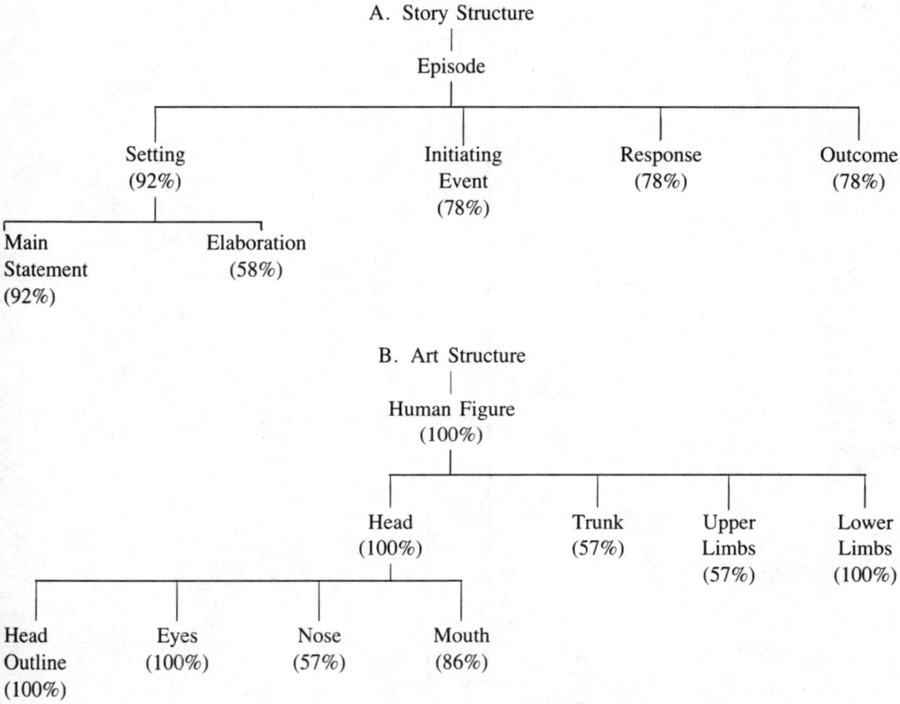

Figure 6. Four-year-old story and drawing structures.[11]

on'' of an additional terminal episode. Similarly, one might expect that there would be a doubling of the complexity of their human figures in the drawing task and the addition of a background scene whose complexity was roughly equivalent to that which their figures displayed earlier. In fact, both of these expectations would appear to be confirmed, at least in a rough fashion. The quantitative data are presented in Fig. 10.

General Conclusion

We began the present chapter with a quote from John Flavell in which he suggested that children's development in different domains proceeds quite independently, and that there is, as a consequence, very little horizontal structure in their development. Although Flavell himself has modified this position, we suggested that it remains the dominant view in North America today, and one which deserves to be reconsidered.

The theoretical reasons for reconsidering this view were presented in the first two sections. In the first section, it was pointed out that the supposedly massive evidence against the notion of age-related structure in children's thought was, in fact, only evidence against the notion that certain particular forms of structure automati-

[11]Figures in parentheses indicate proportion of subjects generating the type of element in question.

A. Story Structure

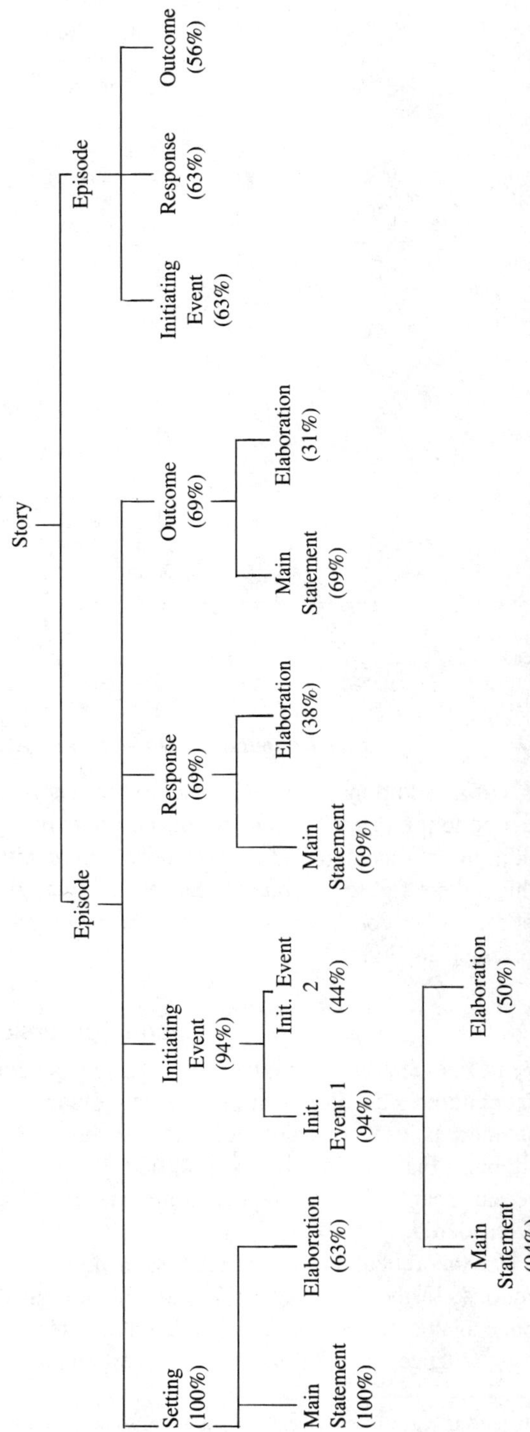

B. Art Structure

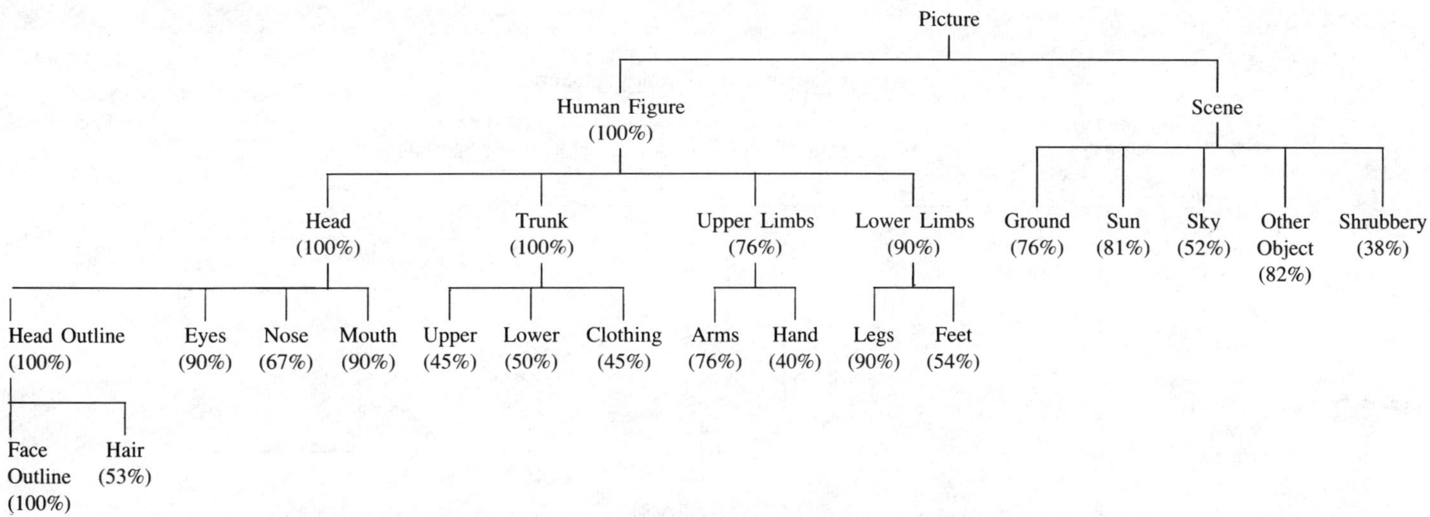

Figure 7. Six-year-old story and drawing structures.

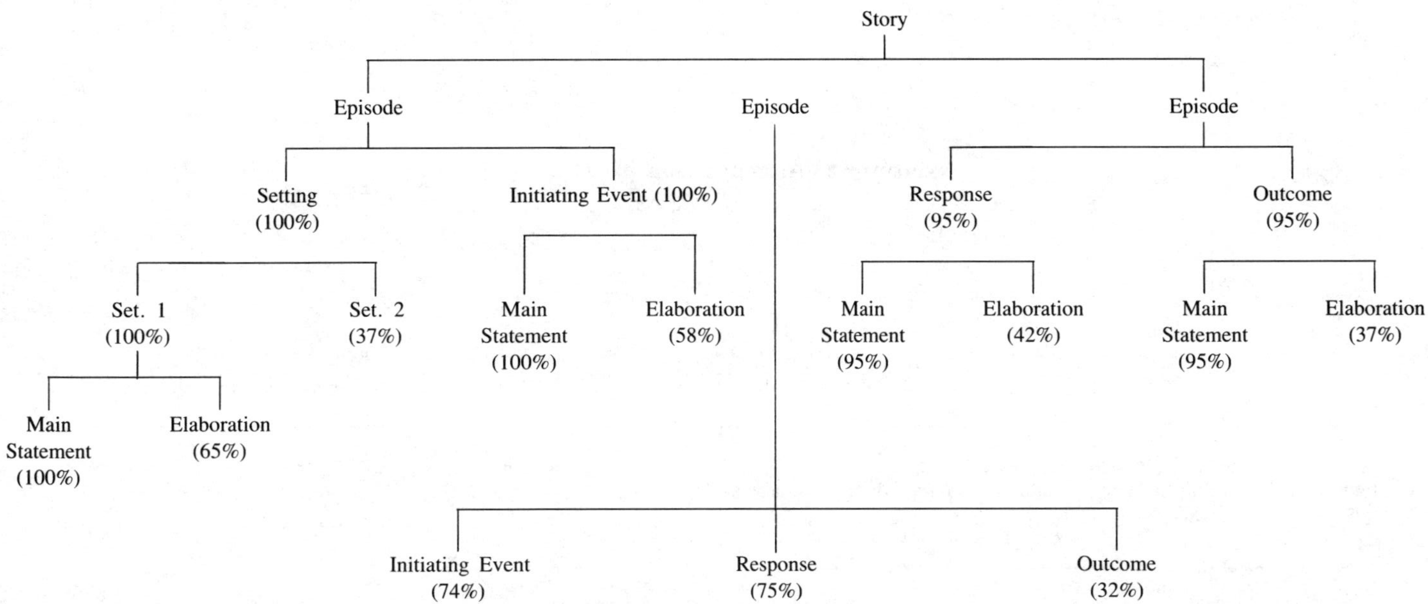

Figure 8. Eight-year-old story structure.

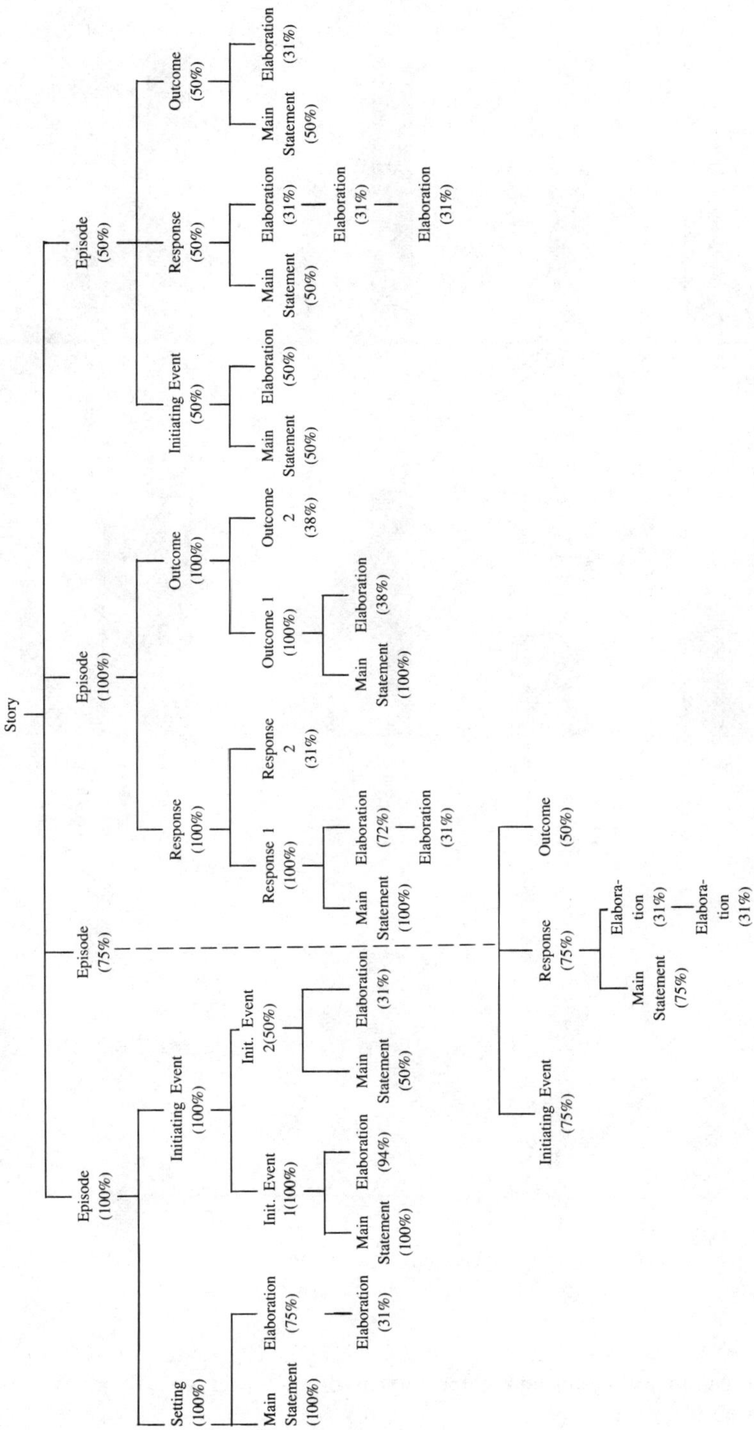

Figure 9. Ten-year-old story structure.

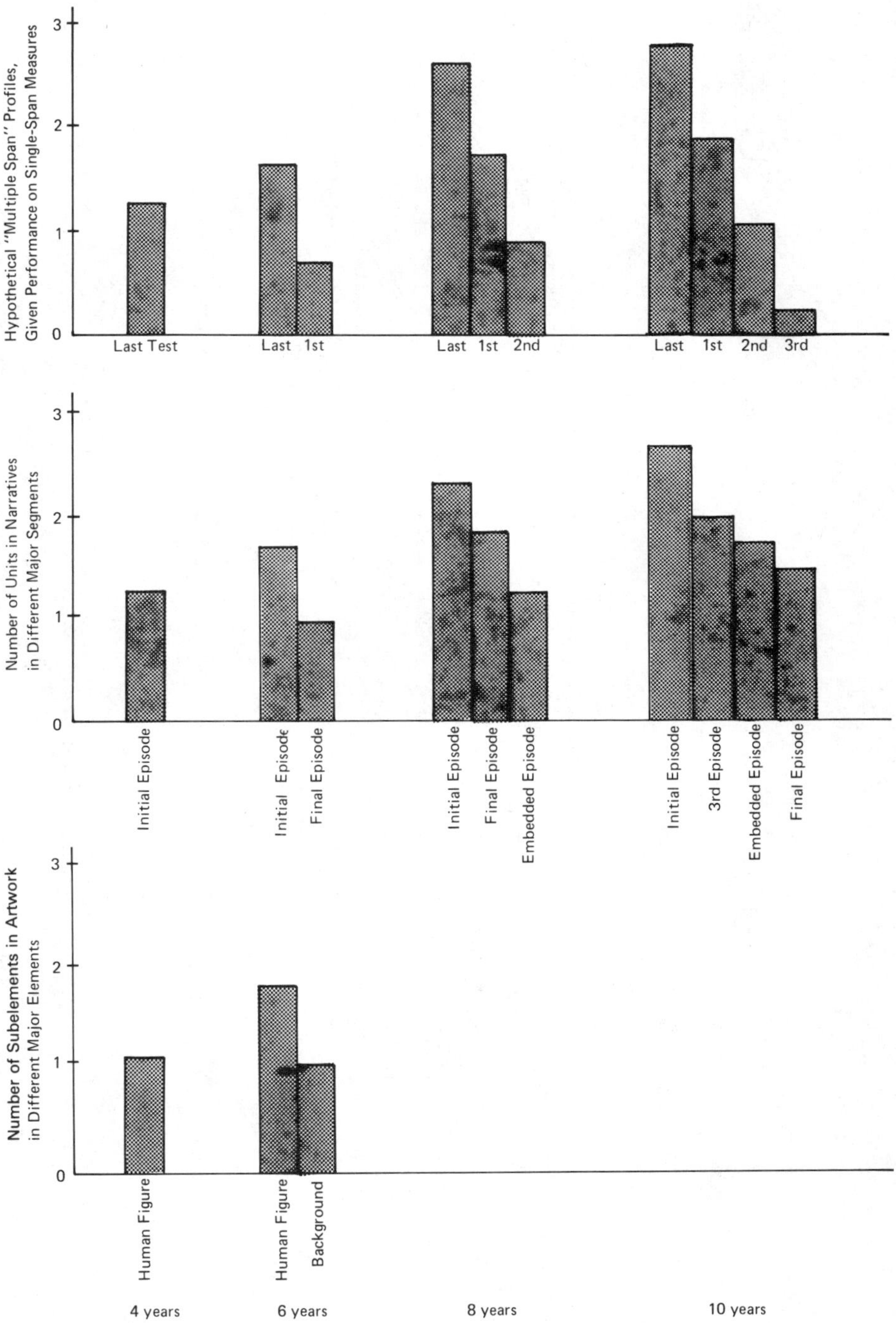

Figure 10. Number of major and minor units at different ages, on multiple-span, story, and drawing tasks (Study 5).

cally lead to the acquisition of certain particular concepts or abilities. Once it is realized that any given concept or ability can be acquired by a variety of different routes or mastered at a variety of different levels, then it becomes apparent that the various tests which are currently in existence in fact confirm the existence of a lawful and age-related progression. Moreover, it also becomes apparent that this pattern is actually very similar to the one which was initially postulated by Piaget. In the case of conservation, for example, the tests which are passed at the age of 3 or 4 can be solved by the application of global or intuitive thought. The tests which are passed at the age of 5 or 6 can be solved by the application of unidimensional thought, $X = f(Y)$. The tests which are passed at the age of 7 or 8 can be solved by the application of bidimensional thought $X = f(Y \text{ or } Z)$. Finally, the tests which are passed at the age 9 or 10 require a more fully developed or elaborated form of bidimensional thought in which compensation often plays a major role, $X = f(Y-Z)$.

Of course, the data on décalages are not the only ones which argue against the presence of the sort of global structures which were postulated by Piaget. In addition, there are the data on low intertask correlations and training effects. While these data do argue against the most extreme or "wholistic" form of the horizontal structure hypothesis, however, they do not argue against a somewhat more liberal view, in which the core of Piaget's thinking is preserved, and in which development is seen as moving through a sequence of universal stages, but in a fashion which can be modulated by individual differences in task-specific experience or ability.

One such liberalized view was presented in the second section. According to this view, children progress in different domains relatively independently, working out their answers to problems, as it were, from first principles. While they do not apply the same "structure-of-the-whole" to different tasks, however, they very often end up constructing answers which do have the same structure, because their problem solving is subject to the same age-related constraint in working memory. In effect, then, children often bump their heads up against the same age-related ceiling, thus reaching structurally similar conclusions across structurally similar tasks.

According to this neo-Piagetian view, one would expect to find a good deal of developmental asynchrony as well as synchrony at the level of individual performance, because of the presence of the above-mentioned variability in task-specific experience or ability. However, provided that a set of tasks all gave children equal access to the information which was of relevance, one should also expect a good deal of horizontal structure. This should be particularly true at the group level, where individual differences of a nonsystematic sort should tend to cancel each other out, and where the resultant performance should be a good indication of children's modal developmental level.

In the third section, three studies were reported which were designed to test this hypothesis. In the first two studies, two different sets of tasks were described in which task format and structure were carefully controlled, while task content was systematically varied. It was shown that the results in each case were the same. In spite of the fact that a moderate degree of asynchrony was present at the individual level, at the group level there was virtually perfect synchrony across tasks, with no significant difference being present in the mean scores and with the age-related progression conforming to what would be expected from a working-memory analysis. Recent evidence from other laboratories was also reviewed, and it was suggested that the same pattern is present there as well, with the exception of single anomalous task, namely, probability judgment. Finally, it was shown that the same pattern can even appear in tasks in which the problem formats are not rigidly controlled, provided once again that the tasks in question are ones with which children are reasonably familiar. In fact, in the case of the story-telling and picture-drawing tasks, our exploratory analyses suggested that a substantial amount of horizontal structure may be present in the way in which children organize the individual elements of their productions, as well as in the way in which they coordinate these individual elements into an overall whole.

Taking the various analyses and studies which were reported together, the following conclusions may be drawn. First, it would be a serious mistake to abandon the belief in the presence of horizontal structure for a view which holds that development in each domain proceeds completely independently, with performance being determined entirely by experiential or domain-specific factors. What the data appear to call for, rather, is a more balanced view in which children's problem solving in different domains is seen as proceeding *relatively* independently but as being subject to strong age-related constraints of some sort. Second, a likely candidate for such a constraint is working memory, since working memory would be expected to produce a high degree of horizontal structure at the group level and a substantial (thought significantly lower) degree of such structure at the individual level. Another piece of evidence which argues in favor of working memory as an important factor is the absolute *number* of elements which children can handle, both within and across chunks, at the ages of 4, 6, 8, and 10. In all cases, these numbers appear to agree with the values that would be predicted by working-memory analysis, as well as the values that are actually obtained when working memory measures are administered.

Having stated these two conclusions, however, it must be acknowledged that a number of issues remain unresolved and are worthy of further investigation.

Synchrony in Group as Opposed to Individual Data

The first issue concerns the difference in the pattern between the individual and group data. As was demonstrated in each one of our studies, the evidence for horizontal structure is a good deal stronger at the group level than it is at the individual level. Two questions which naturally arise, therefore, are (1) Why is this so?, and (2) Have we attributed too much significance to the group data? After all, it could be argued that these data reflect mere ''statistical averages'' not the behavior of any individual subject.

Our own answer to these questions has been hinted at several times in the chapter. However, it is worthwhile to reiterate this answer, and to make it more explicit. Our theory postulates the existence of two sorts of factors: (a) specific factors such as those which accrue with particular experience and/or talent; and (b) a general factor, which is not unique to any task and which constrains subjects' problem solving "across the board." From this theoretical perspective, what the group data indicate is the action of the latter factor, when specific effects have been removed due to averaging. What the individual data indicate is the action of both factors.

While the evidence of horizontal structure is a good deal weaker at the individual level, then, our theory is one which predicts that this should be the case. Although our theory is not necessarily the *only* one which entails such a prediction, it is important to realize that it is at least *one* such theory and that there do exist theories which would generate different predictions. If one were to adopt an extreme form of developmental determinism, for example, one would have to predict that few deviations from modal functioning should exist, even at the individual level. Similarly, if one were to adopt an extreme form of experiential determinism, one would have to predict that little synchrony should be present in the individual data *or* the group data. It is because the present theory adopts an intermediate position, that it can account for the general pattern of data at both levels. And it is because the theory has a working-memory component that it can make predictions about this pattern which are quantitative as well as qualitative.

Higher Degree of Asynchrony at the Final Two Substages

A second issue which is worthy of discussion concerns the tendency for asynchronies in individual performance to be more frequent at substages 2 and 3 than at substage 1. The data indicating this to be the case are not strong. For Study 3 for example, the difference was only on the order of 15%. Nevertheless, the trend did appear to be a consistent one across the three studies and thus warrants some attempt at explanation.

The tentative explanation we would advance draws once again on the work of Pascual-Leone (1974). At Level 1, all the dimensions which children had to notice were ones which were reasonably salient and to which attention would be expected to be called in the course of day-to-day experience. By contrast, at Level 2, the second dimensions were all *less* salient. One might therefore expect that within-subject variation in experience, coupled with variation in the ability to detect salient versus unsalient dimensions, might lead to greater variation in the extent to which this hidden or less obvious dimension was noticed and integrated with the first. These differences would not show up at the group level, since they would tend to cancel each other out (e.g., for every child who had more experience on task A, there would be another who had less). However they would show up at the individual level and generate precisely the sort of deviations which were actually observed.

Parallels Within and Across Structural Domains

A third question which is raised by the present data concerns the relationship between development in different structural domains. For the first set of tasks that were reported, it was clear that—although the surface content varied—there was an underlying structural progression that did not. In each case, what was observed was the same sort of progression that Piaget first described, namely, a progression from predimensional thought, to unidimensional thought, to bidimensional thought without much integration, to bidimensional thought with integration via compensation.

For the story-telling and picture-drawing tasks, however, the situation was quite different. First of all, the progression within each task could not be predicted on the basis of Piaget's formulae. Rather, it had to be induced. Second, the cross-task parallel was not nearly as clear, either between the two tasks themselves or between either of the two tasks and the previous five tasks that were described.

The interesting finding, of course, was that the empirical pattern of results *was* the same. Children went through the same number of qualitative shifts on each of the two tasks and did so at the same ages. Moreover, there was the same absence of any significant task effect and evidence of performance in accordance with the expectation generated by a working-memory analysis. The obvious question which these empirical results raise, then, is whether there is any more general structural commonality among all the tasks, or whether the only parallel is in the load which they place on working memory.

We believe that all the tasks did share a more general set of structural properties. The only way to demonstrate this, however, is to conduct a detailed developmental analysis of the precursors of each of the competencies which emerge at the age of 6 and which were assessed by the present battery of tasks. Unfortunately, a full analysis of this sort is beyond the scope of the present chapter. Since such an analysis has been presented elsewhere (Case, 1985), however, the general thrust of this analysis may be summarized, and an indication of the more general parallel may at least be suggested.

By about the age of 4 months, children for the first time become capable of coordinating the executive structures which they have consolidated for monitoring complex sensory events. As a consequence, for the first time, they become capable of engaging in thought which is sensorimotor, i.e., thought in which the *first-order* relationships between their own actions and the responses of external objects can be apprehended. At first, only one such relationship can be dealt with at a time (4–8 months). Later, two such relationships can be dealt with in sequence (8–13 months). Finally, two such relations can be coordinated in an "on-line" or "integrated" fashion (13–20 months). The result is that children can apprehend the workings of a simple *sensorimotor system* in which the relationship between two external objects may be manipulated as a function of the child's sensorimotor activity on each.

At about the age of 20 months, a second major qualitative shift takes place, and children for the first time become capable of coordinating entire sensorimotor systems. The result is they become capable of apprehending and manipulating *second-order* relationships, or relationships between relationships. At first they can only focus on one such second-order relationship at a time (20–27 months). Later, they become capable of focusing on two such relationships in sequence (27 months–3½ years). Finally, they become capable of coordinating two such entities in a parallel, or integrated fashion (3½–5 years). The result that children can apprehend the workings of second-order *relational systems.*

Finally, during the age range from 4 to 10, children go through a parallel set of developments at a third epistemic level. First they become capable of coordinating two second-order relational systems and apprehending the third-order relationships which emerge as a result. Next, they become capable of focusing on two such higher-order relationships in a sequential or global fashion. Finally, they become capable of integrating two such higher-order relationships in an "on line" or "parallel" fashion.

Table 5 presents the general characteristics of each of the substages at this third epistemic level and the way in which these characteristics were instantiated by the various tasks in our studies. It also presents a more general set of labels for the various substages. As may be seen, what all the Level 0 tasks shared in common is not really that they require a form of thinking that is "predimensional" but rather that they require subjects to work within a second-order relational system. The nature of this system varies considerably from task to task. In the case of the figure-drawing task, the system is one of geometric relations, of the sort that can be produced by appropriate sensorimotor control of a pencil and paper. In the case of the story-telling task, the system is one of episodic relations, of the sort which can be captured by verbal description of recurrent sequences of action. In the case of the balance-beam and shadow-projection tasks, the system is one of cause-and-effect relations, of the sort which emerge when certain objects (or movable object parts) are brought into contact with each other, in an appropriate fashion. Finally, in the case of the birthday-party and reward-sharing tasks, the system is one of social relations, in which certain actions of one person have a regular effect on the reactions of another person. While the content varies considerably from task to task, then, we would suggest that the level of relationship which children have to consider does not, nor does the complexity of the system within which these relationships have to be embedded.

As one moves from the Level 0 to the Level 1 tasks, the new feature which all the tasks entail is the requirement for coordinating two of the relational systems which were dealt with at the previous level and for apprehending the new (and higher-order) relationship which emerges as a consequence. The most obvious example of this sort of coordination is that which is involved in the tasks that were used in the first two studies, i.e., the tasks where children had to coordinate their system for comparing numbers with their system of establishing cause-and-effect relationships. However, the story-telling and the figure-drawing tasks require coordinations which are formally similar. In the case of the drawing task, the coordination which is required is between the geometric relations that exist within the elements of a figure and those that exist within the elements of the background. In the case of the story-telling task, the coordination is between two different sorts of episode. In all the

Table 5. General Characteristics of Each Substage Plus Specific Manifestations

		Specific Manifestations			
Substage	**General Characteristics**	**Picture Drawing**	**Story Telling**	**Physical Cognition (balance beam, shadows)**	**Social Cognition (birthday party, rewards)**
0. Operational consolidation (3½–5 years)	System of 2nd order relations is consolidated; product can be treated as a single unit	Child can draw figures consisting of several parts and corresponding (spatial) relations; result can be treated as a single object (e.g., "a man")	Child can describe sequence of several events, and corresponding (temporal) relations; result can be recognized as a single "episode"	Child can understand operation of a mechanical device consisting of several parts, with corresponding (mechanical) relations; result is ability to predict result of particular action, or variation along a single (polar) dimension	Child can understand operation of a system consisting of several people or objects and appropriate (social) relations; result is ability to predict result of a particular social action or variation along a single (polar) dimension
Major Qualitative Shift					
1. Operational coordination (5–7 years)	Units of previous stage coordinated, often in a nested fashion; result is	Child can coordinate relations among isolated pairs of objects (e.g., ground & man, sky & sun); higher order unit	Child can coordinate episode leading to problem with episode leading to resolution; higher-order unit	Child can coordinate polar variable-evaluation with counting (also a 2nd order relational system); higher-order	Child can coordinate polar variable evaluation with counting; higher-order unit which emerges is a continu-

	emergence of a new kind of unit of a higher order	which emerges is the "scene"	which emerges is the plot or "theme"	unit which emerges is a continuous (physical) variable	ous (social) variable
2. Bifocal coordination (7–9 years)	Two higher-order units now considered and related, albeit in a somewhat global fashion	Child can differentiate and globally coordinate foreground and background relations; result is first global indication of depth (e.g., sky comes down behind foreground figures)	Child can focus on plot and subplot and resolve both successfully, albeit in sequence	Child can focus on 2 quantitative physical variables and make global (nonquantitative) integration	Child can focus on 2 quantitative social variables and make a global (nonquantitative) integration
3. Elaborated bifocal coordination (9–11 years)	Two higher order units considered and related, via fully elaborated procedure or rule. Result is consolidation of 3rd order relational system, and readiness for next major qualitative shift	Child can relate objects in foreground and background, via sequence of intermediate objects or cues; result is unified and realistic scene with clear and continuous appearance of depth	Child can develop solution in which plot and subplot are resolved in parallel or brought to well-developed conclusion; result is elaborated and well unified story	Child can focus on two quantitative physical variables, and make precise trade-off between them via addition or subtraction; result is understanding of quantitative (physical) compensation	Child can focus on two quantitative social variables and make a precise trade-off between them via addition or subtraction; result is understanding of quantitative (social) compensation

various tasks, then, there is the same underlying requirement for moving to a new epistemic level and constructing an entity of a higher sort via the coordination of two lower-order entities.

Finally, as one moves from the tasks at Level 1 through those of Level 3, two new requirements are added: (a) an increase in the breadth of focus from one to two higher-order elements, and (b) the introduction of some additional elaboration, thus in effect constructing an integrated *system* of third-order relations.

In summary, we would hypothesize that all seven of the tasks considered in the present chapter did share a common set of structural requirements over and above those for a certain level of working memory. These requirements were for apprehending some sort of third-order relationship (of which a quantitative dimension is one example) and then assembling these relationships into a third-order relational *system*. It is this fundamental structural property, we believe, which best characterizes the capability of children between the ages of 5 and 10. And it is this same capability which is responsible for producing the horizontal structure that is observed across problem-solving situations, where the information is familiar and task-complexity is controlled.

References

Brainerd, C. J. (1978). The stages question in cognitive-developmental theory. *The Behavioral and Brain Sciences, 1,* 173–182.

Case, R. (1978). Intellectual development from birth to adulthood: A neo-Piagetian interpretation. In R. S. Siegler (Ed.), *Children's thinking: What develops?* Hillsdale, NJ. Erlbaum.

Case, R. (1985). *Intellectual development: Birth to adulthood.* New York: Academic Press.

Case, R., Kurland, M., & Goldberg, J. (1982). Operational efficiency and the growth of short term memory. *Journal of Experimental Child Psychology, 33,* 386–404.

Chandler, M. J., Mangione, P., & Moss, P. (1977, June). Developmental changes in the structure of children's inferences about intelligence in their peers. Paper presented at the annual meeting of the Jean Piaget Society, Philadelphia.

Damon, W. (1973). *The child's conception of justice as related to logical thought.* Unpublished doctoral dissertation, University of California.

DeMerssman, S. (1976). *A developmental investigation of children's moral reasoning and behavior in hypothetical and practical situations.* Unpublished doctoral dissertation, University of California, Berkeley.

Dennis, S. (1984, April). *Stages in the development of children's drawing.* Paper presented at the annual meeting of the American Educational Research Association, New Orleans.

Fischer, K. W. (1980). A theory of cognitive development: The control and construction of hierarchies of skills. *Psychological Review, 87,* 477–531.

Flavell, J. H. (1980, October). *Structures, sequences, and stages in cognitive development.* Paper presented at the Minnesota Symposium on child psychology. Minneapolis, MN.

Flavell, J. H. (1982). On cognitive development. *Child Development, 53,* 1–11.

Flavell, J. H., Bolkin, P. T., Fry, L. C., Wright, J. W., & Jarvis, P. E. (1968). *The development of role-taking and communication skills in children.* New York: Wiley.

Gelman, R. (1972). The nature and development of early number concepts. In H. W. Reese (Ed.), *Advances in child development and behavior* (Vol. 7). New York: Academic Press.

Gelman, R., & Weinberg, D. H. (1972). The relationship between liquid conservation and compensation. *Child Development, 43*, 371–383.

Gelman, R. (1979). Why we will continue to read Piaget. *The Genetic Epistemologist, 8*(4), 1–3.

Halford, G. S. (1980). Toward a redefinition of cognitive developmental stages. In J. Kirby & J. B. Biggs (Eds.), *Cognition, development, and instruction*. New York: Academic Press.

Liu, P. (1981). *An investigation of the relationship between qualitative and quantitative advances in the cognitive development of preschool children*. Unpublished doctoral dissertation, University of Toronto (OISE).

Marini, Z. (1981). *The relationship between social and non-social cognition in elementary school children*. Unpublished master's thesis, University of Toronto (OISE).

Marini, Z. (1984). *The development of social and physical cognition in childhood and adolescence*. Unpublished doctoral dissertation, University of Toronto (OISE).

McKeough, A. (1982). *The development of complexity in children's narratives*. Unpublished master's thesis, University of Toronto (OISE).

McKeough, A. (1984, April). *Developmental stages in children's narrative composition*. Paper presented at the annual meeting of the American Educational Research Association, New Orleans.

Noelting, G. (1982). *Le développement cognitif et le mécanisme de l'Équilibration*. Chicoutimi, Canada: Gaëtan Morin.

Pascual-Leone, J. (1969). *Cognitive development and cognitive style*. Unpublished doctoral dissertation, University of Geneva, Switzerland.

Pascual-Leone, J. (1974, August). *A neo-Piagetian process-structural model of Witkin's psychological differentiation*. Paper presented at the Second International Conference of the Association for Cross-Cultural Psychology. Kingston, Ontario.

Piaget, J. (1952). *The child's conception of number*. London: Routledge & Kegan Paul. (First published in French in 1942),

Siegler, R. S. (1976). Three aspects of cognitive development. *Cognitive Psychology*, 1976, *4*, 481–520.

Siegler, R. S. (1978). The origins of scientific reasoning. In R. S. Siegler (Ed.), *Children's thinking: What develops?* Hillsdale, NJ: Erlbaum.

Siegler, R. S. (1981). Developmental sequences within and between concepts. *Monographs of the Society for Research in Child Development, 81*, (Serial No. 189).

Wilkening, F., & Anderson, N. H. (1980). Comparison of two rule assessment methodologies for studying cognitive development. Technical Report #94. Centre for Human Information Processing, University of California, La Jolla, California.

2

*Similarities Between Developmental Sequences at Different Age Periods**

Pierre Mounoud
Faculté de Psychologie et des Sciences de l'Education Université de Genève, Switzerland

Introduction

We could be forgiven for thinking that Brainerd's (1978) well-documented article, "The Stage Question in Cognitive Developmental Theory," together with the 44 open peer commentaries arising from it, would have closed the debate on this subject for several years. Apparently this is not the conclusion arrived at by the organizers of the present workshop on human development, and neither is it the conclusion to which I have come. I experienced a certain uneasiness reading his article. It is as if the stage problem was never really dealt with. Although the article is concerned with precise experimental criteria and results, these criteria and results are not situated within a broader context. In particular, it should have been necessary to clarify the epistemological options that led Piaget to pose developmental problems in the terms we know. I feel we should take a step backward to establish a certain distance from the stage problem in psychology. For Piaget, for example, the stage problem is omnipresent and the criteria he tried to clarify a posteriori are already to be found in the premises of his theory. Consequently, these are not necessarily the criteria to set us on the right track. It is as if Brainerd, in his article, had omitted redefining the object of study of developmental psychology in favor of a discussion too exclusively centered on criteria for stages as they were established by Piaget. It is for this reason that I would like to use a different route to approach the problem concerning us today.

Preliminary Questions

The object of psychology is the study of behaviors. These may be defined as the means of exchange between the subject and his or her environment. More specifically, developmental psychology is concerned with the problem of *the acquisition of new behaviors,* in other words, with new modes of interaction or exchange

---

*I would like to thank Claude-Alain Hauert for very helpful comments. This paper was prepared for the Workshop on Stage and Structure in Human Development, Tel Aviv University, October 1–7, 1983, and is translated into English by Peter Coles.

between the subject and his environment. Based on this, I would like to raise a certain number of questions:

- Are there steps in the acquisition of a new behavior?
- Are these steps the result of an arbitrary partitioning or do they manifest qualitative changes?
- Do these steps have characteristics common to the acquisition of a more general class of behaviors?
- Is the acquisition of a new behavior carried out by passing from simple to complex, from elementary to composite, from noncoordination to coordination, from local to global, etc., as it is more or less explicitly assumed?
- Does the child's acquisition of a new behavior depend principally on:
    - the characteristics of the environment, i.e., confrontation with a new task or situation and the manner in which it is confronted?
    - previously acquired behaviors?
    - the characteristics or resources of the organism, such as age and degree of maturity of the organism?
- Do the characteristics or resources of the organism only define prior or prerequisite conditions, or do they also determine the order and sequence of this construction (as seems to be manifested by the laws of maturation of the nervous system: cephalo–caudal, proximo–distal and antero–posterior development)?

For me, it seems desirable to take, as a point of departure, the problems posed by the acquisition of a new behavior—in our case, learning to read—in order to see whether the problems encountered are peculiar to this behavior or whether they may be found more generally in the acquisition of other behaviors, whether at the same or at other periods in development. We will therefore try to elucidate the necessary conditions for the acquisition of new behaviors and the principal steps in these acquisitions. In this indirect manner we hope to be able to make some remarks on the general problem of stages.

Before entering into details, it will once again be useful to make one or two general remarks.

Justification of an Approach

In my opinion, developmental psychologists would do well to give up an approach to behavior which is *exclusively* analytical or fragmentary in favor of a reconsideration of the major acquisitions of childhood: reaching, walking, speech, language, writing, reading, arithmetic, etc.—behaviors which fascinated psychologists at the turn of the century. If we look critically at the experimental situations which have been devised over the past few decades, we will notice that they present extreme simplifications and impoverishments of reality. Striving to be more experimental, they turn out to be more and more fragmentary. This does not imply any opposition on my part to strict experimental approaches, but we should be conscious of their limitations as well as of their advantages. Above all, it is necessary to *place studies of specific problems within broader contexts*. Without the broader contexts, these simplifications discard an important part of the general problems to be solved by the child and denature our very conception of development. This position has been adopted by some authors over the past few years, and I am among them. Psycholinguists in particular have recently been led to "recognize the limited, even inappropriate, nature of the recourse to isolated phrases to study both the emergence and development as well as the functioning of linguistic operations in the adult" (Hupet & Kreit, 1983).

Among the child's major acquisitions during his or her development, certain of them, such as reaching and speech, are acquired more or less "naturally," i.e., without systematic adult intervention, whereas others, such as reading, writing, and arithmetic, require the intervention of relatively well-defined pedagogical practices. Now, in their concern to control the role of the environment and, in particular, pedagogical practices, developmental psychologists have often avoided the study of behaviors learned at school, preferring to leave this to educational psychologists (division of labor, you understand!). We are all convinced that there are no spontaneous nor natural acquisitions which would occur without environmental intervention; we are convinced to such an extent that psychologists (and I am one of them) have preferred to make developmental studies of those behaviors where the precise role played by the environment is unknown, (i.e., not subjected to strict school learning) rather than studying behaviors where the role played by the milieu (school, in particular) in their acquisition is more precisely defined. Other factors have also contributed to this state of affairs,—for example, the prejudices and criticisms relating to school, with the more or less explicit idea that school plays a rather negative role in the cognitive development of the child—as if it impeded his "natural" capacities or at least did not sufficiently develop his potential. Of course I agree that schooling may, in some cases, impede the creative and imaginative potential of the child. But it is also important to recognize the valid basis of certain pedagogical practices which are the result of knowledge acquired over many years of experience. This is particularly the case regarding the ages at which certain activities are taught. The teaching of reading and writing, which takes place in our cultures when the child is 6 years old seems to me precisely to be founded on a profound knowledge of the child's competencies.

The questions I would like to raise are: How do developmental psychologists justify the age of six years to begin teaching reading and writing in an alphabetical system? Is this choice to teach a new behavior simply a convention which does not take account of the competences of the child? If it does take these competencies into account, are these principally cognitive competencies? Should the maturation of the nervous system be considered as a determinant condition? And if we talk about intellectual maturation, what exactly does this entail? Similarly, how do psychologists account for this "disposition" or "readiness" to learn a new behavior, specifically this "readiness to read" (cf. Breckinridge & Vincent, 1965, cited by Siegler & Klahr, 1982)?

From my point of view, this question of disposition—of prerequisites—for the acquisition of a specific behavior is at the very heart of the stage problem. Siegler and Klahr (1982) introduced their article entitled "When Do Children Learn?" in a similar way. Effectively, they considered that the concepts of "readiness," "critical periods," "stages," and "matching" ("an appropriate match between the circumstances that a child encounters and the schemata that he has already assimilated into his repertoire," according to Hunt, 1961, cited by Siegler & Klahr, 1982, p. 127), constitute different ways of approaching one and the same problem—learning new behaviors. I would like to point out that for me the problem is taking account not only of assimilated schemata but also the internal maturation or transformation of instruments of knowledge (or cognitive capacities, or cognitive competencies) where the influence of the milieu would only be nonspecific (Mounoud, 1979/1981).

The Acquisition of a New Behavior: Learning to Read

Boundaries of the Problem: Teaching in the Strict and in the Broad Sense

I would like to show that *there exist conditions related to the development of cognitive competencies in the child which are able to justify teaching reading in the strict sense, within an alphabetical system, around 6 years of age.* I do not wish to imply that this instruction may not be preceded by other teaching but that, in this case, it is not teaching alphabetical reading in the strict sense but rather other activities more or less directly related to reading which we will call *reading in the broad sense.* This, of course, includes many activities developed by the child from his earliest years (recognition of visually and auditorily perceived forms, development of spatial and temporal organizing capacities through a diversity of activities, such as constructing games, singing, dancing, etc.). Some authors even talk of games *without any relationship* to reading itself (Alegria & Morais, 1979). As can be see, the notion of prerequisites is very imprecise. It can be understood in its narrow sense or, at the other extreme, in a very broad sense, where recursively it may include the quasi-totality of the child's previous acquisitions.

The same applies to alphabetic writing, the teaching of which *in the strict sense* generally begins at about the same age but is preceded by a whole series of activities—especially graphic activities—which the child has been practicing for several years and which may be considered an integral part of learning to write *in the broad sense.* Some French researchers (Auzias, Casati, Cellier, Delaye, & Verleure, 1977) who wrote the book *Writing at Five Years?* came to the following conclusions:

> We can envisage an intitiation to writing at *around six years of age.* In fact, it is only at age 5;9 that the majority of children in the first grade manage to copy legibly and with relative ease—even those who are given daily practice in copying a short phrase. As a result, it is *pointless, even deleterious,* not to respect the *normal rhythm* of child development. . . . On the other hand, exercises *in preparation* for writing seem to us to be *indispensible* in this first grade. By this we mean all those language activities, or rather those occasions for language, all the motor and graphic activities (e.g., painting,

> drawing, picture stories, etc.) which prepare the child for story telling, for symbolic representation, for handling instruments and which apply his eye and hand to the creation of forms. (pp. 40, 131)

Nevertheless, Auzias et al. (1977) do not think it would be very useful to "systematically initiate the first-grade child in "pictographics" even in simplified form, to familiarize him with conventional (figurative) transition signs for translating language. Our writing system is not ideographic but phonographic, using alphabetic representation; the signs in this system (letters) have absolutely no direct symbolic value expressed by their form" (p. 164).

Prerequisites for Learning to Read

As far as reading is concerned, the question of prerequisites or readiness to learn was reformulated in an interesting way by Liberman and his colleagues. (Liberman, Shankweiler, Fischer, & Carter, 1974; Shankweiler & Liberman, 1976; Liberman, Shankweiler, Liberman, Fowler, & Fischer, 1977) and was taken up again and developed by Alegria and Morais (1979). At the beginning of their article, Alegria and Morais review the principal causes which have been implicated in reading difficulties: letter perception, the perception of sounds, and letter-sound association. According to these authors, reading difficulties are not automatically associated with the defective functioning of one or more of these abilities. They point to a number of studies which show that children with reading difficulties do not have difficulties in perceiving letters and sounds any more than in audiovisual integration (grapho–acoustic association) or spatiotemporal association. These capacities certainly constitute prerequisites for learning to read—in the broad sense they are necessary conditions; in the same way, a certain level of mastery of spoken language is a prior condition. But these prerequisites are all satisfied in the 6-year-old child, including those who present difficulties in learning to read.

On the other hand, according to these authors, learning to read within an alphabet system does presuppose the *capacity for explicit analysis (or segmentation) of speech in terms of phonemes*. Now this capacity appears in the child at around 6 years of age. It increases rapidly at the onset of reading instruction and seems to be an important *accelerator* of this ability. It is present in a small percentage (17%) of 6-year-olds after three months of primary schooling but reaches a high percentage (70%) by the beginning of the second year of primary school. The question raised by these authors is whether the improvement of this ability is a direct result of a normal process of intellectual maturation or whether it is mainly the result of training in reading. Morais and Alegria provide an answer to this question on the basis of results obtained in a number of studies: One study compared the performance of children of the same age (Mean age of 6;6 years) in tasks involving phoneme segmentation and phoneme fusion after 3 or 6 months of primary schooling. After 3 months of primary school, they obtained scores of 17% of successful phoneme segmentation and 29% of successful phoneme fusion; after 6 months, they obtained scores of 37% and 57%, respectively, for each of these tasks. The authors therefore conclude that instruction in reading or schooling has a *net accelerator effect* on the capacity for phonemic segmentation. A second study concerning phonemic segmentation ability in illiterate and literate Portuguese adults (Morais, Cary, Alegria, & Bertelson, 1979) shows that the performance of adult illiterates is similar to that of Belgian children in the first year of primary school (6;6 years), following 3 months of schooling. The performance of the literate adults was equivalent to that of children starting their second year of primary schooling.

The first conclusion which Alegria and Morais (1979) draw from these studies I find a little disconcerting: "Simple (intellectual) maturation does not lead man to an awareness of the existence of phonetic units in speech" (p. 265). I must admit that I do not understand how intellectual maturation, in the absence of specific exercises, could ever lead to these units. The second conclusion is that conscious analysis of speech on a phonetic level is not a prerequisite for learning to read, meaning by prerequisite, "a necessary and preliminary competence." On the other hand, they evoke an *underlying or potential* capacity which would be *necessary* for the phonetic analysis of speech, but which would require confrontation with a problem demanding this analysis—such as reading—in order to emerge. This would be a good example of what I would be inclined to call a *prerequisite in the strict sense*. The authors finally point out that we do not know what the (intellectual) maturation process for this potential capacity may be, a capacity which could, according to them, be present well before 6 years.

At this point two essential questions arise: What is this underlying capacity not defined by Alegria and Morais? Is this underlying capacity present before 6 years? I am going to use these questions as a basis for formulating some hypotheses which will then serve as a guiding thread for the remainder of my discussion:

1. This underlying capacity is a *general capacity for segmenting a whole into abstract units*.
2. This capacity is the corollary of *the general capacity for integrating elements into a whole* whose meaning and properties are not reducible to the sum of the properties of its constituent elements. In fact, in order to be able to divide a whole into parts, one must be able to define it and to compose it.
3. The units resulting from segmentation are not of the same nature and do not have the same status as the constituent elements of the whole.
4. The general capacities for segmentation and integration are not constructed in the way Piaget conceived them to be but are brought into play at different stages in development, for example, at around 6 years of age with respect to the conceptual elaboration of behaviors (2 to 10 years), around age 6 months for the "perceptual" elaboration of behaviors (0 to 2 years) and around 14 years for the "semiotic" elaboration of behaviors (see previous discussion).
5. These capacities only appear in relation to the realities with which the child has been confronted in the years or months preceding their implementation (and only if the situations or tasks proposed or encountered demand it).

As a function of these hypotheses, I will first try to show that the capacity underlying the *phonemic segmentation of a word* is this general capacity for segmenting a whole into abstract units, which would not appear until around age 6 years. Next, I will examine several kinds of segmentation, emphasizing the particular nature of the one appearing at 6 years. Then I will analyze the different kinds of sentence *segmentation*, once again trying to show the radical change produced at around 6 years. In this way, I hope to show that the prerequisite revealed with respect to learning to read is not specific to this particular behavior, but also intervenes at the same age in other areas such as sentence comprehension. Then, still generalizing, I will try to show that this same underlying capacity for segmenting a whole, and its corollary, the capacity for integrating elements into a whole, also constitutes prerequisites allowing the child to define simple physical tools as homogeneous wholes and not as sets of juxtaposed elements.

This approach aims to demonstrate the steps in the acquisition of a new behavior (reading, sentence comprehension) and to show that these steps are determined by important and general qualitative changes in the organisation of the child's behaviors.

Having demonstrated the existence of steps (phases or levels) in the acquisition of new behaviors principally related to new processing capacities becoming functional at certain ages, we will mention briefly and finally how the appearance of new coding capacities (at birth, at 2 years, and at 10 years) determines the existence of the major general stages.

Regarding the Capacities for Segmentation

Word Segmentation

The purpose of the previous examination of the problems of prerequisites for learning to read allowed us to take a more general look at segmentation or the analysis of activities as prerequisites to a reading activity. Activities of segmenting or fractioning speech are carried out by children younger than 6 years, especially syllabic segmentation which is achieved by 46% of 4-year-olds (4;10) (Liberman et al., 1974). Therefore, we are not dealing with a general impossibility for children under 6 years to fragment or decompose an auditorily (or visually) perceived continuum from a temporal (or spatial) point of view. These capacities for syllabic segmentation would explain why Japanese children learn, without systematic instruction, to read the *katakana* before entering school (Sakamoto & Makita, 1973). It would also explain how Rozin and Gleitman (1977) managed to teach children who experienced difficulties with the alphabet to read syllabic writing without any notable effects on their capacity to read with the alphabetic system.

Syllabic segmentation seems to me to be possible, because it is based upon units (elements or segments) which can have a reality of their own and may sometimes have their own meanings for the child, independent of the totalities into which they may be placed. In contrast, phonemic segmentation is based on *units,* which have been described as *abstract* or *formal* and which have no existence or meaning independent of the whole of which they are a part. They can only result from breaking this totality into parts and have no existence outside these totalities. Expressed differently, the phoneme would only exist as a *part* of a whole and would not be accessible to 3-, 4-, or 5-year-olds, whereas the syllable could exist on its own, independent of any larger entity which might include it as a segment. This does not mean, however, that the syllable has the same signification or status when it is identified as an isolated entity as compared to when it is a part of a whole. These distinctions are, of course, only relative and correspond to what may be considered the subject's point of view at different steps in development. Liberman has shown that children can break words into syllables from the age of 4, while a more recent study by Bellefroid and Ferreiro (1979) shows that the syllable becomes a *part* of a word (with a defined position with respect to other parts) from the age of 6 years (amazing coincidence!). As a part of a word, the syllable no longer has its own meaning and is thus *defined with respect to the ensemble of word parts,* in particular, by its relative position. Thus, during development, the syllable can have two fundamentally different statuses: (a) before 6 years it would have the status of *independent unit* (inseparable from meaning) which may be regrouped or juxtaposed with other syllables; (b) after 6 years it would have the status of *part of a word,* but essentially defined in terms of its relative position.

We have shown that segmenting activities were performed by children under 6 years, but that segmentations performed before and after 6 years did not have the same status. From this point of view, it seems possible to consider the capacity for segmenting a whole into abstract units as the major prerequisite for learning to read in the strict sense.

Sentence Segmentation

In another branch of psycholinguistics concerned with the comprehension of sentences (or phrases) (cf. Bronckart, Kail, & Noizet, 1983, for a review), different strategies have been revealed for segmentation or fractioning. These strategies for segmenting sentences are based on different indices: semantics, positionals, and formals (morphosyntactics). The passage from pragmatic strategies (based on the semantic characteristics of lexemes or on their proximity) to formal strategies (based on the relative positions of lexemes or on morphosyntax as well as semantic characteristics) also takes place around 6 years of age, with, of course, important variations depending on the kinds of expressions used (Sinclair & Ferreiro, 1970).

Now the question is to know what it is that distinguishes these different varieties of segmentation or these different levels of analysis of words and phrases (syllabic and phonemic; pragmatic and formal) and, above all, whether they are specific to language. Is it possible that clearly distinct capacities exist for segmentation and composition before and after the age of 6? If so, to what competencies would they belong? As far as phonemes are concerned, we should point out that Gleitman and Rozin (1977) consider the segmentation of speech into phonemes as being extremely abstract. These segments would only be very indirectly represented in the sound wave. In contrast, syllabic segmentation could be described as more concrete. Thus we would have a passage from concrete to abstract segmentations, or

Table 1. Different Varieties of Segmentation of Words and Sentences.

Elaboration of the Word or Verbal Sign		
Syllabic segmentation ("concrete"): The syllable is defined by its semantic characteristics	or	Syllabic pseudo-segmentation: The syllable is an independent unit which may be found in relationships of proximity with other syllables
6 Years		
Phonemic segmentation ("abstract"): Phones are defined by their relative positions in the acoustic wave	or	True syllabic segmentation: The syllable is a part of the word it is defined by its relationships with the ensemble of constituent parts of the word
Elaboration of the Sentence		
Pragmatic segmentation: Based on the semantic characteristics of lexemes or on their spatial proximity		
6 Years		
Formal segmentation: Based on the relative positions of lexemes or on morphosyntax		

from pragmatic to formal. These qualifiers have a familiar ring for developmental psychologists, although they are usually used to distinguish distinct stages, whereas here they characterize periods or steps within a stage.

I think that these two categories of segmentation are radically different, as I have already suggested, and that they depend on important qualitative changes at the level of cognitive instruments or cognitive capacities. For me, these changes would not occur before 6 years of age as far as the developmental stage between 2 and 10 years is concerned. They would depend on a process of maturation of the neural system that is relatively insensitive to environmental influences.

For certain psychologists (Boutet, Gauthier, & Saint-Pierre, 1983) this transition characterizes the passage from linguistic activities to metalinguistic activities defined as "the *capacity to consider language as an object* upon which logical operations may be performed, such as segmentation, grouping, and serial positioning" (Bellefroid & Ferreiro, 1979, p. 33).

In Karmiloff-Smith's (1979*a* and *b*) studies on determinants in language acquisition, this cutoff at 6 years would correspond to the passage from surface markers to an organization based on deep structures (formal procedures capable of fulfilling different functions).

Using examples borrowed from psycholinguistics, I have tried to show how, after 6 years of age, new means for segmenting words or sentences can be explained by a general capacity for segmenting a whole into abstract units being brought into play. I would now like briefly to show how this segmenting capacity manifests itself in relation to the way the child is able to define simple objects, such as tools, in problem-solving situations. This example will also allow us to demonstrate the capacity for integrating elements into a whole (the corollary of the former). Finally, we will mention two other research areas where the age of 6 years is also typical of a change in organization comparable to those we have defined.

The Construction of Tools: Two Levels of Analysis

Regarding this passage from a pragmatic or concrete organization to one which is formal or abstract, occurring at around six years, I would like briefly to summarize the results of some research carried out on the construction of simple tools by 4- to 8-year-old children (Mounoud, 1970). In these studies I demonstrated two levels of analysis and resolution of problem situations which were radically different. These two levels seem to me to bear a close correspondence to the two levels of organization I just discussed regarding the comprehension of words and sentences.

A *primary level* of analyzing problem situations and of defining tools, typical of 4- and 5-year-old children, is based on a decomposition of the problem into tasks or elementary actions/properties: for example, in tasks involving reaching, reaching around, pushing, and seizing. These actions are used to define or qualify different segments or pieces of a tool. Each segment thus has its own property (or constituting function): reaching, reaching around, pushing, taking, grasping, hooking, etc. This kind of segmentation could easily be called pragmatic or semantic as have the first strategies for understanding phrases (or sentences), also typical of 4- and 5-year-olds.

At the *second level,* tools are defined by a general function or by global transformation progressively specified by the relationships between different parts of the tool and of the situation. These parts only have meaning in relation to the whole, and these relationships are referenced to a signification for the entire tool. An example of one of these tools would be one designed to move a wood block while avoiding obstacles or to remove a block with a hook on it from a jar. The defining properties of the tools are at the level of relationships such as spacing, length, inclination, or curvature of different constituent parts. *These parts are no longer defined in isolation by means of specific properties, but, rather, they are defined by their mutual relationships.* Such a conception could perfectly well be described as morphological, formal, or abstract, and could therefore correspond to the formal or abstract capacities for analyzing and segmenting words and phrases. The second level is also characteristic 7- and 8-year-old children.

Segmentation or composition of tools is possible at both levels, but they are completely different in nature. It is as if, for the 4- and 5-year-old children, *the tool is gradually defined by juxtaposition of segments or pieces,* each one having a defined property or direct relationship with the child's different actions (pragmatico-semantic aspect). This is reminiscent of the figural collections defined by Inhelder and Piaget (1959) in the realm of classifications. In contrast, at the second level for 7- and 8-year-olds, the tool is defined by a global transformation relative to certain constraints and conditions inherent in the situations. *The tool is defined as a whole composed of parts* (which only have meanings with respect to the whole) for which only the structural relationships between parts give it its function.

The features of the tool, of the situation, and of the actions taken into consideration by 4- and 5-year-olds or 7- and 8-year-olds do not have the same status, despite how they might appear or, especially, despite our adult observer's point of view. Both cases might deal with "length," for example, but what the 4- and 5-year-old calls long or short will not have the same meaning as that for the 7- and 8-year-old. In the first case, with 4- and 5-year-olds, we would be in the presence of what Piaget called preconcepts, where the object and what it signifies are not clearly dissociated. For these children objects or instruments are characterized by isolated properties which may be juxtaposed to make up what we might call "amalgams," in reference to the notion of "amalgamated predicate" defined by Wermus (1976, 1977). Object properties are directly dependent on the meaning of actions performed on them or for which they substitute. Objects represent or stand for actions. They are a kind of transposition of them, an analogue translation, a *substitute*. Their definition will depend on the presence or absence of this or that segment to which a particular meaning is attached and where the whole is not taken into account.

With 7- and 8-year-old children, the instruments become the *support* for meanings attributed to the whole, which are no longer relative to such and such particular actions but to one or several transformation(s) of the whole. The instruments are no longer defined by the presence or absence of this or that isolated characteristic but uniquely by the relationships between their different parts.

Once again the age of 6 years constitutes the transition between these two levels of organization. It is at around 6 years that the integration of previously isolated and juxtaposed properties takes place, which eventually give rise to wholistic meanings enabling transformations to be considered. I noticed that when the children succeeded in defining an instrument by means of a holistic property, they momentarily no longer succeeded in constructing or modifying an instrument. They were only satisfied by the discovery of an instrument which had the whole set of anticipated characteristics.

Before 6 years it would be possible to say that objects do not exist "conceptually" for the subject as wholes. Their sole conceptual existence would be linked to the current or previous actions associated with them and for which they act as a kind of extension or substitute. The objects would only have partial, local and momentary "conceptual" identity. We will not discuss here the different forms of perceptual identity.

From (about) the age of 6, objects become identifiable in a stable and global way, and they have acquired a global qualitative identity (to use Piagetian terminology) without the relationships between parts of the object or between different objects yet having been mastered. Their identity no longer depends upon current contingencies of the action. In other words, the objects become permanent. This permanence, in terms of conceptual elaboration, is still limited by the degree of organization of the relationships the child is capable of mastering, both between the constituent parts of the object and between different objects.

It is noteworthy that we have found this critical age of 6 years in our research on the development of the self-image in the child from 3 to 11 years of age (Mounoud

& Vinter, 1981b, in press). This research studied the precision and stability of the child's image of his or her own face using a distorting mirror. In particular, we are studying the way in which children are affected by their initial confrontations with distortions of their face. Our findings were that, at age 6, children had a precise and faithful representation of themselves.

Similarly, in our research on the planning and control of movements, the study of visuo-manual tracking of periodic signals in 3- to 9-year-old children (Mounoud, 1982; Mounoud, Hauert, Mayer, Gachoud, Guyon, & Gottret 1983; Mounoud, Viviani, Hauert, & Guyon, in press) the age of 6 years marks the passage from *local control* to *global control* of movement, which we interpret as the child's capacity for anticipating the to-and-fro' movements of the target and their arm movements as a totality and not locally, step by step, in a way which then permits the resolution of the problem of coincidence between target movements and their own movements.

From the outset of this text I have tried progressively, using different contents (word, phrase, instrument, face, movement) to define the articulation between two levels of behavioral organization or object elaboration, that is, two qualitatively distinct levels of organization of exchange between subject and environment. These two levels would characterize an *invariant developmental sequence* which defines the internal dynamics of each stage. In this way the principal articulation in the elaboration of new behaviors within each stage would be determined fundamentally by general segmenting and integrating capacities becoming functional. The changes that occur are changes in *processing capacities*.

Our analysis so far fits into a general conception of the psychological development of child behavior which should be outlined briefly before concluding. Within this conception, stages are defined by changes in *coding capacities*, whereas changes in processing capacities determine the steps, levels, or revolutions within each major stage.

Model for the Construction of Representations

The construction of ''new'' structures is performed, according to Piaget, by the process of reflective abstraction or convergent reconstruction with overtaking. This process brings about the construction of structures which might be qualified as formal, in the sense that they are more or less independent of the contents of the objects to which they are applied.

As I have already written elsewhere (Mounoud, 1979/1981), Piaget's psychogenetic theory regarding the construction of structures appears to me to be extremely vulnerable to criticism. From my point of view, what the child constructs are not structures but *representations or internal organizations of contents*. The child arrives at them by means of structures, which I feel are more economically considered as preformed. Now it is possible to reinterpret the process of reflective abstraction as a *process of construction of representations* and no longer as a process of construction of logical structures (internalization of general coordinations of action). This is what I have been trying to do in my analysis of the child's construction of simple tools (Mounoud, 1970) and which I have progressively generalized and transposed to the sensorimotor period (Mounoud, 1971, 1973, 1976/1982, 1983).

Piaget distinguished two phases or types of reflection in the process of reflective abstraction, which are repeated at each developmental stage, giving rise to new structures. The first is a primary reflection in the physical sense of the term, consisting of transposing the elements (action schemata or operations) of the previous structure onto another plane. The second reflection is in the psychological or mental sense, and it consists of a "*reshaping through thought* of previous material presented in its raw or immediate state" (Piaget, 1967, p. 366).

Within the perspective I have been developing over several years, these two phases consist of the translation, by means of a new code, of the different contents with which the child interacts (with objects, people, or his own body). It is necessary to distinguish several types of codes giving rise to representations on different levels. These can be found in Table 2.

It is the appearance of new coding capacities, which I have suggested calling perceptual at birth, conceptual around 2 years, and semiotic or formal around 10 years, which determine the stages of cognitive development of the child. These new coding capacities constrain the child to reorganize, to redetermine his behaviors, and not, as Piaget insisted, the achievement or the closing off of a new structure. An early revolution consists of dissociating the properties of actions and of objects

Table 2. Codes and Representations on Different Levels.

Stages	Levels of Representations	Type of Organization
Birth	Sensorial representations (preformed) linked with preformed structures + new coding capacities (the perceptual code) ⇒ Construction of perceptual representations	Sensorimotor Organization (preformed)
18–24 Months	Perceptual representations (constructed) + new coding capacities (the conceptual code) ⇒ Construction of conceptual representations	Perceptivomotor Organization (constructed)
9–11 Years	Conceptual representations (constructed) + new coding capacities (the semiotic code) ⇒ Construction of semiotic representations	Conceptuomotor Organization (constructed)
16–18 Years	Semiotic representations (constructed)	Semioticomotor Organization (constructed)

(represented in the previous organization) in order to constitute new elementary representations. The principal characteristics of these new representations are that they are partial, local, and juxtaposed. They would also be of the analogous type. This first revolution would correspond to the first physical type of reflection described by Piaget). The reorganization comes to an end with the integration of these elementary representations into *new or global representations* which are momentarily rigid and nondecomposable, nonsegmentable. The second revolution consists precisely in the decomposition of these total representations into parts or units in such a way as to establish relationships between their components. It is not necessary to explain the correspondence between these two revolutions and the two levels of functioning or elaboration of objects described in the first part of this text.

Illustration of the Model in the Stage of Perceptuomotor Organization

It is possible to demonstrate this developmental sequence during the first two years of life. A transition comparable to that defined around age 6 occurs at around 6 to 9 months. It is at this moment that children are capable of defining objects in a singular or individualized way on the basis of the collection of properties accessible to them.

Let us briefly describe this sequence using data concerning knowledge manifested by the baby regarding his body, in particular, his arms and his hands. This analysis will draw mainly on the example of reaching for objects. It is precisely around 24 to 32 weeks that the baby is capable of correct organization of his arm and hand movements in order to seize objects within his reach. The movement is usually described as visually triggered or ''ballistic,'' i.e., programmed before its execution, the hand opening during the course of the trajectory, closing again on the object. Before and after this age, it seems possible to distinguish two levels of organization comparable to those defined earlier.

Before 24 weeks or, more precisely, between 2 and 5 months, infants present a whole series of behaviors showing the way in which they are progressively discovering the different properties of their hands and arms. Hand opening and closing, wrist rotation, extension, and flexion of the elbow, reciprocal grasping of the two hands, etc. Through these different behaviors the baby discovers and progressively controls the properties of different arm segments and will eventually be able to order these movements sequentially to carry out the complex act of reaching. This brings us to the 24-to-32-week level which White, Castle, and Held (1964) considered the ''top level'' of reaching.

After 32 weeks, this nondecomposable, nonmodulable collective activity constituting reaching gives rise to an extremely complex reorganization which consists of controlling the different parts or units of this complex activity to adapt to the different conditions under which the activity is supposed to be able to be used. This second level, which could be qualified as late evolution, has been remarkably described by Halverson (1931). We could also consider that some situations de-

Table 3. Elaboration of New Representations.

Stage 1	Stage 2	Stage 3	Steps	Process
0–1 month	1½–3 years	10–11 years	Initial global representations: Syncretic	
				Sampling of object and action properties: By means of new code
1–4 months	3–5 years	11–13 years	New elementary representations: Separated and juxtaposed	
				Coordination–integration of elementary representations: And establishment of correspondence with objects and situations
4–8 months	5–7 years	13–15 years	New total representations nondecomposable: Rigid with global relationship between them	
				Decomposition–analysis of new total representations in their components: And establishment of correspondence with objects dimensions
8–14 months	7–9 years	15–16 years	New total representations partly decomposable: With partial relationship between them and their components	
				Composition–synthesis of the components of new total representations
14–18 months	9–10 years	16–18 years	New complete representations fully decomposable: With complete relationship between them and with their components	

signed to study the development of object permanence, such as reaching for an object hidden behind a screen or a detour situation, equally characterize this late evolution (Bruner, 1970; Lockman, 1980). Adaptations to the reach as a function of different orientations of the object to be seized, studied by Ashmead, Lockman, and

Bushnell (1980; Lockman & Ashmead, 1982), as well as adaptations to the grasp as a function of the weights of objects which I have studied in the past (Mounoud, 1973; Mounoud & Bower, 1974) are other examples of the second level of organization beginning around 6 to 8 months and ending at around 16 months.

I would have liked to demonstrate in greater detail the existence of this "developmental sequence" at this period of development. This has been done in two recent articles (Mounoud, 1983; Mounoud & Vinter, 1981a).

In order to get a better understanding of the significance of this sequence, it is necessary to take into consideration the behaviors manifested by the baby during the first weeks of life. Several authors (Bower, Broughton, & Moore, 1970*a* and *b*; Trevarthen, Hubley, & Sheeran, 1975; von Hofsten, 1982) have demonstrated that the children are, during the first days of life, capable of a kind of reaching which is quite surprising: They are able to project an arm in the direction of a visually perceived moving object. The newborn's reaching behaviors manifest a processing of certain categories of information related to the situation, his or her actions, and the objects. These data clearly show (if it was necessary to do so!) that there exist different levels of organization and control of the child's behavior. The passage from one organization to another, or what amounts to the same thing, the construction of a new organization of behaviors (i.e., interactions or exchanges between the subject and his environment) defines a stage. The developmental sequence which I have tried to set out defines, in gross terms, the fundamental steps in these successive reorganizations.

Conclusion

The conception presented here is in favor of describing developmental stages that are common to all areas of knowledge. I am aware that this point of view goes against some current conceptions which reject the idea of unified stage encompassing all areas of knowledge (cf. especially Bates, 1979). This conception ascribes a more important role to the process of neural maturation and the biological substrates of behavior which determine the origin of the steps in this developmental sequence. The maturation of the neural system itself depends on the nature of the interactions of the organism with the environment but in a nonspecific way these interactions at the most being able to accelerate or slow down the process. On the other hand, the environment plays a specific role in cognitive development in relation to the contents of experience and the exchanges of subjects with their environment, and more precisely, the realities with which they have been confronted. These experiences will directly determine the contents of the representations elaborated.

References

Alegria, J., & Morais, J. (1979). Le développement de l'habileté d'analyse phonétique consciente de la parole et l'apprentissage de la lecture [The development of conscious phonetical analysis of words and reading proficiency]. *Archives de Psychologie, 47,* 251–270.

Ashmead, D. H., Lockman, J. J., & Bushnell, E. W. (1980, April). *The development of anticipatory hand orientation during infancy*. Paper presented at the International Conference on Infant Studies, New Haven, CN.

Auzias, M., Casati, I., Cellier, C., Delaye, R., & Verleure, F. (1977). *Ecrire à cinq ans ?* [Writing at 5 years?]. Paris: Presses Universitaires de France.

Bates, E. (1979). *The emergence of symbols: Cognition and communication in infancy*. New York: Academic Press.

Bellefroid de B., & Ferreiro, E. (1979). La segmentation du mot chez l'enfant [Word segmentation in children]. *Archives de Psychologie, 47,* 1–35.

Boutet, J., Gauthier, F., & Saint-Pierre, M. (1983). Savoir dire sur la phrase [Explicit knowledge of the structure of sentences]. *Archives de Psychologie, 51,* 205–228.

Bower, T. G. R., Broughton, J. M., & Moore, M. K. (1970a). The coordination of visual and tactual input in infants. *Perception and Psychophysics, 8,* 51–53.

Bower, T. G. R., Broughton, J. M., & Moore, M. K. (1970b). Demonstration of intention in the reaching behavior of neonate human. *Nature, 228,* 679–681.

Brainerd, C. J. (1978). The stage question in cognitive-developmental theory. *The Behavioral and Brain Sciences, 2,* 173–213.

Bronckart, J.-P., Kail, M., & Noizet, G. (1983). *Psycholinguistique de l'enfant. Recherches sur l'accuisition du langage*. [Child psycholinguistic. Researches on the acquisition of language] Neuchâtel and Paris: Delachaux et Niestlé.

Bruner, J. S. (1970). The growth and structure of skill. In K. Connolly (Ed.), *Mechanisms of motor skill development*. New York: Academic Press.

Gleitman, L. R., & Rozin, P. (1977). The structure and acquisition of reading. 1: Relations between orthographies and the structure of language. In A. S. Reber & D. L. Scarborough (Eds.), *Toward a psychology of reading*. New York: Wiley.

Halverson, H. M. (1931). An experimental study of prehension in infants by means of systematic cinema records. *Genetic Psychology Monographs 12,* 107–285.

Hupet, M., & Kreit, B. (1983). L'articulation d'informations "connues" et "nouvelles" dans le langage d'enfants de 3 à 12 ans [The articulation of "given" and "new" information in the speech of 3 to 12 year-olds]. *Archives de Psychologie, 51,* 189–204.

Inhelder, B., & Piaget, J. (1959). *La genèse des structures logiques élémentaires* [The genesis of elementary logical structures]. Neuchâtel and Paris: Delachaux et Niestlé.

Karmiloff-Smith, A. (1979a). *A functional approach to child language. Linguistic theories*. Cambridge: Cambridge University Press.

Karmiloff-Smith, A. (1979b). Micro- and macrodevelopmental changes in language acquisition and other representational systems. *Cognitive Science, 3,* 91–118.

Liberman, I. Y., Shankweiler, D., Fischer, F. W., & Carter, B. (1974). Reading and the awareness of linguistic segments. *Journal of Experimental Child Psychology, 18,* 201–212.

Liberman, I. Y., Shankweiler, D., Liberman, A. M., Fowler, C., & Fischer, F. W. (1977). Phonetic segmentation and reading in the beginning reader. In A. S. Reber & D. L. Scarborough (Eds.), *Toward a psychology of reading*. New York: Wiley.

Lockman, J. J. (1980). *The development of detour knowledge during infancy*. Unpublished doctoral dissertation, University of Minnesota.

Lockman, J. J., & Ashmead, D. H. (1982). Discontinuities in the development of manual behavior. In L. P. Lipsitt (Ed.), *Advances in infancy research* (Vol. 2) Norwood, NJ: Ablex.

Morais, J., Cary, L., Alegria, J., & Bertelson, P. (1979). Does awareness of speech as a sequence of phones arise spontaneously? *Cognition, 7,* 323–332.

Mounoud, P. (1970). *Structuration de l'instrument chez l'enfant* [Structuration of tools in children]. Neuchâtel and Paris: Delachaux et Niestlé.

Mounoud, P. (1971). Développement des systèmes de représentation et de traitement chez l'enfant [The development of representational and processing systems in children]. *Bulletin de Psychologie, XXV, 5–7,* 261–272. Trans. in B. Inhelder & H. Chipman (Eds.), *Piaget reader*. New York: Springer Verlag, 1976.

Mounoud, P. (1973–1974). Les conservations physiques chez le bébé [Physical conservations in infants]. *Bulletin de Psychologie,* XXVII, *13–14,* 722–728.

Mounoud, P. (1976). Les révolutions psychologiques de l'enfant [Psychological revolutions during childhood]. *Archives de Psychologie, 44,* 103–114. Trans. in T. G. Bever (Ed.), *Regressions in development: Basic phenomena and theoretical alternatives.* New York: Erlbaum, 1982.

Mounoud, P. (1979). Développement cognitif: Construction de structures nouvelles ou construction d'organisations internes [Cognitive development: Construction of new structures or construction of internal organizations]. *Bulletin de Psychologie, 33,* 107–118. Trans. in I. E. Sigel, D. M. Brodzinsky, & R. M. Golinkoff (Eds.), *New directions in piagetian theory and practice.* Hillsdale, NJ: Erlbaum, 1981.

Mounoud, P. (1982, October). *Visuo-manual tracking in children from 3 to 9 years old.* Paper presented at the Symposium on the Development of Action at the Institute of Child Development, University of Minnesota, Minneapolis.

Mounoud, P. (1983). L'évolution des conduites de préhension comme illustration d'un modèle du développement [The evolution of reaching behaviors: supporting evidence for a developmental model]. In S. de Schönen (Ed.), *Le développement dans la première année.* Paris: Presses Universitaires de France.

Mounoud, P., & Bower, T. G. R. (1974). Conservation of weight in infants. *Cognition, 3,* 29–40.

Mounoud, P., & Hauert, C.-A. (1982). Development of sensorimotor organization in children: Grasping and lifting objects. In G. E. Forman (Ed.), *Action and thought: From sensorimotor schemes to symbolic operations.* New York: Academic Press.

Mounoud, P., Hauert, C.-A., Mayer, E., Gachoud, J.-P., Guyon, J., & Gottret, G. (1983). Visuo-manual tracking strategies in three to five-year-old child. *Archives de Psychologie, 51,* 23–33.

Mounoud, P., & Vinter, A. (1981a). Representation and sensorimotor development. In G. Butterworth (Ed.), *Infancy and epistemology: An evaluation of Piaget's theory.* Brighton, England: Harvester Press.

Mounoud, P., & Vinter, A. (1981b). Le développement de l'image de soi chez l'enfant de 3 à 11 ans [Development of self-image in 3 to 11 years-old children]. In P. Mounoud et A. Vinter (Eds.), *La reconnaissance de son image chez l'enfant et l'animal.* Neuchâtel and Paris: Delachaux et Niestlé.

Mounoud, P., & Vinter, A (in press). Self-image in children and teenagers as illustration of a theoretical development model. In L. C. R. Restaino, V. Shulman, & L. Butler (Eds.), *The future of Piagetian theory: The neo-Piagetians.* New York: Plenum Press.

Mounoud, P., Viviani, P., Hauert, C.-A., & Guyon, J. (in press). Development of visuo-manual tracking in 5- to 9-year-old children. *Journal of Experimental Child Psychology.*

Piaget, J. (1967). *Biologie et connaissance* [Biology and knowledge]. Paris: Gallimard.

Rozin, P., & Gleitman, L. R. (1977). The structure and acquisition of reading 2: The reading process and the acquisition of the alphabetic principle. In A. S. Reber & D. L. Scarborough (Eds.). *Toward a psychology of reading.* New York: Wiley.

Sakamoto, T., & Makita, K. (1973). Japan. In J. Downing (Ed.), *Comparative reading: Cross-national studies of behavior and process in reading and writing.* New York: Macmillan.

Siegler, R. S., & Klahr, D. (1982). When do children learn? The relationship between existing knowledge and the acquisition of new knowledge. In R. Glaser (Ed.), *Advances in instructional psychology* (vol. 2). Hillsdale, NJ: Erlbaum.

Sinclair, H., & Ferreiro, E. (1970). Etude génétique de la compréhension, production et répétition des phrases au mode passif [Developmental study of the understanding, production and repetition of passive sentences]. *Archives de Psychologie, 40,* 1–42.

Shankweiler, D., & Liberman, I. Y. (1976). Exploring the relations between reading and speech. Haskins Laboratories Status Report on Speech Research, SR-45/46, 1–16.

Trevarthen, C., Hubley, P., & Sheeran, L. (1975). Les activités innées du nourrisson [The inborn activities in young infants]. *La Recherche, 56,* 447–458.

von Hofsten, C. (1982). Eye-hand coordination in the newborn. *Developmental Psychology. 18,* 450–461.

Wermus, H. (1976). Essai de représentation de certaines activités cognitives à l'aide des prédicats avec composantes contextuelles [The representation of cognitive activities by predicats with contextual components] *Archives de Psychologie, 44,* 205–221.

Wermus, H. (1977). Essai de développement d'une prélogique à partir des foncteurs partiels [The role of partial fonctors in the development of prelogical structures]. *Archives de Psychologie, 45,* 85–100.

White, B. L., Castle, P., & Held, R. (1964). Observation on the development of visually directed reaching. *Child Development, 35,* 349–364.

3

*Structure and Process: Developmental Psychology as Looking in the Mirror**

Sidney Strauss
Tali Ephron-Wertheim
School of Education
Tel Aviv University

In today's English-language cognitive developmental literature, one gets the very distinct feeling that Piaget's structuralist model is being replaced by information-processing approaches. If the sheer number of articles inspired by one approach or the other is a guide to this, then a comparison of the experimental literature in developmental psychology over the past several years with the same literature from the 1960s through the mid-1970s surely indicates this. Shifts in what occupies the attention of researchers is part of the normal business of competition between alternative approaches to a field and, when they occur, what often happens is that the approaches are seen in either/or, us/them terms. In addition to this adversarial relation, different models of cognitive development, such as structuralist (Piagetian) and qualitative information-processing (Siegler's) approaches, give expression to different tasks and experimental methodologies. These methodologies include different kinds of apparatus and procedures, or sometimes the same apparatus is used but different procedures are employed for administering them.

It has been claimed that schools of thought in psychology can be best characterized as competing tasks rather than competing theories of cognition and development (Fodor, 1968). Piaget's structuralist approach is a clear exception since he had a theory (some say a too-grand theory) of the mind and its development. On a note similar to Fodor's, Glick (1983) and Karmiloff-Smith (1981) argue that in the present state of cognitive developmental psychology, we are overly involved in taskology; i.e., we fret over endless issues of task context and are even developing theories of tasks. Much recent work in the developmental literature certainly has the quality of people ruminating over how changes in tasks affect children's solutions.

---

*This chapter is based on part of an MA thesis submitted by Tali Ephron-Wertheim in 1982 to the Tel Aviv University School of Education. The thesis was guided by Sidney Strauss. We would like to thank Susan Carey, Iris Levin, Robert Siegler, and Friedrich Wilkening for very helpful comments on an earlier draft of part of our paper. This chapter was written while author Strauss was a Visiting Scientist in the Department of Psychology at the Massachusetts Institute of Technology during his sabbatical year. Thanks are extended to the department for the services and help provided. This is Working Paper No. 28 of the Tel Aviv University Study Group on Human Development and Education.

What is often missing in these studies are the relations between these tasks and what they might be illuminating, such as aspects of theories of the mind and its workings, models of development, and their educational meanings.

An unfortunate link between the us/them, either/or view and taskology is that often, barely beneath the surface, there are claims that a demonstration that children can solve tasks from a methodology consistent with an information-processing viewpoint shows that children are information processors and that a structuralist account cannot adequately capture their understanding of the world. We shall challenge this view. Perhaps because of the steady movement of information-processing approaches, the structuralist account is less all or none. Even in Geneva, the heartland of structuralist orthodoxy, recent attempts have been made to reconcile the structuralist approach and the information-processing approaches, as can be seen in the latest issues of the *Études,* although it should be noted that attempts were made earlier (Karmiloff-Smith & Inhelder, 1974).

We report here some findings from a study that was conducted for three purposes, only one of which is discussed in this chapter. That purpose was to empirically compare and contrast Piaget's structuralist theory and Siegler's qualitative information-processing model of cognitive development via (you guessed it!)—a task. We wanted one that was common to both Piaget and Siegler and chose the balance scale task. Children were given tasks with the procedures used by Piaget and by Siegler, and we expected to find that when given the former tasks, they would behave in accordance with structuralist expectations and when given the latter tasks, these same children would behave as qualitative information processors. The implications of this for the either/or view of cognitive development will be discussed later; however, we want to point out from the outset that we take the position that children's behaviors are constrained by the methodology one chooses, and it is in this sense that, as in the title of our chapter, some of developmental psychology is like looking in a mirror—we see reflections of ourselves.

We begin our comparison with Piaget, who views children as active explorers of their world, constructing forms of knowledge about it. Piaget has argued that these forms of knowledge are conceptual products that are constrained by the structure of logic children have constructed.

An example of this and how one can assess children's intellectual stages comes from Inhelder and Piaget's (1958) study of how children solve balance scale tasks. Here, they administered two kinds of tasks. The first task was to place, say, a 5-g weight at a certain position on one side of the balance scale. The children were then given a 15-g weight and were asked to place it on the other side in such a way that it would balance the arm. The arm was a ruler so that the children could calculate the distances at which they were placing the weights. The second task was similar to the first, except that the experimenter placed a basket with a certain number of dolls (in the case of the present experiment, we used sweets) on one side of the arm and presented children with a second basket with fewer or more dolls in it, and then asked the children to place the basket on the other side of the arm in such a way that the arm would be balanced. In the second task there were no markings on the arm so that the placement of the basket could be only approximate and could not be calculated exactly as in the first task.

Notice that in these tasks, the children are allowed to manipulate the positioning of the weights/baskets, test whether or not the arm balanced and, if not, they could alter the position to wherever they wanted until the arm balanced. Inhelder and Piaget (1958) offered a classification system of children's developmental stages that was based on strategies children used to balance the arm and the structure of logic thought to underlie them; e.g., the INRC structure of formal operations that allows, in Piaget's view, proportional reasoning which is necessary for solving balance scale tasks correctly. We shall detail Piaget's stage criteria in the Methods section.

In contrast to this view is Siegler's (1976, 1978, 1981, 1983; Richards & Siegler, 1982) model of cognitive development which portrays the child as a qualitative information processor. The picture drawn by Siegler is that of the child who is rule-governed. These rules can be described in the form of binary decision trees, although they can be described otherwise (see Siegler, 1983, in response to Strauss & Levin, 1981).

Siegler's methodology can involve a forced-choice procedure in which children do not receive feedback about the correctness of their choices. In such cases, the pan balance scale is presented to the children with blocks under both arms, and weights are placed on both sides of the arm. The children are asked to predict if the right arm will go down, the left arm will go down, or if they will be balanced. Since the blocks prevent the arms from moving, they cannot see the effects of the arrangement on the balance scale.

Siegler (1976) predicted and corroborated empirically the existence of four models of rules that children used when solving tasks of this sort. The models are found in Fig. 1. The first model has weight as the dominant variable for the child. If the number of weights are the same on both sides, Rule 1 children predict that the arm will balance, and if there are more weights on one side they will predict that that side will drop, regardless of the distances of the weights from the fulcrum. Rule 2 is like the first except that now the children take the distance into account, but only under conditions where the weights are equal on both sides. Rule 3 has the children understanding that they must take into account both the weight and the distance, but they do not have a consistent rule for their relation and so they guess in certain (conflict) cases. Rule 4 is the mature rule that allows children to multiply the weight by the distance on both sides of the arm and to judge that if the numbers arrived at are the same on both sides, the arms will balance, whereas if the numbers on one side is larger, that side will drop.

Siegler assessed these models by administering six kinds of tasks that are presented in Table 1. We shall use the example of the conflict-weight task to describe how to read this table. It turns out that the correct judgment on this task is that the side with more weight should tip down. If we multiply the weights times the distances on each side and compare them, the product on the left side is greater: On the left side we multiply 2 weights $\times$ 2 pegs (distance) $=$ 4 plus 2 weights $\times$ 3 pegs $=$ 6, giving a grand total of 10, whereas on the right side we multiply 2 weights $\times$ 4

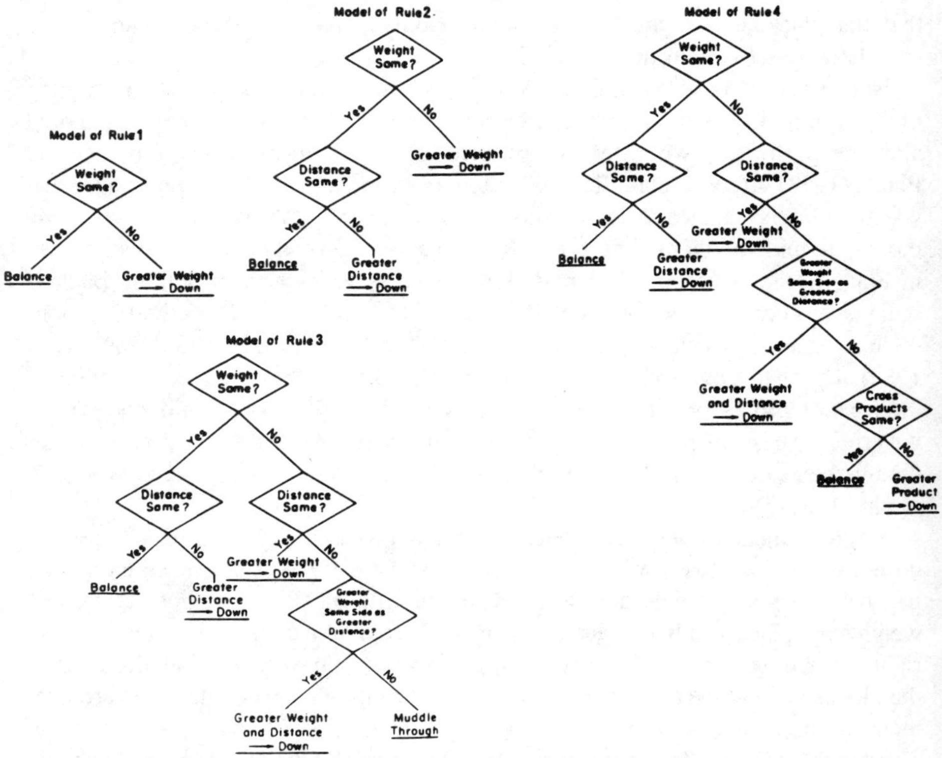

Figure 1. Rules for Solving Balance Scale Tasks

pegs and arrive at the product of 8. Since the side with the larger product falls, the left side falls in this example. The Rule 1 child compares the weights and correctly judges that the left side will fall. The Rule 2 child does the same since the number of weights on the two sides is not equal. The Rule 3 child guesses since s/he does not have a rule that adequately handles the conflict situations. And finally, the Rule 4 child solves the task correctly in the numerical manner described already.

Siegler presented children four tasks per kind so that children were administered 24 tasks in all. In order for children to be assessed as having used any one model, they must have produced the predicted judgment on 20 of the 24 tasks which gave rise to the patterns found in Table 1.

In the present study, children were given balance scale tasks via both Piaget's and Siegler's methodologies. Our first hypothesis was that we would be able to assess the development of children's reasoning according to Piaget's categories when Piaget's methodology was used and that these same children would evidence rule-governed behavior when presented balance scale tasks via Siegler's methodology. One expression of this hypothesis is that we should find age-related changes in rule and stage assessment so that younger children should evidence lower, and older children should evidence higher, rules and stages.

Table 1. Predictions for Percentage of Correct Answers and Error Patterns for Children Using Different Rules.

	Rule			
Problem-Type	**1**	**2**	**3**	**4**
Balance 	100	100	100	100
Weight 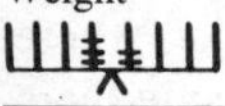	100	100	100	100
Distance 	0 (Should say "Balance")	100	100	100
Conflict-Weight 	100	100	33 (Chance Responding)	100
Conflict-Distance 	0 (Should say "Right Down")	0 (Should say "Right Down")	33 (Responding)	100
Conflict-Balance 	0 (Should say "Right Down")	0 (Should say "Right Down")	33 (Chance Responding)	100

The methodologies were counterbalanced between these methodologies while the order of task presentation within these methodologies was random. This counterbalanced design allowed us to investigate the effect of order of presentation. We reasoned that a forced-choice procedure without feedback (as is possible within Siegler's approach for the balance scale task), if presented first, should not influence children's stage levels as assessed by Piaget's methodology. There is no reason to believe that a procedure without feedback would lead to learning that could influence the stages at which children were assessed.

On the other hand, first presenting Piaget's methodology that requires exploration could influence the rules at which children were assessed. In other words, we are arguing here that children may learn about the balance scale and its functioning via Piaget's methodology, although it is clear that Piaget would argue that the limits of what they can learn are constrained by the logic of their mental structures. This learning could raise the level of the rule at which a child is assessed, and if this turns out to be the case, it could be claimed that Siegler's assessment procedure underestimates children's potential rule level. Hence, our second hypothesis was that children who received Siegler's methodology before Piaget's would be assessed as using a lower-level rule than children who received Siegler's methodology after Piaget's.

For reasons not relevant to the purpose of the present chapter, we presented three types of tasks to the children in our study: (a) small number tasks, where the number of weights on both sides of the arm was four or less; (b) large number tasks, where there were at least six weights on both sides of the arm; and (c) mixed tasks, where on one side of the arm there were four or less weights and on the other side there were at least five weights.

A general methodological point is that on certain tasks, Siegler (1976) placed weights on more than one peg on either or both sides of the arm, and it was our feeling that this could be confusing to the child. In addition, it causes a certain asymmetry between some of the tasks that Siegler presented. For example, as one can see in Table 1, on the distance and conflict-balance tasks the weights on each side are on one peg only, whereas on the other tasks the weights are on more than one peg on at least one side. We were not interested to test this problem by systematically varying the number of weights per peg and, as a consequence, in *all* of our tasks, the weights were placed on only one peg on each side of the arm.

Method

All subjects were administered two types of tasks that measured: (a) Siegler's rules, and (b) Piaget's structures. The order of presentation between Siegler's and Piaget's tasks was counterbalanced. Within each type, the tasks were administered randomly with one exception, to be noted subsequently. Each child was tested individually in one session. Session lasted between 20 and 30 minutes.

There was a total of 80 middle-class children in the experiment, half boys and half girls. The children were divided into five age groups with 16 children per age group. The age ranges and mean age (years, months) per group was: (1) 2,8–3,2; M = 2,11; (2) 3,6–4,0; M = 3,9; (3) 5,6–6,0; M = 5,10; (4) 7,6–8,0; M = 7,7; and (5) 11,6–12,0; M = 11,7. For reading ease we shall refer to these groups as consisting of 3-, 4-, 6-, 8-, and 12-year-olds, respectively. Each child was randomly assigned to one of the two orders of presentation in such a way that per age group, there were 8 children who received one of the orders each and of these 8 children, 4 were boys and 4 were girls.

Tasks Assessing Siegler's Rules

Sixteen tasks were administered in random order. The tasks, presented in Fig. 2, were similar to those in Siegler (1976) in that all six of the categories he used were found in the present study. The tasks were divided into three types: small number, large number, and mixed. There were six small number tasks, six large number

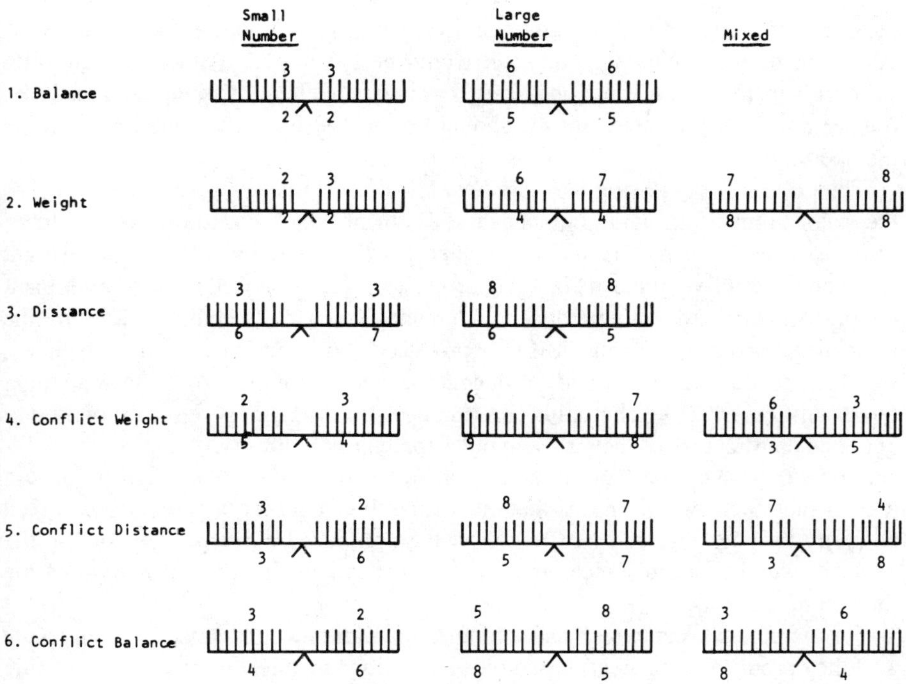

Figure 2. Tasks Used to Assess Children's Rule Use

tasks, and four mixed tasks. The two tasks missing from the mixed tasks were impossible to construct. The balance tasks as well as the distance tasks require the same number of weights to be placed at equal distances, but one cannot have equal weights in a mixed task since the small number side has less than five weights and the large number side has at least five weights.

The tasks in Fig. 2 represent the scale with 10 pegs on each arm. The number below the arm represents the peg number on which the weight was hung and the number above the peg represents the number of weights placed on that peg. We were forced to use 10 pegs versus 4 pegs in Siegler's tasks because of our decision to include small number tasks. For example, in the conflict-distance task for small numbers, we placed 3 weights on the third peg (the product being 9) on the left side, and in order to meet the conditions of the conflict-distance task, we needed to place fewer weights farther out so that the product on the right side would exceed 9. We could have placed either one weight on the tenth peg or two weights on the fifth, sixth, etc., peg. In any of these cases the number of pegs exceeds four. If the reader works out other cases, it will be seen that, except under the condition where the first peg is used, there will be a need for more than four pegs.

The children were told that all the weights weighed the same. They were shown the blocks under the arm and were told that the arm would not move, but that they were to tell the experimenter what would happen were she to remove the blocks: Would the right side go down, the left side go down, or would they balance? After the children produced their judgments, they were asked to justify them: "Why do you think this side would go down/it would be balanced?" Although requesting children's justifications is not necessary for analyzing children's judgment patterns, Siegler has done the same, so this should not be considered a deviation from his methodology.

The criteria for assessing the rule that children were using were as follows. For the small number and large number tasks (both of which had six tasks), children were assessed as using: (a) Rule 1, if they produced judgments based on weight alone on at least five of the six tasks; (b) Rule 2, if they produced a correct judgment on the distance task and produced judgments based on weight on four of the remaining five tasks; (c) Rule 3, if they produced correct judgments on the balance, weight, and distance tasks and produced one correct judgment on the remaining three conflict tasks. We chose this criterion because Rule 3 children are supposed to guess on conflict tasks. The guess involves three possibilities (left-side down, right-side down, balance) so there is a 33% chance that the children would produce a correct judgment, which means that one of the three conflict tasks could be solved correctly by guessing; and (d) Rule 4, if they produced correct judgments on the balance, weight, and distance tasks and correct judgments on at least two of the three conflict tasks.

For the mixed tasks (which had four tasks), children were assessed as using Rule 1, if they produced judgments based on weight alone on three of the four tasks. Rule 2 use could not be assessed for mixed tasks, because there was no task measuring the distance variable for reasons we already explained. Rule 3 use was determined when children produced a correct judgment on the weight task and one of the three conflict tasks. Rule 4 use determination was the same as that for Rule 3, except that children had to produce correct judgments on at least two of the three conflict tasks.

Tasks Assessing Piaget's Structures

As in the Inhelder and Piaget (1958) attempt to measure proportional reasoning, we administered two kinds of tasks. The first kind will be called the quantitative tasks since they can be solved by numerical means; i.e., by multiplying the weight times the distance on both sides of the arm and comparing the numbers. The six tasks we administered are found in Fig. 3. The scale used in this task was the one we used in Siegler's tasks. The weights used in the present tasks were clearly labeled (i.e., had the following written on them): 5 grams, 10 grams, 15 grams, and 20 grams. The experimenter placed one of the weights on a certain peg on one side of the arm, gave the children another weight, and asked them to place it on a peg on the other side so that the scale would be balanced. The children were told that they could keep on moving their weight until the scale balanced. We shall take the first task to illustrate how to read Fig. 3. The experimenter placed a 15-g weight on the third peg of the right arm and gave the children a 5-g weight with the described instructions to balance the arm and to explain why it balanced in the end.

The second kind of task will be called the qualitative tasks because they could not be solved by quantitative methods, but the children could make qualitative approx-

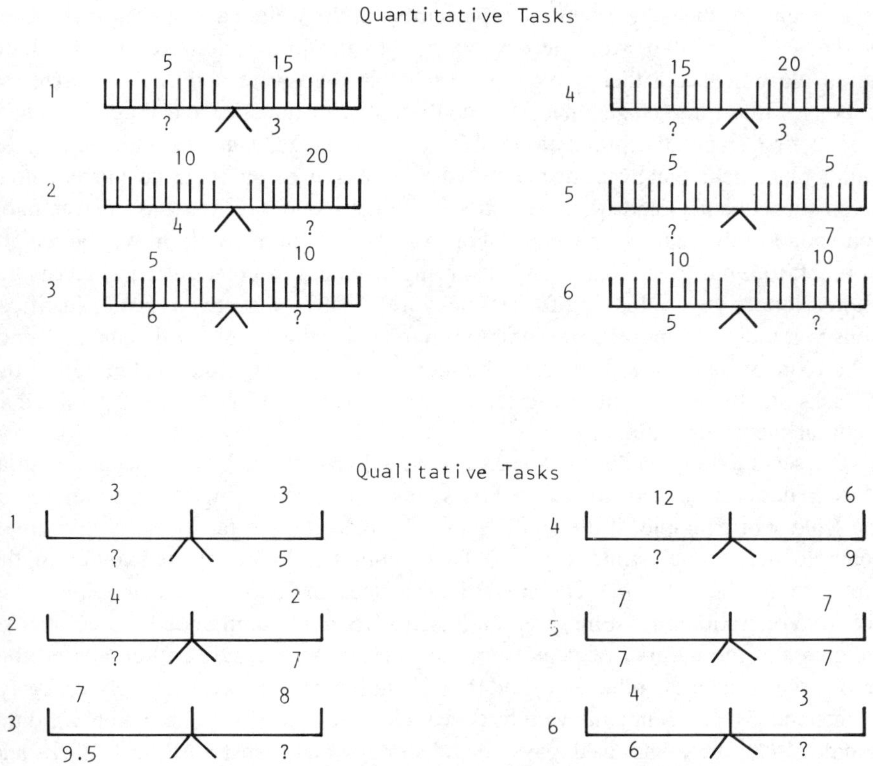

Figure 3. Tasks Used to Assess Piagetian Stages

imations to reach the correct solution. The six tasks measuring qualitative proportions are also illustrated in Fig. 3. The scale used was a yardstick with the numbers painted over on the side facing the children but not the experimenter. The yardstick had a hole in the center, and a metal projection on a plastic triangle (that served as a base) fit into the hole, allowing the yardstick to swing freely.

The experimenter presented a basket that had a certain number of pieces of candy in it and placed it at a certain point on the scale. She then gave the children a basket with pieces of candy and asked them to place it on the other arm in such a way that the arm would balance. They were also told that they could keep on moving their basket until the apparatus balanced. An illustration of how to read Fig. 3 can be given with the first task. The experimenter placed a basket with three pieces of candy on the right arm at a position equivalent to the position of the fifth peg in the first (quantitative) task and gave the children a second basket that also had three pieces of candy in it with the instructions to balance the scale and to explain why they thought the scale was balanced at the point it balanced.

The criterion for placing the children in one of the stage categories was complex, in part, because Inhelder and Piaget (1958) did not give clear indications as to how they categorized children's responses into stages. We used two criteria for stage assessment: (a) the strategy of initial placement of the weight and subsequent moves of the weight, if they were necessary; and (b) the justificatons offered for their initial placement and their moves of the weight. We were able to categorize the responses into three stages that were consistent with Inhelder and Piaget's stages.

The first stage, the preoperational or intuitive stage, was characterized by an initial placement that was inaccurate (the definition of an accurate placement is given subsequently) and where the strategy of subsequent moves was unsystematic and ended in the scale's imbalance. For example, children would move the weight from left to right, to left again, and at varying distances from the fulcrum, giving the impression they were looking for a "lucky hit." There were two types of justifications that indicated intuitive reasoning: references to the weight or distance, but not to both on the same task. Children who met the strategy criterion on at least 9 of the 12 tasks and the justification criterion on at least 9 of the 12 tasks were classified as being at the preoperational stage.

The second, concrete operations, stage was characterized by an inaccurate initial placement and a subsequent systematic series of moves leading to the balancing of the scale. For example, if the children initially placed it too far from the fulcrum, a concrete operational strategy would be to consistently move the weight in the direction of the fulcrum. There were two types of justifications that indicated concrete operational reasoning: (a) additivity, where children added the number of weights and the number of pegs from the fulcrum on one side and compared that sum to the sum on the other side; and (b) qualitative proportions where children say, for instance, that "Since the weight is heavier, I have to put it closer in toward the middle." Children who used a systematic strategy on at least 9 of the 12 tasks and produced the criterion justifications on at least 9 of the 12 tasks were classified as being concrete operational.

The third, formal operations, stage was characterized by an accurate initial placement of the weight on at least 9 of the 12 tasks and a numerical justification on at least 4 of the 6 quantitative tasks. An accurate initial placement was defined as a placement on the correct peg (or the place where the peg would have been in the qualitative tasks) or on one peg to either side of the correct peg (or the place where these pegs would have been in the qualitative tasks). A numerical justification included reference to the numerical relations between the weights; e.g., for both the qualitative and quantitative tasks: "Since this weight is 10 grams and the other is 5 grams, it's twice as much as the other one so we'll have to put it at half the distance."

Results and Discussion

The first and second hypotheses, respectively, were that more older than younger children would be found at higher stages (according to Piaget) and would use higher rules (according to Siegler) and that there would be an effect of order presentation on Siegler's rules only. A 5(Age—3,4,6,8,12) × 2(Orders of Presentation) ANOVA for repeated measures was run separately on each of the types of Siegler's tasks and on Piaget's tasks. The first hypothesis, on age effects, was confirmed and the second hypothesis, on order effects, was confirmed as predicted. The data on age effects are found in Table 2, and the results for order of presentation are found in Table 3.

On Siegler's three tasks, there were significant main effects for age and order of presentation, respectively: for small number, $F(4,60)=37.8$, $p<.001$ and $F(1,60)=3.65$, $p<.017$; for large number, $F(4,60)=54.1$, $p<.001$ and $F(1,60)=5.40$, $p<.002$; for mixed, $F(4,60)=39.4$, $p<.001$ and $F(1,60)=3.4$, $p<.02$. On the Piagetian tasks there were main effects for age but not for order of presentation: $F(4,60)=177.9$, $p<.001$ and $F(1,60)=1.4$, $p>.24$, respectively. The interactions were not significant. Using Tukey's Honest Significant Differences Test (HSD), the age differences for Piaget's tasks were found between children from age groups 1,2,3, and 4 together, versus 5. For Siegler's tasks, the age differences for both small-number and large-number tasks were between age groups 1 and 2 together, 3 and 4 together, and 5. The age differences for the mixed tasks were very complex and for reasons of space we will not present them here.

In sum, there were age effects for all of the types of tasks and order effects for the Siegler tasks but not Piaget's tasks. The order effects were such that children who were given Sigler's tasks before Piaget's (orders 1 and 3) were at lower-level rules than those who were given Siegler's tasks after Piaget's (orders 2 and 4).

In addition, it is of importance to determine the relations between being at a stage and having rules *within* each child. The relevant data are found in Table 4. In order to determine the nature of the relations, we had to compare a child's stage level with his rule level. Each child had three rules (one for each of the three task types: large number, small number, and mixed), and we need criteria to reduce them to one rule to make the comparison with stage level. It was decided that if a child was assessed on at least two of the three task types to have the same rule (including Rule 0, which occurred in cases where a child did not conform to any of the rules on a task type), s/he was assessed to be at that rule. In contrast, if a child was assessed on all three tasks to be at a different rule, s/he was assigned to have Rule X.

The data in Table 4 indicate that there is a match in four places between stage and rule assessment: (a) 17 children were unclassifiable (0) for both; (b) 13 children were at the intuitive stage and at Rule 1; (c) 4 children were at the concrete operations stage and at Rule 2; and (d) 7 children were at the formal operations stage and at Rule 4. If we remove the 11 children at Rule X from the analysis, 59% (41/69) of the children were at equivalent stages and rules.

If we take the remaining children who were at nonmatching stages and rules, we do not see any clear pattern of subtle differences of one technique over the other. Children at the concrete operations stage were assessed at Rules 1 (13 children) and 3 (5 children), and this suggests that Siegler's procedure may allow more subtle distinctions than Piaget's. However, we also see that Rule 3 children are assessed at both the concrete-operations and formal-operations stages, which suggests that in this case Piaget's procedure may allow more subtle distinctions.

The main purpose of our study was to compare and contrast Piaget's structuralist

Table 2. Frequency of Children Using Rules and at Stages Across Ages

	Rules																		
	Small Number					Large Number					Mixed					Stage			
Age	0[b]	1	2	3	4	0	1	2	3	4	0	1	2	3	4	0[c]	1	2	3
3	11	4	0	1	0	15	0	1	0	0	11	5	—[a]	0	0	16	0	0	0
4	7	8	1	0	0	8	8	0	0	0	4	12	—	0	0	6	10	0	0
5	0	8	4	4	0	1	10	3	2	0	0	12	—	3	1	0	1	15	0
8	0	8	2	6	0	0	7	5	4	0	0	13	—	3	0	0	3	13	0
12	0	0	1	8	7	0	0	3	8	5	0	2	—	6	8	0	0	2	14

[a]Children could not be assessed as using Rule 2 in the mixed tasks.
[b,c]0 represents unclassifiable rule or stage.

Table 3. Mean Scores Across Ages on Siegler's and Piaget's Tasks for Different Orders of Presentation

Order of Presentation	Small Number Ages					Large Number Ages					Mixed Ages					Piaget Ages				
	3	4	6	8	12	3	4	6	8	12	3	4	6	8	12	3	4	6	8	12
Siegler—Piaget	0.10	0.50	1.35	1.60	3.10	0	0.10	1.10	1.85	2.85	0.25	0.50	1.35	1.35	2.60	0.00	0.40	1.90	1.80	2.90
Piaget—Siegler	0.75	0.75	2.10	2.10	3.60	0.25	0.85	1.60	1.70	3.35	0.35	1.00	1.75	1.25	3.85	0.00	0.90	2.00	1.90	2.90

Score range for Siegler's tasks 0–4.
Score range for Piaget's tasks 0–3.

Table 4. Frequency of Rule and Stage Use Within Children.

	Rules					
Stage	**0**	**1**	**2**	**3**	**4**	***x***
0	17	3	0	0	0	2
Intuitive	1	13	0	0	0	0
Concrete Operations	0	13	4	5	0	8
Formal Operations	0	0	1	5	7	1

and Siegler's qualitative information processing models via the balance scale task. The findings were that children, when given tasks with Piaget's methodology, behave as if they have a structure of reasoning that fits Piaget's description (Inhelder & Piaget, 1958) and that *these very same children,* when given tasks with Siegler's (1976) methodology, behave as if they are using rules as described by Siegler. In other words, it appears that we can bring out the structuralist side of children or the information-processing aspect of their reasoning by changing our methodology by increments noticeable only by those who have an interest in noticing these increments.

An initial reaction to these data could be that all we did here was show that both Piaget and Siegler were good experimenters in the sense that their work is replicable and that, in fact, we have shown nothing more than that. If this were the case, there would be no general significance for this finding. In contrast to this potential reaction, we believe there are several points of general importance here for developmental theory. One is that the experimental methodologies we use to assess the development of reasoning constrain, even produce, the phenomenon we wanted to assess in the first place (see Strauss & Levin, 1981). In fact, one gets the feeling that children have many other ways to process and structure knowledge and that they'll expose them to us only at the moment someone invents a psychological model and corresponding methods to test for them.

Be that as it may, we are still in a position where we have found that the same children look as if they have both structures and rules. This phenomenon poses more than the usual dilemmas for trying to figure out how to make sense out of data. Whether we want them or not, these findings force us to consider how Piaget's and Siegler's models are related. If one accepts this, then the question before us is: Is there an in-principle reason why the structuralist approach and Siegler's rule assessment approach are incompatible?

Before answering this question, it would be well to briefly specify how Piaget and Siegler stand on three issues that bear on this question. For the first, we begin with the basics, namely, that neither Piaget's nor Siegler's accounts are descriptions of children's behaviors or performances. They are inferences from these performances. The units of analysis they use when making these inferences are quite different and, in part, set the stage for the content they examine. Piaget searches for cognitive structures that are claimed to be organized wholes with elements and relations between them, whereas Siegler searches for rules that are thought to be descriptions of knowledge states as well as the process that children go through when solving various problems. So the point of the first issue is that Piaget's and Siegler's units of analysis are different.

The second issue pertains to the formalisms used by theoreticians and experimenters to describe the units to be tested. Piaget has used models of logic and mathematical models as descriptions of the cognitive structures posited to give rise to behavior, although it should be mentioned that he did not seem particularly wed to these formalisms. In fact, just prior to his death he called for abandoning these models and stated the need to find other formalisms that could better account for his psychogenetic model. Siegler (1983) has pointed out that there are a number of ways to describe the basic unit he is out to measure. Among them are decision trees, flow diagrams, production systems, and algebraic rules.

The third area of issues is the level of what is being tested. Piaget's search is for general, formal cognitive structures in the sense that they are thought to be *relatively* domain- and content-independent. In this view, Piaget's approach is somewhat removed from the data and can be argued to be a broad stroke, general model located at a deep level of cognitive reality. In contrast, Siegler's search is for specific rules that are domain- and content-specific. This puts his rules closer to children's performances, thus allowing a precision and reliability which, we should add, are quite remarkable for developmental psychology.

In the event that you, the reader, may have lost his/her way, the reason we sketched some differences between Piaget's and Siegler's approaches was to provide some background material for the question: Are the two approaches incompatible in principle? As you may have guessed, there are three answers to the question: yes, no, and maybe. As we shall see, these answers involve either denial or cooption.

Let's begin with the first answer to the question which is that, yes, there is an in-principle reason why the structuralist and the rule assessment approaches are incompatible. The main strategy used here is to do what we often do when approaches are confronted: We deny like crazy. Here one can deny the existence of entities such as general cognitive structures as defined by Piaget or the significance of rules as defined and measured by Siegler.

One basis for denying general cognitive structures a la Piaget is the allegation that there is little empirical evidence for what we would expect were there such structures; i.e., consistency across task domains and invariant sequences within task domains. In the present volume, Levin makes this point and Case et al. and Turiel and Davidson argue against it. The reason for questioning the significance of rules a la Siegler is that they are too close to children's performances to have any general significance. That is, they are strategies derived from yet are at the edge of children's performances. Here the idea is that developmental psychology ought to get more removed from rather surface phenomena.

A second reason that can be offered for the incompatibility of the two approaches is that, in the example found in the present study, Piaget and Siegler are out after different animals. Piaget is searching for children's structures and Siegler is attempting to find their rules. And no less important, their tasks are reflections of what they are aiming to find. If this is considered unobjectionable, then there is little point in discussing whose task or way of testing is better, which is an aspect of taskology. (We should note two exceptions to this idea. The first is when two tasks have been constructed for the same ostensible specific purpose, and the second is when the same data have alternate explanations, as in the debate between Siegler, 1976, 1983, and Wilkening and Anderson, 1982, or Klahr & Siegler's 1978 production system analysis of Siegler's 1976 data.)

If one accepts this, then the arena of debate when comparing these two models of cognitive development shifts to questions that strike us as more edifying than taskology questions, namely, what *are* they out to find? What *are* the theoretical positions held by the experimenters? What *are* the relations between their experimental findings and those positions of theory? In other words, in this view, theories of tasks should be replaced by theories of mind where among other things, epistemological, developmental, psychological, and educational issues form the focus of debate.

The second answer to the question we are asking is that, no, there is no in-principle reason why the structuralist and rule assessment approaches are incompatible. Two approaches can be taken to this answer. The first is the *depth approach.* Here the argument is that there is something deeper than Piaget and Siegler that gives rise to both. When analyzing what this deep entity could be, not only is it unclear what could be deeper than both Piaget and Siegler, it's not even clear to us where we might want to start searching for the answer. This we will leave to those wiser than we. The second approach is that of the *ecumenical rabbi* who, not unlike the one from *Fiddler on the Roof,* listened to the claims and counterclaims of a feuding couple, and after hearing them out, proudly declared, ''You're right and you're right.'' The rabbi's student heard this pronouncement and pointed out, in distress, that the husband's and wife's arguments contradicted each other, so both couldn't be right. To this the rabbi responded after some thought, ''You're right too!''

Although this is a way to deal with marriage difficulties, it's not clear if this approach bridges the gap or papers over the crack between competing models of cognitive development. What is clear is that it has the potential to be an eclectic approach that removes real differences that may exist between Piaget and Siegler. If the ecumenical approach is followed, it is not enough just to proclaim conjugal bliss; one must also show how two seemingly different positions can provide adequate, nonconflicting accounts of cognitive development.

In this vein, one variant of the ecumenical approach is the *levels approach.* The point here would be that Piaget and Siegler are measuring the same thing, but what they are measuring is at different levels of specificity. As we have shown, Piaget is in search of general cognitive structures and Siegler is attempting to describe children's knowledge states and the processes of cognitive functioning. Now the distinctions between structure, conceptual knowledge, and process has had a long airing (for recent treatments of its main features, see Beilin, 1983, and Newell, 1972, and both Karmiloff-Smith and Mounoud in this volume). We do not intend to spend much time on it here, but we would like to note that the rules that concern Siegler are very similar to what children use as justifications for their judgments on Piagetian tasks. Piaget's argument was, of course, that it is the child's cognitive structure that produces the judgments and justifications, whereas Siegler concluded that rules can be determined from patterns of children's judgments. Notice that the levels approach involves co-option because the processing of information is done within constraints set by the structure. In other words, this approach suggests that Siegler's model of information processing should be viewed as being limited by the cognitive structures outlined by Piaget, and it is in this sense that Siegler gets co-opted by Piaget.

If the answer to our question is that there is no in-principle reason why the structuralist and rule assessment approaches are incompatible as described in these two approaches, then we implicitly (and in the case of this chapter, explicitly) argue that children's behavior can be described by both models. If this is accepted, then the either/or view of cognitive development ought to be questioned (Strauss, 1981). What this means is that the debate between advocates of different models of cognitive development ought to turn on questions about what makes one model interesting about how it informs the aforementioned epistemological, developmental, psychological, and educational issues.

The third answer to our question is that maybe there is (and maybe there isn't) an in-principle reason why the structuralist and rule assessment approaches are incompatible. Here the argument usually goes as follows: There may be such things as general cognitive structures that develop, but we are not yet in a position to know how to test for them, and so we are better off, for the moment, concentrating on the development of specific content. The study of the development of specific concepts is what characterizes Siegler's work. This is a pragmatic route since it doesn't rule out the possibility of the development of general cognitive structures but, rather, postpones our search for them until we better understand how specific content develops. This approach suggests that the structuralist and information-processing approaches may be compatible (Strauss, 1981), but it should not be used as a platitude or a cop-out, quicky solution for a thorny problem.

The second main finding in our study was that there was an order-of-presentation effect. It was found that solutions on Piaget's tasks were unaffected by whether Siegler's tasks came before or after Piaget's tasks, but solutions on Siegler's tasks were affected by the order of presentation. The children in the group where Piaget's tasks were administered before Siegler's tasks were assessed as using higher-level rules than children in the group where Siegler's tasks were administered first. This suggests that Siegler's technique may underestimate the level of children's rule use. Since order-of-presentation effects are generally rather spurious findings, even when predicted, as in this study, we feel that this part of our study is in need of replication. Given this caveat, we now pursue interpretations of this finding.

One explanation of these findings is that Piaget's tasks could be seen as a training procedure (for solving Siegler's tasks) that sensitizes the children to the distance variable of the balance scale. In our Piagetian tasks, children were given unequal weights on both sides of the arm and were asked to manipulate them such that the arm would balance. The only manipulation possible was to change the distance of the weights from the fulcrum. It is conceivable that this experience allowed the children to learn about the role of the distance variable.

Siegler (1976) reported using a condition in which children were told that they could put weights on the arms to see how the balance scale works. In that condition, children were allowed to manipulate both the number of weights and their distances from the fulcrum, and the results were that children were assessed to be at the same level of rule use as children who were simply given Siegler's tasks without any prior experimental manipulations. An explanation of the differences between our findings and Siegler's may be that children in Siegler's experiment almost surely changed both the weights and distances, and they may have had difficulties separating out the effects of each variable. In the present study, only the distance variable was highlighted, as mentioned earlier, and that might have cued the children to the importance of that variable.

What comes out of this work is the idea that via our experimental methodologies we can produce many behaviors that accord with the theoretical model that inspired those methodologies. Now this idea has its limitations since we cannot produce *any* behavior, but there is no doubt that we seriously constrain, if not produce, the very behaviors we set out to test. That is the case for both Piaget and Siegler. We then suggested that this finding has implications for the either/or view of developmental psychology. The reader may be thinking that we didn't have to conduct a research study to make the point and, at a certain level, that is probably correct. Nevertheless, the data are convincing, and what we are up to should be clear. We believe the argument here revolves around what empirical data one can take as evidence that we are looking at a structure, a rule, or something else, for that matter. Further research on this issue is currently being conducted by Pnina Frankel, a graduate student at Tel-Aviv University, and it is our hope that we will have more to say on this topic in the near future.

References

Beilin, H. (1983). The new functionalism and Piaget's program. In E. K. Scholnick (Ed.), *New trends in conceptual representation.* Hillsdale, NJ: Erlbaum.

Fodor, J. (1968). *Psychological explanations: An introduction to the philosophy of psychology.* New York: Random House.

Glick, J. (1983). Piaget, Vygotsky, and Werner. In S. Wapner & B. Kaplan (Eds.), *Toward a holistic developmental psychology.* Hillsdale, NJ: Erlbaum.

Inhelder, B., & Piaget, J. (1958). *The growth of logical thinking from childhood to adolescence.* New York: Basic Books.

Karmiloff-Smith, A. (1981). Getting developmental differences or studying child development? *Cognition, 10,* 151–158.

Karmiloff-Smith, A., & Inhelder, B. (1974). If you want to get ahead get a theory. *Cognition, 3,* 195–212.

Klahr, D., & Siegler, R. S. (1978). The representation of children's knowledge. In H. W. Reese & L. P. Lipsitt (Eds.), *Advances in child development* (Vol. 12). New York: Academic Press.

Newell, A. (1972). A note on process-structure distinctions in developmental psychology. In S. Farnham-Diggory (Ed.), *Information processing in children.* New York: Academic Press.

Piaget, J. (1970). Piaget's theory. In P. H. Mussen (Ed.), *Carmichael's manual of child psychology.* New York: Wiley.

Richards, D. D. & Siegler, R. S. (1982). U-shaped behavioral curves: It's not whether you're right or wrong, it's why. In S. Strauss (Ed.), *U-shaped behavioral growth.* New York: Academic Press.

Siegler, R. S. (1976). Three aspects of cognitive development. *Cognitive Psychology, 8,* 481–520.

Siegler, R. S. (1978). The origins of scientific reasoning. In R. S. Siegler (Ed.), *Children's thinking: What develops?* Hillsdale, NJ: Erlbaum.

Siegler, R. S. (1981). Developmental sequences within and between sequences. *Monographs of the Society for Research in Child Development, 46* (Whole No. 189).

Siegler, R. S. (1983). Five generalizations about cognitive development. *American Psychologist, 38,* 263–277.

Strauss, S. (1981).Cognitive development in school and out. *Cognition, 10,* 295–300.

Strauss, S., & Levin, I. (1981). Commentary on Siegler's Developmental sequences within and between sequences. *Monographs of the Society for Research in Child Development, 46,* 75–80.

Wilkening, F., & Anderson, N. (1982). Comparison of two rule-assessment methodologies for studying cognitive development and knowledge structure. *Psychological Bulletin, 92,* 215–237.

4

*The Nature of Children's and Adults' Concepts of Time, Speed, and Distance and Their Sequence in Development: Analysis via Circular Motion**

Iris Levin
Helen Simons
School of Education
Tel Aviv University

> Have you observed a complete revolution of a millstone? Each point in the extreme circumference of the stone describes a large circle in the very same time in which a point nearer the center describes a small circle; the velocity of the outer circle is therefore greater than that of the inner circle. You cannot say that the motion of the latter was interrupted by more moments of rest, for the whole moving body, i.e., the millstone is one coherent body. (. . .) reply is based on the twelfth proposition, which is to the effect that the perception of the senses cannot be trusted, and thus only the evidence of the intellect is admissible.—Moses Maimonides, *The Guide for the Perplexed,* part I, ch. 73, 12th century AD

The present chapter focuses on two major issues: the problem of assessing sequences in development and the nature of understanding of basic physical concepts in children and adults. These issues will be addressed through a study analyzing adults' and children's understanding of the concepts of time, speed, and distance via their interpretation of circular movements.

Sequence in the Development of Time, Speed, and Distance

The general question of sequence in the development of various concepts, skills, or operations is of major concern to cognitive developmentalists (e.g., Berenthal, 1981; Campbell & Richie, 1983; Fischer & Bullock, 1981; Flavell, 1982a, 1982b; Flavell & Wohlwill, 1969; Maccoby, 1983; Siegler, 1981). It is generally assumed that finding a consistent sequence among different achievements can provide a clue to the functional or even causal relations among the achievements concerned and, as a consequence, to the fundamental question of how development proceeds.

*Thanks are extended to Esther Dromi, Sidney Strauss, and Friedrich Wilkening for their most helpful comments.

However, conclusions to be drawn from an empirically established sequence are ambiguous on at least two grounds. First, the same sequence can emerge because of different relations between achievements—ranging from hierarchical inclusion of the former in the latter achievement to both achievements being dependent on sequentially ordered maturational factors. Second, a sequence can emerge between achievements because of differences in information-processing requirements—irrelevant to the competence per se—between the tasks used to assess them (e.g., Flavell, 1982b). Even when the mastery of different achievements is assessed by means of what seems to be the same task, the question of task equivalence in terms of information processing must still be seriously addressed. This issue will be illustrated here in a study analyzing the sequence in the development of the time, speed, and distance concepts.

The question of the sequence in development of time vs. speed was originally investigated by Piaget (1969). He related that he was asked by Albert Einstein whether a developmental study might reveal which of the two concepts—time or speed—is the more basic one in human cognition. Einstein probably would have liked to hear that these concepts are basic to the same extent in children's conceptualization, because each is dependent on the other, according to his theory of relativity. In contrast, within the classical Newtonian mechanics, which relativity claimed to place in a broader framework, speed is defined as the ratio of distance (or output) to time ($v = d/t$) and thus is derived from them.

One might expect that the question of which of two concepts is more basic in cognition would be addressed through studying the origins of these concepts, probably through infants' perception. However, for Piaget, understanding a concept often meant being able to apply it correctly in interpreting simple events, despite interfering cues. In the context of this definition of conceptual understanding, Piaget expected that for the two concepts, duration and speed, neither would be psychologically a more basic concept than the other. He claimed that, on the one hand, duration and distance must be understood before speed, because speed is derived from the relationship between them. On the other hand, speed has to be understood before time, because the measurement of time requires considering a distance covered (or output produced) at a constant speed. Piaget resolved this vicious circle (i.e., that each of these concepts is comprehensible only through an understanding of the other) by assuming a gradual progression of alternate advances in understanding each of them.

To examine this developmental process, Piaget assumed or informally ascertained which cues young children use in judgments concerning each concept. Consequently, particular tasks were constructed to assess the understanding of each concept.

Since the concept of time was expected to be confused primarily with the cue of distance, children's understanding of time was assessed by asking them to compare the time that two cars covered different distances. Similarly, since the concept of speed was expected to be judged on the basis of the cue of passing, children's understanding of speed was assessed by asking them to compare the speeds of two cars running in tunnels, so that the event of one car passing the other was not seen. From his empirical investigations Piaget concluded that the concepts of speed and time are constructed gradually, by and large synchronically, and that both concepts are mastered approximately by the age of 9 to 10 years.

However, as pointed out by Siegler and Richards (1979), Piaget's conclusion that time and speed develop more or less synchronically was methodologically ill-founded. Different tasks had been used to assess the comprehension of different concepts with no attempt to match them in terms of their information processing demands. Hence, Siegler and Richards (1979) and later Acredolo and Schmid (1981) tested the sequence question by assessing comprehension of the different concepts via the same task.

The task chosen for both studies presented children with a series of races between two cars running along parallel, visible, linear tracks. The children were asked to compare the cars' performances in terms of running time, speed, and distance covered. Surprisingly, the children were able to compare speed at an earlier age than distance and distance earlier than time. Moreover, the discrepancy between ages of mastery of the concepts was quite impressive—speed was understood by the ages of 8 or 9 years, whereas duration was comprehended only some time between the ages of 12 years and adulthood.

However, it is our suggestion that the sequence obtained was dictated by the structure of the particular task used to assess the concepts' comprehension. More specifically, we claim that a child with an incomplete understanding of the concepts of speed and distance could compare correctly the races in these terms, whereas a child with a similarly incomplete understanding of duration could not succeed in comparing correctly the races according to time. Thus, duration should have emerged in this particular task as the most difficult concept—as, in fact, happened to be the case.

In order to understand why an incomplete understanding was sufficient for correct performance on speed and distance but not on time, we should consider the major cues children use in judging each concept prior to mastery.

Generally speaking, Piaget and others viewed the mastery of time, speed, and distances as developing in three major stages. (e.g., Acredolo & Schmid, 1981; Lovell, 1959; Lovell, Kellet, & Moorhouse, 1962; Lovell & Slater, 1960; Piaget, 1969, 1970; Za'rour & Khuri, 1977). During the first stage, all three concepts are judged according to the same single cue of end positions. When presented with two cars running on parallel tracks, children tend to choose the car which stopped further ahead as having run faster, further, and for a longer duration than its counterpart. Their confusion of speed, distance, and time with end-position cue is thus revealed.

In the second stage, a differentiation emerges among the bases of judgment of the three concepts. When asked to compare the speeds of two cars, children use the cues of passing, catching up, or "perceptual blurriness" (see Kuhn, 1977). Hence, the car which is catching up with the other, passing it, or simply looking more blurred is judged as being faster. When asked to compare the distances covered, children use the cue of displacement; i.e., the straight-line distance between starting position and ending position, irrespective of the path traversed. Hence, if two cars, one running on a linear track and the other on a zigzag one, both start and finish the race side by side, then they are judged to have covered the same distance. When asked to compare durations, children in this stage refer primarily to the cue of distance, judging the car which covered a longer distance to have run for more time.

Now it may be expected that when children in this second stage are presented with two cars running along parallel, visible, and linear tracks, as in Siegler and Richards's (1979) and Acredolo and Schmid's (1981) studies, they will compare speed and distance correctly, but not time. To elaborate, they will be able to compare speeds correctly on the basis of passing, catching up, or "perceptual blurriness." They will be able to compare distances correctly, even if they use displacement as a cue, because displacement and distance are identical for parallel straight tracks. At the same time, children may fail to judge durations correctly, because the existence of different distances traversed by the cars provides an opportunity to confuse durations with distances.

To sum up, there is an interplay between the structure of the task chosen and the cues children in the second stage attend to in judging the different concepts. This interplay should lead to frequent "false positive" assessments of comprehension of speed and distance, whereas comprehension of time is assessed more stringently. Therefore, even if the three concepts do develop more or less synchronically, as assumed by Piaget, it is to be expected that time should emerge the latest on the task used by the later studies. This analysis leads to the conclusion that in order to assess the developmental sequence of different concepts in a "fair test," it is not enough to use the same single task.

To support our conclusion that the structure of the task previously used dictated the sequence obtained, we tested the sequence with a different task that according to the previously described analysis of these concepts' development should lead to a different sequence than the one previously established by Siegler and Richards (1979) and Acredolo and Schmid (1981). On our task we expect distance and time to be compared correctly at an earlier age than speed. If these expectations are empirically confirmed, a comparison of our results with those of the former studies would support the general contention that a sequence obtained between concepts is a function of the interaction between the nature of children's understanding of the concepts and the structure of the task.

The Circular Motion Task

The task used incorporated features of two tasks used by Piaget in assessing children's comprehension of speed; i.e., cars moving circularly and dolls attached to a bar that moved in an arc. In our task (see Fig. 1) two magnets were attached to the same disc, one closer to the central axis and the other to the perimeter. The disc rotated continuously for five rotations and the subjects were required to compare the durations, speeds, and distances of the magnet's movement.

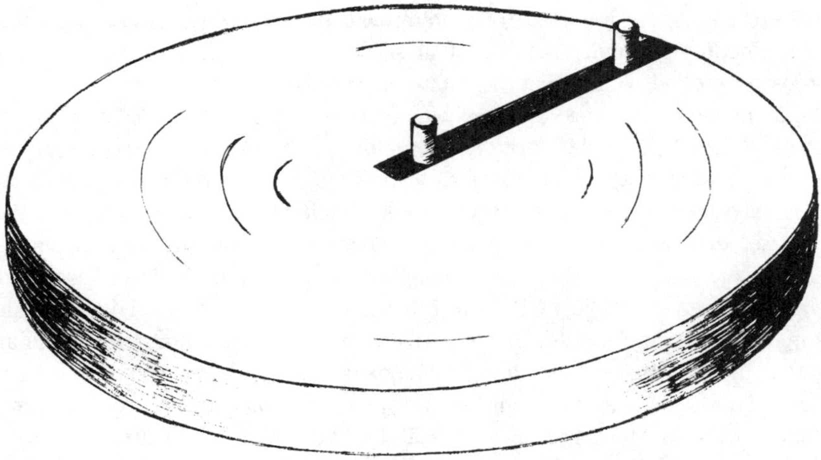

Figure 1. The circular motion task with 2 magnets attached to a disc

The rotational movement in this task opens up the possibility of conceiving speed in two different frameworks, in terms of distance per time (linear speed) and in terms of number of rotations or angles per time (rotational or angular speed). Thus, comparing the speed of the outer magnet to the speed of its inner counterpart could lead to two different correct responses. The correct response in linear terms is that the outer magnet moved faster than the inner one (since it covered a longer distance in the same time). The correct response in rotational terms is that both magnets moved at the same speed (since they completed the same number of rotations (or angles) in the same time).

In the same vein, comparing the distances covered by the two magnets can be conceived in two frameworks and lead to two different correct responses. The correct response in linear terms is that the outer magnet covered a longer distance (since its circle has a longer radius or circumference). The correct response in rotational terms, however, is that both magnets covered the same distance (in number of rotations).

Comparing the time that the two magnets moved can lead, in the present context, to only one correct response. The correct response is that the magnets moved for the same time (since they started moving and stopped moving simultaneously, or since they both moved as long as the disc moved).

To sum up, the movements presented could be analyzed correctly in two different ways. A linear analysis would lead to the conclusion that the two magnets moved for the same time, but the outer magnet moved faster and covered a longer distance. A rotational analysis would lead to the different conclusion that the two megnets moved at the same speed for the same amount of time and covered the same distance.

However, the responses of "same time," "same speed," or "same distance" could be reached through different ways of reasoning as well. The circular movements presented here involved four interrelated features which could lead subjects with an incomplete understanding of these concepts to arrive to the "same" responses on grounds other than the rotational analysis.

First, in the present task, the two magnets stayed side by side throughout the movement, the distance between them remaining fixed. Thus, a subject who judges an object to be moving faster, further, or for a longer duration than its counterpart, if it finishes ahead or passes the other, would conclude here that the magnets moved for the same time, at the same speed, or covered the same distance. Second, the magnets were attached to the same moving disc and could be conceived as constituting a single moving object. Therefore a subject who is convinced that a single moving object cannot involve different distances, speeds, or durations would again come to the conclusion that the two magnets moved in the same way in each respect. Third, the magnets could be grasped as being attached to a disc which rotated them so that they were passive and did not move at all. Thus, a subject who conceives speed, time, or distance in terms of effort or work could conclude here that the magnets did not move at all, thus having the same "zero" speed, duration, or distance. Fourth, the magnets completed five circles, thus finishing their movement more or less in the same spot that they started it. If a subject analyzes the distance covered in terms of the spatial discrepancy between starting point and ending point, he or she might conclude that the magnets did not move at all, covering no distance in no speed and in no time.

Since in our task, different judgments (e.g., both move at the same speed vs. the outer magnet moved faster) could stem from different correct analyses of the movements (rotational vs. linear) and the same judgment (e.g., both move at the same speed) could rise from either correct or incorrect conceptions, the subject's understanding could not be inferred from the judgment per se. It had to be based on the judgment and its explanation.

We expected the comparison of the movements in terms of speed to follow three major developmental phases. First, speed is intuitively conceived in terms of effort or work and is judged by the cue of which object moves "harder," finished ahead, or passes the other. In this phase the subject will tend to conclude that the speeds of the magnets are equal and explain it in these "intuitive" terms. Second, speed is conceived linearly in terms of distance per time. In this phase the subject will conclude that the outer magnet moves faster and will explain it by relating speed to distance. Thirdly, speed will be conceived as a dimension that can be conceived in different frameworks: linearly and rotationally.

Similarly, we expect the comparison of the movements in terms of distance to follow the same three phases. However, the comprehension of linear distance should precede the comprehension of linear speed, because speed in this case can be derived almost only from the comprehension of the interrelation of linear distance, linear speed, and time ($v = d/t$). The subject who understands that the outer magnet covered a greater distance than its inner counterpart, can deduce that it moved faster. In contrast, the comprehension of linear distance must not include speed.

The concept of time is expected to follow two phases. In both, the subject judges the durations of the two magnets to be equal, but explains his or her conclusion differently. In the first phase the explanation refers to the aforementioned "intuitive" terms (e.g., both magnets being attached to a single line) and in the second, to the correct cues of starting times and ending times.

In comparison with the findings of previous studies which assessed the concepts' comprehension in a linear presentation (Acredolo & Schmid, 1981; Siegler & Richards, 1979), we expect to find comprehension of the concepts to emerge at a later age, and perhaps even only among adults with the appropriate specialization. Moreover, speed which was found by others to be the easiest is expected to be the most difficult concept in our study, because a correct linear analysis of speed involves a similar correct analysis of distance plus its integration with time, whereas distance can be comprehended without its relation to speed.

Method

Subjects

The age range examined started from the age at which speed had been claimed to be mastered, i.e., 8 to 9 years, and terminated at adulthood. Two hundred thirty-eight males and females participated in the study: third graders (8–9 years); sixth graders (11–12 years); ninth graders (14–15 years); twelfth graders (17–18 years), and university students (19–44 years, most in their twenties). The twelfth graders and university students were composed of two groups—to be called hereafter, physics/mathematics majors and humanities majors. Thus the sample was composed of seven groups, with 34 subjects in each.

Among the twelfth graders, the physics/mathematics majors took advanced high-school courses in physics and/or mathematics, whereas the humanities majors took advanced courses in foreign languages, history, sociology, etc. Among the university students the physics/mathematics majors were students of physics, engineering, or computer sciences; the humanities majors were students of psychology, education, law, literature, the Bible, etc. All the physics/mathematics majors at the university had at least an introductory course in physics at the school of engineering of Tel Aviv University, and none of the humanities majors had any university training in physics.

All but one of the groups were equally divided by sex. Among the physics/mathematics majors at the university 24 out of the 34 subjects were men, because women of this group were hard to find. All the subjects, children and adults alike, volunteered to participate in the study.

The Instrument

The event to be interpreted by the subjects involved two objects moving in parallel circular paths. The objects were two cylindrical black magnets, 1 cm in diameter, and 2 cm in height. They were attached to a yellow circular disc (originally a phonograph turntable) which was 17 cm in diameter. One magnet, its top colored blue, was placed 1 cm from the center of the disc, and the other magnet, with a white top, was placed at the outer edge. A straight red line running from the center to the outer edge of the disc connected the magnets. The instrument is drawn in Fig. 1.

The session started with the disc being rotated five revolutions at 33 rpm. The velocity was indicated by a pointer which pointed at "33" written on a scale of three numbers, 16, 33, and 45, with no unit mentioned. The subjects could see the scale, but the interviewers ignored it completely.

The Interview

Each subject was interviewed individually by two experimenters in a quiet vacant room in the subject's school or university. One experimenter manipulated the disc and interviewed the subject, while the other operated a tape-recorder and intervened whenever the interviewer seemed to have skipped a question or misunderstood an answer. Before operating the disc the subject was asked to watch the two magnets, because he or she would be asked a few questions about them. The disc was then rotated and the interview began.

The interview was semistructured and composed of five sections:

1. *General reference.* The subject was asked to relate to the time that the two objects moved, to the distances they covered, and to their speeds. The purpose of this question was to obtain data on the way the three concepts—time, distance, and speed—are described or measured with minimal constraining or guiding instructions. The order of mentioning the three concepts was counterbalanced as much as possible across subjects.
2. *Comparison.* The subject was asked to compare the times, distances, and speeds of the two moving objects with the following question: "Did they move for the same time (speed or distance), or did one object move for a longer time (faster or further)?" If the subject answered that one of the objects moved for a longer time (faster or further), he was further inquired which of the two objects did so. We asked about each concept in a separate question.
3. *Explanation.* The subject was asked to explain how he or she came to his or her conclusions about each of the concepts; e.g., "How did you determine that the objects moved at the same speed?"
4. *Confrontation.* The experimenter summed up the three comparisons provided by the subject, confirmed that the subject stood by his/her answers, and posed a question as to whether these answers constituted a logical totality (e.g., "So, is it logical that the two objects travel at the same speed for the same time, and one of them covers a longer distance than the other?") If the subject granted that the answers were illogical, he or she was asked to explain the nature of the problem he or she was having. Whenever interviewees changed their minds about any of the comparisons, they were required to back up the new judgment about the comparison with an explanation and to reexamine whether the three comparisons now went together logically.
5. *Certainty.* The subject was asked to assess how certain his or her final comparisons were on a 1–5-point scale.

Changes of mind about comparisons occurred frequently throughout the interview. When asked, for example, to compare the speeds (Section 2), some subjects changed their prior comparisons of distances. When asked to explain a comparison (Section 3), many subjects changed the comparison itself. Upon being asked the confrontation question (Section 4) and even the certainty question (Section 5), subjects frequently changed their minds time and again.

Subjects were allowed to view the event as many times as they wished. The interview lasted from a few minutes to half an hour. Section 1 (i.e., general reference) was omitted in the interviews of third graders, because in pilot studies it was found to be too vague for them.

Results

The nature of understanding of each of the concepts was assessed in two ways. The first assessment analyzed the units of measure used by the subject when answering the initial unstructured request to describe the movements of the two magnets in terms of the distances they covered, the durations of their movements, and their speeds.

The second assessment analyzed the subjects' answers to the more structured instruction to compare the movements according to time, speed, and distance and to explain their conclusions. Since approximately half of the subjects changed their answers on the comparison question, for one or more of the three concepts, the score for subjects' level of understanding took into account all the judgments and explanations which emerged throughout the interview. Thus, the score ranged from correct to mixed to incorrect answers. It was assumed that when subjects dealt with a more difficult concept, both the category of scale used to describe or estimate it and the frequency of correct comparisons and explanations would be lower.

Scale Categories

The interview started with an open request to refer to the movement displayed in terms of distance covered, running time, and speed of the movement of the two magnets. The responses were sorted into three scale categories:

1. Ratio scales: Describing time in seconds or minutes; describing distance in centimeters, meters, or number of rotations, and describing speed in terms of any distance per time ratio, e.g., number of centimeters per seconds, number of revolutions per minute (rpm), kilometers per hour (kph), etc.
2. Ordinal scales: Comparing the times, distances, or speeds of the two magnets, claiming that one of them moved for a longer time, further or faster, or that they were equal on any of the dimensions.
3. Nominal scales: describing time, distance, or speed in a discrete term reflecting a subjective scale. These descriptions include conventional terms like "short" (for time and distance) and "slow" (for speed). They also included unconven-

tional usages like "reasonable" (for speed) or insisting on not knowing how to describe the dimension concerned.

Table 1 presents the percentages of subjects in each age group according to the scale category they used to describe each concept. Since there were two groups each of the twelfth graders and university students (n = 68) while there was only one group of each lower grade (n = 34), the overall percentages across grades was computed by averaging the means of the age groups, thus equally weighting the different ages.

Three major findings emerge from the data of Table 1. First, in almost every age group, distance and time were described more often by the ratio than by the ordinal scale and the least often by the nominal scale. Second, the opposite was the case for speed: The nominal scale was used more often than the ordinal scale, which was used more often than the ratio scale. Third, the use of the nominal scale for describing each of the concepts decreased systematically with age. Chi-square analyses comparing the use of the different scale categories for the different age groups for each concept separately revealed that the age groups did not differ significantly in descriptions of distance, χ^2 (6) = 9.12, $p<.17$ but did differ significantly in descriptions of time, χ^2 (6) = 13.09, $p<.04$, and of speed, χ^2 (6) = 26.57, $p<.001$.

To examine whether the same trends appeared in adults, particularly those who specialized in physics and/or mathematics, the scale categories used to describe each concept was computed separately for each of the four adult groups: twelfth graders or university students, majoring in humanities or in physics/mathematics. The data is presented in Table 2.

The data in Table 2 reveals unexpected systematic trends: For all three concepts, humanities majors used the ratio scale more often than the ordinal scale, while physic/mathematics majors used the ordinal more often than the ratio scale. For

Table 1. Percentage of Subjects in Each Age Group Using the Scale Categories for Each Concept (*n*=204)

Concept	Scale Category	Grades			University	
		6th	9th	12th	Student	Overall
	Ratio	44	44	43	47	45
Distance	Ordinal	32	41	51	44	42
	Nominal and don't know	24	15	6	9	13
	Ratio	50	41	37	53	45
Time	Ordinal	27	35	53	38	38
	Nominal and don't know	23	24	10	9	17
	Ratio	6	12	18	38	18
Speed	Ordinal	26	35	45	35	35
	Nominal and don't know	68	53	37	27	46

Table 2. Percentage of Subjects in Each Adult Specialization Group Using the Scale Categories for Each Concept (n=136)

		Humanities Majors		Physics/Mathematics Majors	
Concept	Scale Category	12th Graders	University Students	12th Graders	University Students
	Ratio	56	65	29	29
Distance	Ordinal	35	26	68	62
	Nominal and don't know	9	9	3	9
	Ratio	56	73	18	32
Time	Ordinal	38	18	68	59
	Nominal and don't know	6	9	15	9
	Ratio	12	44	23	32
Speed	Ordinal	38	21	53	50
	Nominal and don't know	50	35	24	18

distance and time the nominal scale was rarely used by all adults' groups indiscriminantly, while it was used often for speed, especially among humanities majors. Chi-square analyses comparing the use of the different scale categories in the different adult specialization groups for each concept separately revealed a significant difference between groups for each of the concepts, χ^2 (6) = 17.26, p<.01; χ^2 (6) = 26.57, p<.001; χ^2 (6) = 18.86, p<.01 for distance, time and speed, respectively).

Generally speaking, the data in Tables 1 and 2 indicate that among children and adults the scale category used to describe speed was lower than that used for distance or for time. The tendency to describe speed on a lower level than either distance or time is surprising, if we take into account the fact that the estimation of speed both on the ordinal and the ratio scales can be directly deduced from the estimations of time and distance. If a subject estimated both the distances and the durations on a ratio scale, he could calculate the speeds by dividing distance by duration. If he only qualitatively compared the distances and durations, he could also infer the relative speeds. Thus, the lower-scale category used for speed means that either the subjects did not know the interrelation between time speed and distance ($v=d/t$) or they were unable to access or apply this knowledge when it came to speed estimation. Excerpts taken from interviews illustrate this point. Subject is a university student of literature and the Bible (humanities major)

S: The distance of the outer magnet—two and a half meters, and of the inner magnet—not more than half a meter.
E: What about the time?
S: Approximately 7 seconds.
E: And the speed?
S: How do you want to measure the speed? On what scale? If it's a scale of "fast," "slow," or "very fast," then its a different story.
E: However you understand it . . .

S: Slow . . .

Subject is a student of education (humanities major):

S: It's 5 seconds. The distance is 50 centimeters. What was the third thing?
E: The speed.
S: The speed—how would you like me to express it? In kph?
E: As you like.
S: Slow. . . . I don't know. . . .

Subject is a 12th grader, history (humanities major):

S: The time . . . maybe 15 seconds.
E: The distance?
S: Two rotations, may be two and a half rotations.
E: OK, and the speed?
S: According to what? What does it mean?
E: However you see it.
S: I don't understand. What does it mean? Speed in seconds? I already told you how much time it took.

Comparisons and Explanations

The nature and level of understanding of each of the concepts was assessed by taking into account all the comparative judgments and explanations offered for that concept throughout the interview.

Speed

The answers on speed were categorized into four levels of understanding: correct and complete responses; correct but incomplete responses; mixed responses (i.e., hesitating or vacillating between correct and incorrect answers); and incorrect responses.

Most of the incorrect responses on speed were of claiming that the two magnets moved in the same speed. These "equal speed" comparisons were explained in different ways; the major ones will be presented here. Many subjects explained the "same speed" response by referring to the fact that the magnets moved on the same disc and thus their speed was the same as that of the disc. Some of them even mentioned that, were it two runners rather than two objects attached to a disc, their conclusion as to their speed might have been different. Other subjects referred to the fact that the magnets moved side by side, neither passing each other. Few of them added that if the speeds would have been different, the line connecting the magnets would have bent rather than remaining straight throughout the movement. Some subjects insisted that the magnets were not really moving at all; only the disc moved. Consequently, they concluded that the magnets either had no speed at all because they were still, or they concluded that the magnets had the speed of the disc.

Incorrect responses on speed also involved claims that the inner magnet moved faster than the outer one. These "inner faster" comparisons were explained by referring to the fact that the inner circle was shorter, and so the magnet completed it earlier–faster. Some subjects further claimed that the inner magnet completed more rotations than the outer one. Both explanations were based on inferences contradicting the perceptual givens, because it could be easily seen that the magnets completed each circle simultaneously and did not differ in the number of rotations completed.

Few incorrect responses on speed involved the claim that the outer magnet moved faster. Though such a response might have been based on a correct linear analysis of speed, its incorrectness was revealed through its explanation. Some subjects explained the "outer faster" response by stating that the outer magnet overtook the inner one or that the outer pulled the inner one along the way. These incorrect explanations mentioned neither the difference in distance nor the equality of time. To sum up, incorrect responses on speed included all possible comparative judgments: equal speed, inner–faster, outer–faster, or even claiming that the magnets had no speed of their own. None of the explanations integrated correctly the distance and duration relations involved.

Two types of answers were categorized as correct (though incomplete) responses, one referring to speed in linear terms, the other in rotational terms. The linear correct response involved the claim that the outer magnet moved faster, because it covered a longer distance—with or without mentioning that it did so in the same amount of time. The rotational correct response, which appeared very rarely, was to claim that the magnets moved at the same speed because they completed the same number of rotations—with or without mentioning that they did so in the same time. These responses were considered incomplete, because the subjects did not suggest the possibility that the speeds could be considered in both of the two frameworks, linear and rotational.

A correct and complete response was of providing the two frameworks and explaining that the outer magnet moved faster in linear speed and that both magnets moved at the same speed in rotational or angular terms.

Examples taken from adult interviews illustrate incorrect, correct–incomplete, and correct–complete responses on speed, respectively.

Incorrect speed responses. Subject is a university student of psychology and education (humanities major):

E: How did you determine that they moved at the same speed?
S: Because both are placed on a rotating disc.
E: And that means. . . ?
S: That the speed of the disc is the speed they're moving.

Subject is a university student of mathematics (physics/mathematics major):

E: How did you determine that the speed was equal?
S: They rotated on the same body. They had the same speed. They had the speed of this thing (disc). If they didn't have the same speed, it (the disc) would break.

Subject is a university student of social sciences (humanities major):

E: How did you determine that they moved in the same speed?
S: Because they did it together. They moved together on a single line.

Subject is a university student of literature and education (humanities major):

S: Perhaps the outer magnet moved faster.
E: How did you determine that?
S: I'm trying to reconstruct what I learned in physics. . . . It seems to me that the object at the end moves faster.
E: Why?
S: Its weight. It is lighter.

Subject is a university student of humanities (humanities major):

E: How did you determine that the outer magnet moved slower?
S: I know that if the inner magnet is in the middle, it makes more rotations in comparison to the outer magnet. So the inner one moves faster.

Correct speed responses. Subject is a university student of economics and business administration (humanities major):

E: How did you determine that the speed of the outer magnet was faster than the speed of the inner one?
S: Since it goes a longer distance in the same time.

Subject is a university student of education (humanities major):

S: If the outer one goes a greater distance in the same time, you divide the distance by time, so the speed of the outer one is greater.

Complete and correct speed responses. Subject is a university student of industrial engineering (physics/mathematics major):

E: How did you determine that the outer magnet had a higher linear speed?
S: According to its radius. The longer the radius, the higher is the linear speed.
E: And how did you determine that the rotational speed was equal?
S: They are both on the same surface. And each point on this surface has the same rotational speed.

Subject is a university student of industrial engineering (physics/mathematics major):

S: Now this is my problem: If I say that they moved for the same time and one of them covered a longer distance, it necessarily follows that it moved faster. Now, we should distinguish which speed we are talking about: tangential or rotational. . . . The greater speed is the tangential.
E: What about the rotational speed?
S: I have to think about it. . . . The rotational speed is equal. The rotational speed

tells us what the angle is that the radius opens in a time unit, and the angle was the same (for the two objects).

The quantitative data pertaining to the level of speed comparisons are presented in the lower part of Table 3.

Two major findings stand out in the speed data. First, overall, the majority of subjects do not analyze correctly the movements in terms of their relative speeds; they either err altogether or hesitate or vacillate between correct and incorrect responses. Even among university students, almost half of them err on speed. Second, with age there is an improvement in understanding the speed relations: Correct responses increase and incorrect responses decrease. To carry out a chi-square analysis correct and complete responses were combined with correct responses. The analysis showed the difference between age groups to be significant, χ^2 (8) = 52.59, $p<.001$.

To analyze the extent to which comprehension of speed in adults was related to their specialization, a comparison was drawn between the four adult groups. The lower part of Table 4 presents the percentages of subjects according to the score on speed comparison in each adult specialization group.

The interesting finding in the speed part of Table 4 refers to adults without specialized training in physics and/or mathematics. The majority of them, 90% of the twelfth graders and 70% of university students, were either completely wrong about the speed comparison or vacillated between right and wrong answers, often being unable to decide between the two. Specialization in physics/mathematics improved performance, though even students who had university training in mechanics were still not unanimously correct. Among physics/mathematics majors, university students who had additional courses in mechanics at the university level performed better than the twelfth graders. Among the humanities majors the improvement from the twelfth grade to university is harder to explain, because no

Table 3. Percentage of Subjects According to Level of Comparison for Each Concept by Age Group (*n*=238).

Concept	Level of Comparison	Grades: 3rd	6th	9th	12th	University Students
Distance	Correct	26	61	59	69	76
	Mixed	18	35	38	28	18
	Incorrect	56	3	3	3	6
Time	Correct	0	23	50	81	75
	Partial (mixed)	12	6	15	1	6
	Incorrect	88	71	35	18	19
Speed	Correct and complete	0	0	0	12	22
	Correct	9	6	21	21	32
	Mixed	9	15	17	26	19
	Incorrect	82	79	62	41	27

Table 4. Percentage of Subjects According to Level of Comparison for Each Concept by Adult Specialization Group (n=136).

		Humanities Majors		Physics/Mathematics Majors	
Concept	Level of Comparison	12th Graders	University Students	12th Graders	University Students
Distance	Correct	53	65	85	88
	Mixed	44	23	12	12
	Incorrect	3	12	3	0
Time	Correct	71	62	91	88
	Partial (mixed)	0	9	3	3
	Incorrect	29	29	6	9
Speed	Correct and complete	0	3	23	41
	Correct	9	27	32	38
	Mixed	26	29	27	9
	Incorrect	65	41	18	12

further training in related topic was involved. A chi-square analysis revealed the difference between the adult specialization groups to be significant, χ^2 (9) = 50.23, $p<.001$.

Distance

The responses to the request to compare the magnet's movements according to the distance they covered were categorized into three levels of understanding: correct responses, mixed responses of correct and incorrect answers, and incorrect responses. These categories paralleled the ones used for speed, except that for distance the "complete response" of using two frameworks for distance—in linear and rotational terms—was never mentioned by the subjects.

Incorrect responses on distance mainly involved the claim that the two magnets covered the same distance. The explanations for the "equal distance" responses were similar to the explanations mentioned for the "equal speed" responses. The main explanation was that since the magnets were attached to the same disc, everything about their movement must be equal, including distance. Some subjects referred to the distance between the magnets. Since this distance did not change throughout the movement, the distance they covered must have been equal as well. Other subjects grasped the magnets as not being autonomous in their movement. This led them to the conclusion that the objects did not move at all, and therefore no distance was covered by them, or that their distance was that of the disc. Many subjects who believed the distances were equal for any of these reasons were also impressed by the differences between the length (or size) of the outer and the inner circles. To reconcile the contradiction, some of the subjects came up with the invention that the inner magnet completed more circles than the outer one so that the overall distance was equal.

Rarely, incorrect comparisons on distance involved the claim that the inner magnet covered a longer distance. This type of response was explained mainly by stating that the inner magnet was faster. Occasionally, the response that the outer magnet covered a longer distance was categorized as incorrect. This was done when the explanation was definitely incorrect, such as stating that it was so because the outer magnet was slower.

Correct comparisons on distance were of two kinds, linear and rotational. The vast majority of correct responses were linear, stating that the outer magnet had a longer radius and thus covered a longer circumference, or simply that it had a longer circle to cover. Some subjects explained that the central axis had no distance at all to cover and that the further one is removed from the axis, the longer is its circle. Very few correct responses were of the rotational kind. The subjects who provided them claimed that the distances were equal since the magnets completed the same number of rotations or angles. Excerpts from interviews illustrate incorrect and correct responses on distance.

Incorrect distance responses. Subject is a university student of education and Hebrew literature (humanities major):

S: Logically, it seems that they covered the same distance in the same time.
E: How did you determine that they covered the same distance?
S: Since they were at the same angle.
E: In relation to what?
S: I mean to say that they were on the same line, not at the same angle.

Subject is a 6th grader:

S: The inner magnet makes more circles but they are small, and the outer magnet makes less circles but they are big.
E: So what do you say about the entire distance?
S: That it's equal. . . . The inner one makes many small circles and the outer one makes few big circles, so the overall distance is equal.

Subject is a 12th grader, social studies (humanities major):

S: The outer object seems to have a longer circle, but I think it doesn't.
E: Do you mean to say that they covered the same distance?
S: I don't know if they moved at all. They are fixed. But of course the inner circle is smaller than the outer circle. But they are on the same wheel and it seems to be the same.
E: So you say that. . . ?
S: They actually did not move at all.

Correct distance responses. Subject is a 12th grader majoring in physics/mathematics:

E: How did you determine that the distance of the outer magnet is greater?
S: I see it. This circumference is much greater than this circumference.

Subject is a 12th grader majoring in mathematics:

S: The outer one covers a longer distance because the radius of its circle is longer.

Subject is a 12th grader majoring in physics:

E: How did you determine that the outer one passed a longer distance?
S: Distance is number of rotations times the circumference. Now, the number of rotations they rotated around the center is equal. The formula for a circle's circumference is $2\pi r$. 2π is equal in both cases but the radius of the outer body is longer.

The upper part of Table 3 presents the relevant data for the distance comparison in each age group. Very few subjects from the sixth grade on considered only incorrect answers as a possible response to the distance comparison question. However, approximately one third of them considered both correct and incorrect answers, frequently being unable to decide between them. Even among university students, approximately a quarter of the sample was either unsure or wrong on distance. Overall, performance improved with age, though the level did not change much from the sixth to the twelfth grade.

To carry out a chi-square analysis, the mixed responses were combined with the incorrect responses. The analysis revealed the difference between age groups to be significant, χ^2 (4) = 27.71, p<.001. The effect of specialization on comprehension of distance was analyzed among the four adult specialization groups. The distribution of subjects according to the score on distance comparison in each group is presented in the upper part of Table 4. These data reveal that twelfth graders and university students without specialized training in physics/mathematics performed similarly to children of the sixth and ninth grade. They rarely mentioned only incorrect responses, but approximately a third of them provided both correct and incorrect answers. Similarly to speed, performance improved from the twelfth grade to the university. Those with specialized training performed correctly most of the time in both the twelfth grade and in the university. A chi-square analysis that was carried out combining mixed responses with incorrect responses, showed the difference between the adult groups to be significant, χ^2 (3) = 14.66, p<.002.

Time

The majority of responses on time, except in the third grade, involved the claim that the two magnets moved for the same time. Thus, the distinction between correct and incorrect responses relied mainly on explanations. The responses were categorized into three levels: correct, partial (mostly mixed), and incorrect. Correct responses included an explanation of the equal duration by referring to the fact that the magnets both started and stopped moving together. Some subjects added that neither of the objects stopped while the other continued to move. A partial explanation involved referring either to the fact that the objects stopped together or to the fact that they started together, but not to both. Often such a partial explanation involved referring also to the "irrelevant features" such as the magnet's being on the same line.

Incorrect responses included explanations which referred only to the nontemporal cues. Subjects claimed that the magnets moved for the same time, because they were attached to the same disc or because they moved neck and neck. Incorrect responses, were also of claiming that the outer magnet moved for a longer time because it covered a longer distance; or that the inner magnet moved for more time because it finished its circle faster. These responses occurred mainly among younger subjects. excerpts from interviews illustrate incorrect, partial, and correct responses.

Incorrect time responses: Subject is a university student in arts and education (humanities major):

S: The outer magnet moves faster, goes a longer way in a shorter duration.
E: In a shorter duration?
S: If it moves faster its time becomes shorter.

Subject is a 6th grader:

S: Since the inner one is closer (to the axis), it is moving slower and certainly in less time, because this circle is smaller than the circle of the outer one.

Subject is a 6th grader:

E: What made you think that they moved for the same time?
S: (long pause) I don't know. The wheel is rotating. They are both in the same . . . they are moving together.

Partial (or partial and mixed) time responses. Subject is a university student of machine engineering (physics/mathematics major):

E: How did you determine that they moved for the same time?
S: Since they started moving together.

Subject is a 12th grader majoring in physics/mathematics:

E: What about the time?
S: The time is the same time—since you operated them at the same time and the inner object made more circles.
E: So what do you infer from that, that the inner object made more circles?
S: The overall time is equal. But the time per circle is shorter for the inner circle.

Correct time responses. Subject is a 12th grader majoring in physics/mathematics:

E: How did you determine that the time is equal?
S: You started them at the same time and stopped them at the same time, so the duration is equal.

The middle part of Table 3 presents the percentages of subjects according to the score on time comparison in each age group. These data show that from the third to the ninth grade at least half of the subjects either completely erred or produced mixed correct and incorrect time comparisons. Surprisingly, almost 20% of the adults still erred, either in their comparison or in their explanation. A chi-square analysis was performed combining partial and incorrect responses. The analysis revealed the difference between age groups to be significant, χ^2 (4) = 84.91, $p < .001$.

A further analysis of adult's performance was done by comparing the four adult specialization groups. The middle part of Table 4 presents the relevant data. It shows that, among adults without specialized training, almost a third of the sample erred on time. The errors were almost always in the explanation provided for the "equal time" comparison. Specialization improved performance, but university students did not perform better than twelfth graders. A chi-square analysis combining partial and incorrect responses, revealed the difference between the adult groups to be significant, χ^2 (3) = 11.80, $p < .019$.

Concept Comparison

To compare the relative difficulty of the three concepts—speed, time, and distance—the correct and complete responses on speed were combined with the correct responses thus creating a 3-level scale of correct (scored 2), mixed or partial (scored 1), and incorrect (scored 0) responses for each of the three concepts. The mean scores in each age group, are presented in Table 5.

The data in Table 5 support the conclusion that across a wide age range, within the context of the circular motion task used here, distance was the easiest concept, speed was the most difficult concept, and time was in between. This order of difficulty failed to appear only among the youngest group who erred on time as well as on speed almost unanimously. To analyze whether the same order of difficulty between the three concepts appears in adults with and without specialization in physics/mathematics, a comparison was drawn between the four adult specialization groups. Table 6 presents the mean scores in each group.

These data reveal that even among adults, speed is the most difficult concept. Performance on distance and time was similar, with three of the four groups still

Table 5. Mean Score of Level of Understanding of Distance, Time, and Speed by Age Group[a] (n=238).

Concept	Grades				University Students	Overall
	3rd	6th	9th	12th		
Distance	.71	1.59	1.56	1.66	1.71	1.51
Time	.12	.50	1.15	1.63	1.54	1.16
Speed	.26	.29	.59	.91	1.29	.79

[a]0 = incorrect responses; 1 = mixed (or partial) responses; 2 = correct responses.

Table 6. Mean Score of Level of Understanding of Distance, Time, and Speed by Adult Specialization Group[a] (n=136).

	Humanities Majors		Physics/Mathematics Majors	
Concept	**12th Graders**	**University Students**	**12th Graders**	**University Students**
Distance	1.50	1.53	1.82	1.88
Time	1.41	1.29	1.85	1.79
Speed	.44	.91	1.38	1.68

[a]0 = incorrect responses; 1 = mixed (or partial) responses; 2 = correct responses.

performing on distance somewhat better than on time. Specialization improved performance in each concept.

The general order of difficulty—increasing from distance to time to speed—differs from that found in studies using linear movements in that speed was found here to be the most difficult, whereas previously it had been shown to be the easiest concept. Moreover, many adults (especially without specialization) showed difficulties in analyzing the movements in terms of speed—a concept which was claimed to be mastered by elementary school children.

Discussion

Dependence of Development Sequence on Criteria of Comprehension

One of the major general conclusions of cognitive developmental psychology reached during the past decade has been that the comprehension of mathematical and physical concepts emerges earlier than Piaget led us to believe (Gelman, 1983). Studies keep piling up demonstrating that preschoolers, toddlers, and even infants possess more advanced knowledge and reasoning about number, space, time, and the like than was attributed to them only 10 years ago (e.g., Levin, 1982; Levin, Goldstein, & Zelniker, 1984; Miller, 1983; Strauss & Curtis, 1981, 1983; Starkey, 1983; Wilkening, 1981, 1982).

The major implication of the present work is different thought not contradictory. This study supports the conclusion that adults' comprehension of some physical concepts—particularly speed—is far less advanced than Piaget and others suggested (Acredolo & Schmid, 1981; Lovell & Slater, 1960; Piaget, 1970; Siegler & Richards, 1979; Wilkening, 1982). This conclusion is, however, in line with a few recent studies which have analyzed university students' reasoning about physical concepts such as movement, nonuniform speed, and acceleration (Caramazza, McCloskey, & Green, 1981; Crepault, 1983; diSessa, 1982; Ehri & Muzio, 1974; Hewson, 1983; McCloskey, 1983a, 1983b; Shanon, 1976; Trowbridge & McDermott, 1980, 1981; Whitaker, 1983).

These two conclusions—that precursors of understanding emerge early, and that mature understanding emerges late, if ever—taken together imply that the development of understanding of physical concepts is a very long and complex process and that, more often than not, never reaches a stable end-state concordant with the laws of Newtonian physics. Taking the length of this process into account, it becomes immediately evident that the developmental sequence of different concepts is dependent on the criteria of comprehension selected for each of the concepts concerned.

According to Piaget's analysis, speed and time are interrelated concepts that evolve in an alternating parallel process. He posited that one cannot understand speed as the relation of output to time without understanding each of its constituents. At the same time, one cannot measure time without taking into account the constant speed of the criterial event (e.g., the speed of the hands on a clock or the rate of the sand's flow in a hour glass). Thus he concluded neither of these concepts can be fully understood without an equivalent comprehension of the other. However, Piaget's conclusion is based on particular criteria for understanding speed and time. A different conclusion might be reached, if other criteria of understanding were applied.

Each of these concepts involve different strata of knowledge. The lowest level may involve perception of differences, while the highest level of its conception may be in terms of Newtonian mechanics or even Einsteinian relativity theory. There is no reason to believe that at each and every level of understanding one of these concepts is necessarily a precursor to the development of the other. Moreover, at the higher levels, understanding of one of the concepts necessarily implies some level of understanding of the other, because it involves understanding of their interrelations.

We have argued that the developmental sequence obtained for distance, time, and speed is dependent on the interplay between the subjects' nature of understanding of the concepts concerned and the structure of the task used for assessment. The structure of the task was, however, dealt with until now as consisting only of the features of the display. We now wish to point to the fact that the structure of our task consisted also of the particular interview we held and of the specific criteria we used for analyzing responses. We claim that the sequence we obtained could be an outcome of the effects of these aspects of the task as well. Furthermore, we know it to be the case that by using different criteria for differentiating correct from incorrect performances, we arrived at different sequences.

In the present chapter we categorized responses into correct and incorrect answers according to the comparative judgments and their explanations. The decision to include explanation as a criterion for level of performance was taken halfheartedly because of two major reasons. First, the explanations given for judgments do not necessarily reflect the reasoning that led to their judgments. Explanations sometimes serve to convince the audience in the validity of a judgment rather than reflecting the prior reasoning. Second, an incorrect explanation can stem from verbal limitations rather than from an erroneous reasoning (See Acredolo, 1981, Acredolo, Adams, & Schmid, in press; Brainerd, 1973). However, in the present study we assumed that the same judgment could stem from different reasonings, some correct and others incorrect, and different judgments could be based on correct but different conceptualizations. Thus, the distinction between correct and incorrect responses had to take into account both judgments and explanations.

To investigate the effect of using judgments plus explanations on the relative difficulty of the three concepts concerned—distance, time, and speed—we reanalyzed the data, using judgments only. To do so, we had to decide which judgments were to be considered correct. This was easy for time because correct judgments on time could only be of claiming the two objects to be moving for the same duration. For distance and speed, the decision had to be somewhat arbitrary. Since most of the correct responses on distance and speed according to explanations were of the linear version, we decided to accept as correct judgment the claim that the outer object moved faster and covered a longer distance than its inner counterpart. In addition, speed judgments using explicitly both linear and rotational terms were also considered correct. The sequence obtained in this judgment-based analysis was systematic across groups but different than the sequence obtained when explanations were considered as well. In each age group speed was still the most difficult concept, but time was consistently easier than distance, rather than vice versa. (See details in Levin, 1983.)

It is quite plausible that time emerged as the easiest concept when only judgments were concerned. The display used in the present study consisted of many features (e.g., single disc, no passing) which could lead the subject to judge the movements of the objects to be equal in all respects. Since the time that the two objects moved was indeed equal, while their speeds and distances (when analyzed by judgments alone) were not, a bias toward a response of equality should have raised the frequency of correct responses to the question on time only and thus bring about more "false positive" answers on time than on distance or speed. Thus, this sequence rather than revealing something new about development points to the methodological difficulty of assessing sequences in general. A sequence obtained is even dependent on using or not explanations in assessing comprehension.

The Limitations in Speed and Distance concepts

Before the sequence question can be meaningfully approached, we are obliged to explicate the levels of knowledge involved in the development of each concept. The results of the present study allow us to contribute a few suggestions on this issue, in particular with regard to the limitations of the concepts of distance and speed.

Five major limitations in children's and adults' concepts of speed were suggested by the present data. The first two are related to the issue of measurement. When asked to describe the movements in terms of distance, time, and speed, many subjects including adults, were unable to describe the speed by using an objective scale. Many explicitly expressed their inability to find a satisfactory way to describe the speed, or even more specifically, their inability to find the scale that should be used. Consequently, they either preferred to admit they did not know the answer or fell back to subjective descriptions such as "slow" or "reasonable." We suggest that herein lies one of the basic difficulties many people face when analyzing movements in terms of speed. While they have an internalized objective unit of length and time (e.g., cm and s, respectively) with which they can estimate distances and durations, this is not the case with speed. Since they are unable to represent a unit of speed, they do not know how to estimate the speed of a movement.

It should be noted that in the sixth grade, children learn how to calculate the speed of a movement given its duration and distance, and many of them (but not even all university students) know how to do it. However, the result of this calculation is probably grasped as a general description of the speed of the movement with the units themselves having no definite or clear meaning. The difficulty of deriving an estimation of speed out of estimations of distance and duration and the confusion about the meaning of the derived unit of measure is illustrated in the following example. Student is a graduate student in educational counseling (humanities major):

S: The distance—do you want it in cm or in rotations?
E: As you wish . . . understand . . . .
S: I presume in rotations, 5 rotations. The time is half a second. The speed, according to what?
E: As you understand it . . .
S: I know that the speed of a car is kms per hour. It's the time divided by the distance. I told you the time and the distance. So it's that divided by that . . . I got a smaller number for speed of the inner magnet.
E: So are you saying that the speed of the inner magnet is less or that the number is smaller?
S: Well, its distance is shorter, so the formula . . .

The lack of an internalized speed unit with an objective meaning and the confusion about the meaning of conventional units (such as kph or mph) should raise difficulties in estimating speeds, not only of circular but of linear movements as well. However, an additional problem of measurement stems from the fact that the distance and speed of circular movements can be conceived and measured by two different sets of scales: linear scales and rotational scales. Thus distance can be measured by centimeters or by the number of rotations, and speed by the relation of either to a time unit (e.g., seconds). The linear and rotational scales are nonequivalent, so that two circular movements can cover the same distance in terms of the number of revolutions and different distances in terms of number of centimeters (as was the case in our display), or vice versa. Many children and adults mention one kind of scale for distance and another for speed but seem to be unaware of their inconsistency. Thus, they may conclude that the two movements cover different distances (in centimeters) at the same speed (in terms of number of rotations), or rarely vice versa, a conclusion that for some of them seems embarrassing in light of what they know to be the relation between time, speed, and distance.

An example of an inconsistency between the scales of distance and speed illustrates the point. Subject is a university student of psychology (humanities major):

E: What about the distance?
S: The outer one is between 90 and 100 cm, and the inner one is 10 cm.
E: What about the time?
S: About 10 seconds.
E: And the speed?
S: 33.
E: 33 what?
S: (very long pause) Revolutions per minute (said in English).

The third limitation pertains to the concept of movement itself and to its implication for speed as well as for distance. We suggest that many subjects, including adults, do not seem to be aware of the relativity of movement, (i.e., the fact that an object can be both moving and stationary, depending on the frame of reference). In the present study the two magnets did not change their spatial positions from two points of view: relative to the disc they were attached to and relative to each other. At the same time they did change their positions relative to frames of references other than the disc: the perceiver, the table, etc. Thus, when asked to describe or compare the movements, some tended to conclude that the objects did not move at all or that their movement was merely an extension of the disc's movement and thus it was undifferentiated in any of the terms—time, distance, or speed.

The fourth limitation of the concept of speed, and to some extent, that of distance as well is that some individuals confused them with notions of work, energy, or force. People feel that when they run faster they use more energy and end up exhausted. Similarly, when they want to drive a car faster they put in more energy by pushing the accelerator harder. In the same vein, they feel that when they run further they end up more tired, and in the case of their car, they use up more gas. Perhaps through these experiences and others, speed, and to some extent, distance, become equivalent to some intuitive notions of energy, work, or output. An outcome of this equivalence is the assumption that a disc moved by a single motor consuming a certain amount of energy in its movement cannot be moving at different speeds in two different spots or even cover different distances. Moreover, since the magnets are not moving autonomously under their own power, their energy is that of the disc itself. Hence, the speed of each and its distance is either null or is equal to that of the disc, and thus the two speeds as well as the two distances must be equal to each other.

The confusion of speed with energy or force seems to be extremely resistant to counter suggestions. Subjects, who were embarrassed by their own conclusion that an object moving for the same time and at the same speed as another could cover a longer distance, tended to change their judgment of distance as frequently as they changed their judgment of speed. Many changed correct distance judgments into incorrect judgments, although they had previously been able to explain why the distances must be unequal.

The following excerpt from an interview illustrates this point. Subject is a 12th grader of sociology (humanities major):

S: The distance that this (the outer object) covered is more than this (the inner object) since its diameter is longer.
E: And what about the speed?
S: The speed is the same speed. Oh, no, wait a minute! Something doesn't seem right. If the time is the same time and the speed is the same speed, so it must be the same distance, too!
E: So, what do you say?
S: Now that I am thinking about it, this (the inner object) simply rotated more times. Let's say this one (the outer object) rotated twice, and this one (the inner object) rotated five times, so it's the same distance.

The fifth limitation of the concepts of speed and distance is the tendency to judge them according to end positions, catching up or passing. The object which catches up with and/or passes the other is perceived as being faster or as covering a longer distance. According to Piaget (1970), this kind of reasoning implies ordinal concepts of speed and distance rather than interval ones. This is the case for distance because it involves comparing only end positions rather than beginning and end positions. For speed, it involves comparing spatial positions at one point in time with the spatial positions at another point in time, rather than considering the entire distance and duration for both.

It should be noted that the cues of catching up and passing are indeed reliable cues for speed in many situations and for distance in few situations. Particularly, they are valid cues in a wide variety of parallel track races. Therefore, many adults may consider these cues to be valid for estimating speed and some may even consider them to be valid cues for estimating distance. But in circular movements, inter alia, these cues can lead to errors.

Concept Structure

Up to this point we described different limitations of physical concepts. This description raises the question of how knowledge concerning a concept is organized in children and adults. McCloskey and his co-workers (Caramazza, McCloskey, & Green, 1981; McCloskey, 1983a, 1983b) have interpreted errors exhibited by adults in different problems concerning motion as stemming from well-organized and coherent theories based on erroneous assumptions, assumptions that may be traced in the history of physical theories. In contrast, it is our impression that the adults' as well as children's concept of speed (as well as distance) is a conglomerate of uncoordinated fragments of knowledge derived from different sources. For many subjects, speed is equivalent to the perceptual impression of "fastness" or blurriness; is also equivalent to an intuitive notion of force or energy; is assessed on the basis of catching up and passing; is read from speedometers and from pointers in turntables; is defined by distance per time or by number of revolutions per time, and so on. The limitations on the validity of each of the various cues for speed are unclear. The distinction between circular and linear speeds are vague or nonexistent. The meaning of its unit of measure is unclear, just as it is unclear as to when an object moves or stands still.

Different fragments of knowledge are more readily available to some subjects than to others or at least considered by them to be more reliable bases for judgments. For those who believe that speed is primarily distance per time, the conclusion that the speed of an outer object on a rotating disc is greater is easy to arrive at. For those who believe that speed is essentially energy, the conclusion that the speeds of all spots on the same disc are equal is self-evident. For many subjects over a very wide age range, the two conclusions clash, and they can not resolve the conflict with a feeling of confidence about the correctness of the resolution.

The erroneous notions and confusions adults exhibited in the present context about speed, and to some extent, about distance, are equivalent to those young children showed in Piaget's studies. The surprising similarity between our adults' reasoning and that of children dealing with linear movements leads us to propose that ''childish'' misconceptions about physical concepts are frequently not replaced with correct concepts through learning or development. Experience may provide correct responses which are restricted to well-known limited contexts such as linear motions. When the same concepts are applied to circular motions, ''childish'' misconcepts emerge. This description does not necessarily imply that adults are potentially incapable of functioning at a higher level than children. It is possible that adults might learn more from structured experience than children would. Such an experience should take into account the intuitive misconceptions which appeared in the present study.

References

Acredolo, C. (1981). Acquisition of conservation: A clarification of Piagetian terminology: Some recent findings and an alternative formulation. *Human Development, 24,* 120–137.

Acredolo, C., Adams, A., & Schmid, J. (1984). On the understanding of the relationships between speed, duration and distance. *Child Development, 55,* 2151–2159.

Acredolo, C., & Schmid, J. (1981). The understanding of relative speeds, distances and durations of movement. *Developmental Psychology, 17,* 490–493.

Berenthal, B. J. (1981). The significance of developmental sequences for investigating the what and how of development. In K. W. Fischer (Ed.), *Cognitive development.* San Francisco, CA: Jossey-Bass.

Brainerd, C. J. (1973). Judgments and explanations as criteria for the presence of cognitive structures. *Psychological Bulletin, 79,* 172–179.

Campbell, R. L., & Richie, D. M. (1983). Problems in the theory of developmental sequences: Prerequisites and precursors. *Human Development, 26,* 156–172.

Caramazza, A., McCloskey, M., & Green, B. (1981). Naive beliefs in ''Sophisticated'' subjects: Misconceptions about trajectories of objects. *Cognition, 9,* 117–123.

Crepault, J. (1983). Modeles, raisonnements et notions de temps chez l'adolescent [Models, reasonings and notions of time in the adolescent]. *Cahiers de Psychologie Cognitive, 3,* 387–392.

di Sessa, A. (1982). Unlearning Aristotelian physics: A study of knowledge-based learning. *Cognitive Science, 6,* 37–75.

Ehri, L. C., & Muzio, I. M. (1974). Cognitive style and reasoning about speed. *Journal of Educational Psychology, 66,* 569–571.

Fischer, K. W., & Bullock, D. (1981). Patterns of data: Sequence synchrony and constraint in cognitive development. In K. W. Fischer (Ed.) *Cognitive development.* San Francisco, CA: Jossey-Bass.

Flavell, J. H. (1982a). Structures, stages and sequences in cognitive development. In W. A. Collins (Ed.), *The concept of development: The Minnesota symposia on child psychology.* Hillsdale, NJ: Erlbaum.

Flavell, J. H. (1982b). On cognitive development. *Child Development, 53,* 1–10.

Flavell, J. H., & Wohlwill, J. F. (1969). Formal and functional aspects of cognitive development. In D. Elkind & J. H. Flavell (Eds.) *Studies in cognitive growth: Essays in honor of Jean Piaget.* New-York: Oxford University Press.

Gelman, R. (1983). Recent trends in cognitive development. In J. C. Schierer & A. R. Rogers (Eds.), *The G. Stanley Hall lecture Series* (vol. 3). Washington, DC: American Psychological Association.

Hewson, P. W. (1983). *Microcomputers and conceptual change: The use of microcomputers program to diagnose and remediate an alternative conception of speed.* Paper presented at the Annual Meeting of the American Educational Research Association, Montreal.

Kuhn, T. S. (1977). *The essential tension.* Chicago, IL: University of Chicago Press.

Levin, I. (1977). The development of time concepts in young children. Reasoning about duration. *Child Development, 48,* 435–444.

Levin, I. (1979). Interference of time-related and unrelated cues with duration comparisons of young children: Analysis of Piaget's formulation of the relation of time and speed. *Child Development, 50,* 469–477.

Levin, I. (1982). The nature and development of time concepts in children: The effects of interfering cues. In W. J. Friedman (Ed.) *The developmental psychology of time.* New York: Academic Press.

Levin, I. (1983). The sequence in development of time, speed and distance concepts: task analysis. *Cahiers de psychologie cognitive, 3,* 371–374.

Levin, I., Goldstein, R., & Zelniker, T. (1984) The role of memory and integration in early time concepts. *Journal of Experimental Child Psychology, 37,* 262–270.

Lovell, K. (1959). A follow-up study of some aspects of the work of Piaget and Inhelder on the child's conception of space. *British Journal of Educational Psychology, 29,* 104–117.

Lovell, K., & Slater, A. (1960). The growth of the concept of time: A comparative study. *Child Psychology and Psychiatry, 1,* 179–190.

Lovell, K., Kellett, V. L., & Moorhouse, E. (1962). The growth of the concept of speed: A comparative study. *Child Psychology and Psychiatry, 3,* 101–110.

Maccoby, E. E. (1983). *Socialization and developmental change.* Presidential address, Society for Research in Child Development, Detroit, MI.

McCloskey, M. (1983a). Naive theories of motion. In D. Gentner & A. L. Stevens. *Mental models.* Hillsdale, NJ: Erlbaum Association.

McCloskey, M. (1983b). Intuitive physics. *Scientific American, 248,* 122–130.

Miller, K. F. (1983). *Child as the measurer of all things: Measurement procedures and the development of quantitative concepts.* Paper presented at the Carnegie Symposium on Cognition: Origins of Cognitive Skills, Pittsburgh, PA.

Piaget, J. (1969). *The child's conception of time.* London: Routledge & Kegan Paul.

Piaget, J. (1970). *The child's conception of movement and speed,* New-York: Ballantine Books.

Shanon, B. (1976). Aristotelianism, Newtonianism and the physics of the layman. *Perception, 5,* 241–243.

Siegler, R. S. (1981). Developmental sequences within and between concepts. *Monographs of the Society for Research in Child Development, 46* (No. 2).

Siegler, R. S., & Richards, D. D. (1979). The development of time, speed, and distance concepts. *Developmental Psychology, 15,* 288–298.

Starkey, P. (1983). *Some precursors of early arithmetic competencies.* Paper presented at the Biennial Meeting of the Society for Research in Child Development. Detroit, MI.

Strauss, M. S., & Curtis, L. E. (1981). Infant perception of numerosity. *Child Development*, 52, 1146–1152.

Strauss, M. S., & Curtis, L. E. (1983). *Development of numerical concepts in infancy*. Paper presented at the Carnegie Symposium on Cognition: Origins of Cognitive Skills. Pittsburgh, PA.

Trowbridge, D. E., & McDermott, L. C. (1980). Investigation of student understanding of the concept of velocity in one dimension. *American Journal of Physics, 48,* 1020–1028.

Trowbridge, D. E., & McDermott, L. C. (1981). Investigation of student understanding of the concept of acceleration in one dimension. *American Journal of Physics, 49,* 242–253.

Whitaker, R. J. (1983). Aristotle is not dead: Student understanding of trajectory motion. *American Journal of Physics, 51,* 352–357.

Wilkening, F. (1981). Integrating velocity, time and distance information: A developmental study. *Cognitive Psychology, 13,* 231–247.

Wilkening, F. (1982). Children's knowledge about time, distance and velocity interrelations. In W. J. Friedman (Ed.), *The developmental psychology of time*. New York: Academic Press.

Za'rour, G. I., & Khuri, G. A. (1977). The development of the concept of speed by Jordanian school children in Amman. In P. R. Dasen (Ed.), *Piagetian psychology: Cross-cultural contributions*. New York: Gardner Press.

5

Heterogeneity, Inconsistency, and Asynchrony in the Development of Cognitive Structures

Elliot Turiel
University of California, Berkeley

Philip Davidson
Michigan State University

In recent years it has become increasingly apparent from developmental research that individuals do not display a unitary orientation to the physical or social world. Younger and older children as well as adolescents and adults make judgments that can be characterized as multifaceted and heterogeneous. Moreover, in certain cases individuals manifest judgments that appear inconsistent and even contradictory with each other. For example, a series of studies (summarized in Turiel, 1983a) show that young children have varied understandings of social relations and social systems. A fair amount of research has yielded evidence that the same children display opposing social orientations in their judgments. Children sometimes uphold authority and at other times contradict authority. They are both accepting and rejecting of rules or regulations. Sometimes children are bound to the social context, while at other times they ignore the context. They make judgments that are relativistic, as well as judgments that are universalistic. These research findings do not allow classification of individuals within broad social categories, such as individualistic or collectivistic, egalitarian or hierarchically oriented, autonomous or conformist. All these orientations have been found to *coexist* in that each is reflected in the individual's judgments.

Apparent inconsistencies are also found in the realm of nonsocial cognition. For instance, in some task contexts young children appear to understand numerical concepts of equality, more than, less than, addition, and subtraction; in other contexts, they appear not to. Thus, children as young as 2 years of age understand that adding an element to a set increases its numerosity qualitatively (Mehler & Bever, 1967). Moreover, young children can solve quantitative problems (Starkey, 1983) with sets of less than four elements (i.e., $2 + 1 = ?$, $3 - 1 = ?$). Research also shows that young children do not solve number conservation problems, even with set sizes of 4 or less (Beilen, 1968; Silverman & Briga, 1981), suggesting that they lack these competencies.

Correspondingly, inconsistent judgments are evident in somewhat older children. Children of about 7 or 8 years of age are capable of seriating seven objects by order of size (Inhelder & Piaget, 1964). However, when faced with the problem of seven seriated disks that differ imperceptibly by .2 mm (Piaget, 1980), children of the same age deny they form a series and resort to categorical thinking (A, B, C, D are "small"; E, F, G are "big"). Whereas 6- and 7-year-old children demonstrate an understanding of the notion of unit in that they typically succeed on number conservation tasks, they typically do not succeed on length conservation and measurement tasks, indicating that they have not yet mastered the notion of unit.

Striking examples of heterogeneity and inconsistency can be seen in social actions and in the relations between social judgments and actions. The most dramatic illustrations of inconsistencies between judgment and action come not from developmental research with children but from social/psychological studies with adults (Asch, 1956; Haney, Banks, & Zimbardo, 1973; Milgram, 1974). Consider the example of Milgram's (1974) famous (and infamous) studies on obedience to authority. As is well known, in the basic experimental paradigm subjects were instructed to administer increasing electric shocks, up to a very high voltage, to another person. In one experimental condition the majority of subjects administered shocks to the end of the scale. These findings can be (and often are) taken as evidence of the lack of consistency between moral judgments and behavior, since in all likelihood most of the subjects judged it wrong to inflict harm on another person (Milgram, 1974). This is not the entire story, however. In several other experimental conditions of Milgram's study, the majority of subjects discontinued their participation in the experiment, rejecting the experimenter's instructions to administer shocks to another person. (As discussed in Turiel and Smetana, 1984, those experimental conditions varied in the proximity or salience of either the victim or the authority, relative to the subject.) Therefore, several experimental conditions yielded consistency between moral judgments and behavior. Moreover, the findings of all the experimental conditions, taken together, show that there is heterogeneity in social behaviors.

These examples of heterogeneity and inconsistencies in social behaviors, as well as in social, logical–mathematical and physical concepts, do not point to the existence of a general structure in the individual's thought forming a unitary or homogeneous system. The findings are also incompatible with the proposition that developmental changes involve abrupt, saltatory shifts from one structural system to another. Moreover, it appears that there is no simple deterministic relation between judgment and behavior, because behavior is influenced by the parameters of the task and the situational context (as in the situational variations among the different experimental conditions in Milgram's studies). If the theoretical propositions of stage and structure in development are taken to imply (Flavell, 1971, 1981; Flavell & Wohlwill, 1969) abrupt and synchronized transformations of homogeneous patternings of thought uniformly manifested across tasks and situations, then these propositions receive little support from the types of empirical examples mentioned thus far.

However, the findings of heterogeneity and inconsistency are in concordance with an alternative view of structure in development. The alternative view is that cognitive structure or organization is a feature of the individual's knowledge systems rather than of "the thought" of the individual. This proposition has two central features distinguishing it from the view of structural development as entailing homogeneity of thought and abruptness in change. One feature is that thought is organized in accordance with domains. The second is that development entails qualitative changes, within the domains, that occur through a *gradual* process of the deformation of a system of thinking in conjunction with the reformation of a new system. It is also proposed that the individual's behavior is not strictly determined either by cognitive structures or by situational variations. Thought and action are interrelated in two respects. Children's experiences—their actions and interactions—are a source of conceptual development. In turn, constructed concepts are applied to behavioral decisions and choices.

The perspective taken in this chapter is that heterogeneity in judgments and actions reflects systematic and organized forms of reasoning. Findings of different types of judgments on different types of cognitive tasks do not lend themselves to the interpretation of simple lack of coherence or organization in reasoning. Heterogeneity in individuals' thinking can stem from separate systems of thinking that are organized within delimited domains and thus also reflect coherence in structures of thought. Correspondingly, both consistencies and inconsistencies of judgments in different events or tasks require interpretation and, in themselves, do not constitute a test of the organized nature of thought. In addition, individuals are often posed with multifaceted situations including components from more than one domain, which may require their coordination. This further complicates interpretations of heterogeneity or inconsistency in that patterns of responses to tasks involving the coordination of domains may differ from responses to domain-specific tasks. Consequently, domain coordinations constitute an additional source of inconsistency in judgments as well as a source of conflicts and contradictions. The proposition of domain-specific systems of knowledge, therefore, requires delineating domains, their boundaries, and the criteria by which they are distinguished, and ascertaining how under certain conditions domains are coordinated.

Structure, Development, and Domain Specificity

The main focus of this chapter is on research dealing with conceptual domains and its implications for an understanding of heterogeneity and homogeneity in performance across different cognitive tasks, continuities and discontinuities in development, and inconsistencies and consistencies in thought and action (especially social thought and action). These issues, which have been of much concern to developmentalists, pertain to horizontal (synchronic) organization, vertical (age-related) organization, and the organization of thought and action as it bears upon both horizontal and vertical organization. Whereas there has been a tendency to interpret findings from research on cognitive development as supportive of heterogeneity (over homogeneity), continuity (over discontinuity), and inconsistency (over consistency), we interpret the findings to demonstrate the coexistence of each of these polarities in cognitive organization. In our view, an adequate interpretation of research findings requires explanations and assessments that account for (a) qualitative differences between domains of knowledge (implying heterogeneity); (b) organization within domains (homogeneity); (c) qualitative changes in development within a domain (implying discontinuities); (d) gradual processes of change within a domain (implying *temporal* continuities); and (e) between-domain coordinations (with the potential for multiplicative relations of consistencies and inconsistencies). Before considering these interpretations, we outline the theoretical basis for relating domain specificity to horizontal and vertical dimensions of cognitive organization.

Domains of Knowledge: Horizontal Organization

Turning first to the question of horizontal organization, the assumption is that knowledge is constructed through the interactions of the individual with the environment. It follows that knowledge depends both on the structure of existing schemes, conceptions, ''naive theories,'' etc. (the subjective side), and on the structure of the environment (the objective side). In one sense, the subject's prior structures would appear to have a logical priority during these interactions, because novel information about the environment is not processed directly as a factual given but is itself constructed by the interaction and is thus subject to varying degrees of distortion. On the other hand, normal development entails increasingly successful and stable adaptations to reality, and in that sense, it appears that priority is given to the environment. This differential emphasis is seen in Piaget's work and has been a source of some confusion for interpretations of his position. For instance, in analyzing the constructive processes involved in logico-mathematical thought, Piaget dwells on children's imposition of their own cognitive organization upon reality; raw empirical data is to be transformed, rearranged, or even suppressed (Piaget, 1971). Thus, it may appear that the focus is on the subjective side of interactions. Elsewhere, however, in discussing the characteristics of separate conceptual acquisitions, Piaget attributes their distinctiveness to differences between types of empirical data. For instance, the developmental lag between the acquisition of the concepts of number and measurement is attributed to differences in interactions with discrete objects, on the one hand, and interactions with continuous objects, on the other hand (Piaget, Inhelder, & Szeminska, 1960). The focus here is on the object side of the interaction. A comprehensive view of structural development includes the subject's constructions in interaction with the environment—constructions that are organized within domains corresponding to basic distinctions in the structure of the environment.

With regard to the subjective side, a cognitive structure may be analyzed into its regulations and its elements. We assume that structures are self-regulating; that is, the regulations or controls required for the formation and utilization of knowledge are assumed to be an intrinsic aspect of knowledge itself. An apt metaphor is a biological one in which a knowledge structure is likened to a functioning organ that, although communicating directly or indirectly with other organs and the external environment, yet preserves its own function and structure. This assumption about self-regulation has two immediate theoretical consequences. It means first of all that a given structure actively initiates the assimilation of novel elements and actively accommodates or reinterprets prior elements in the light of new evidence. Thus, a structure reorganizes itself. Second, it means that a structure is bounded: "The structure is preserved or enriched by the interplay of its transformation laws, which never yield results external to the system nor employ elements that are external to it" (Piaget, 1970, p. 5). Thus a second implication of self-regulation is that the range of elements available for assimilation is *delimited* by the requirement that elements are integrated in a conceptually meaningful way with existing knowledge. This consideration implies that a knowledge structure is not all-encompassing.

That is to say, the assimilation process itself selectively includes elements that are appropriate and excludes elements that are not appropriate to the concerns of a given knowledge structure. The alternative thesis, that a single knowledge structure may contain disparate types of elements, requires abandoning the assumption that structures are self-regulating. For, suppose a given structure contains disparate elements (e.g., notions about spatial relations and notions about interpersonal relations), then through self-regulation alone such a structure would inevitably generate inappropriate combinations of elements and therefore inadequate inferences. One might then postulate some additional cognitive component (e.g., an executive control device) which systematically selects and processes the varieties of information in the knowledge structure. But this type of solution is not compatible with the previous assumption that the knowledge structure is self-regulating (since such a regulatory device would not be intrinsic to the substance of knowledge—e.g., neither spatial nor interpersonal in nature—but rather a separate development and cognitive phenomenon). Therefore, if a knowledge structure is self-regulating, then it is bounded from the outset by the concerns of a specific branch of knowledge.

Ultimately, one needs to account not only for regulations within knowledge structures but also for the relations between distinct structures. We propose that it is more productive to think of such a relation as a specific coordination between substantive knowledge domains rather than as a dissolution of domain boundaries resulting in a different, more global domain. The proposition that domains can be conceptualized as entailing relations in coordinations while maintaining their boundaries is similar to Piaget's (1970) notion of a confederation between structures: "In being treated as a substructure, a structure does not lose its own boundaries; the larger structure does not 'annex' the substructure; if anything, we have a confederation, so that the laws of the substructure are not altered but conserved" (p. 14).

Following these propositions, then, a necessary step is to provide a description of how the boundaries of knowledge domains are established. If a structure is not global, then several levels of increasingly restrictive structuration remain to be considered. At the other pole to globality is a view of cognition as fragmented into indefinitely many components. Such a view would be consistent with an interpretation of distinct cognitive structures as corresponding directly to differences in the surface structure of the environment; differences, that is, in specific object properties or in specific, ad-hoc learning contexts. Such a cognitive system, we believe, would be as problematic as the global version at the other pole.

Our hypothesis is that the boundaries of organized knowledge domains reflect differences between types of subject–object interactions. This is consistent with the view that the elements of knowledge do not correspond to elements of raw empirical data; rather, elements themselves are constructed through interaction. The properties of distinct types of interactions determine the assimilation of elements to one domain rather than another; hence, a domain can be defined with reference to a given class of interactions. This definition yields a relatively small number of substantive epistemological categories; we return to a discussion of these in a subsequent section.

Transformations Within Domains: Vertical Organization

If one of the central issues in horizontal organization is that of structural boundaries, one of the central issues for vertical or age-related organization is structural stability. As noted previously, developing systems undergo reorganization. A fundamental guiding principle in this reorganization is the apparent tendency of the system to complete itself by identifying and assigning an inverse for every element. Such a completed system is stable in the sense that, starting with any arbitrary element of thought, it is always possible to return to the identity by combining the given element with other appropriate elements. For example, in the moral domain, the system is stable, if given any element (e.g., some moral transgression) it is possible to combine it with other elements (concerning responsibility, punishment, etc.) so as to conserve, or return to, a conception of what is just.

This definition of stability, it will be noted, refers to a formal property of elements (existence of inverses) rather than the content or quantity of elements. However, a state of stability is not likely to be attained before the content of the structure is extensively elaborated. Given that knowledge is not preformed in the environment, children's initial constructions within any domain are likely to be poorly defined, not precisely related to inverse elements, and thus in need of progressive corrections and refinements. Hence, the early forms of any knowledge domain are relatively unstable. If cognitive development were solely a process of continuously refining the elements of knowledge through progressive discriminations and generalizations, the question of developmental stages would not arise. However, there is evidence that most conceptual systems reach stability not once, but several times.

Developmental change is obviously coincident with change in age or time. However, in our view, developmental stages cannot be defined solely with reference to a time dimension. That is, it is not sufficient to say that children are "in" a certain stage *because* a form of cognitive functioning exists for a relatively long period, or that children are "between" stages *because* their functioning shows variations within a short span. It may be true that the stability of a domain would translate into an observed temporal stability. Nevertheless, there is no necessary connection between the two. It is possible that during certain periods in the lifespan cognitive functioning can be stable (i.e., stagelike) for a relatively short period of time and that transitions can occur over a relatively long period of time (thus appearing continuous on the temporal dimension). A description of stages and their development depends on cognitive criteria and explanation of processes of change in conjunction with measures of time span rather than solely on temporal criteria.

Furthermore, as already noted, the assumption that knowledge is self-regulating implies that the locus of stage development is within distinct knowledge domains. From the domain perspective, homogeneity of cognitive functioning across a large number of domains would not be a sufficient criterion for assuming children are "in" a certain stage. For general homogeneity to be useful as a criterion of stage development, it would have to be related to an explicit theory of interdomain correspondences and coordination. Such a theory would (a) propose that coordination of certain specific domains represents a developmental progression; (b) provide descriptions of qualitatively distinct stages in interdomain coordination and describe mechanisms of transition from one stage of coordination to the next; and (c) derive from these theoretical statements a prediction that the domains in question should show synchronous development. Note that, lacking an appropriate theoretical derivation, nothing prevents the contrary expectation that the development of coordination between two domains, A, B, depends on their being *out* of synchrony (e.g., requires that domain A always be one stage in advance of domain B). In the absence of an explicit theory of the development of interdomain coordinations and specification of the correspondences in the levels of different domains, findings of either heterogeneity or homogeneity across domains are ambiguous with respect to the question of cognitive stages.

In our view, a comprehensive theory of stages in the coordination of knowledge domains is at the present time solely a hypothesis. Fragments of a theory of coordination, covering selected areas of cognitive development, have been proposed. For instance, Piaget (1941/1965) argues that development of the mathematical concepts of ordinality and cardinality is necessarily interdependent with development of intensive order and class concepts; even this rigorous analysis of domain coordination does not, however, predict that these mathematical and logical notions must be acquired in lock-step synchrony. We return to a more detailed consideration of interdomain coordination in subsequent sections. For the present, we note that it is not possible to attempt a description of linkages between cognitive domains, until domains have been isolated and accurately described. Thus, descriptions of within-domain concepts and their development are, at the least, more fundamental than analyses of between-domain developments.

We have proposed that the determination of cognitive stages does not necessarily depend on stability of functioning across time nor on synchrony of functioning across domains. Stage transition involves a change in the relation between the elements and regulations internal to a given knowledge structure. During the consolidation of a structure, both elements and regulations are transformed. Initially, elements are not comprehensively defined and regulations are only partial. An illustrative example comes from research on children's moral judgments (Davidson, Turiel, & Black, 1983). It has been found that moral judgment at age 6 years is limited to fairly familiar issues (the elements); reasoning is systematic but regulated by concrete and categorical concerns with maintaining welfare and avoiding harm. By the age of 10 years, children also apply their moral reasoning to relatively unfamiliar issues, thus indicating a more comprehensive definition of moral elements. The older children's judgments also show a concern with maintaining reciprocity and justice, indicating a more comprehensive conceptual regulation. For 6-year-olds, the concern for welfare and avoidance of harm is well consolidated as a regulatory principle; regulations based on relational issues of justice are not elaborated. The consolidation by the age of 10 years of the more abstract concern for maintaining justice is not independent of children's increasing ability to cognize the notion of welfare per se, thus transforming its status from that of a regulation to that of an element in the novel structure. That is, the more advanced stage of reasoning includes a concern with achieving a just reciprocity between the welfare claims of various individuals. This illustrates how cognitive stage transition can be defined as a transformation in the status of the regulations vis-à-vis the structure: What had been a regulation in a prior structure becomes an element in a subsequent structure. A more general phenomenon of the same type is the transformation of regulations having the form of one-place functions into elements in a subsequent structure whose regulations have the form of functional compositions. Thus, for example, Langer (1982, 1983) traces the origin of representational cognition to the transformation of "first-order" sensorimotor mappings into "second-order" mappings of mappings; these second-order mappings have the properties required for separation of action from meaning, or the initiation of symbolization.

Identification of Domains and Definitional Criteria

The implication of the position we have presented is that different types of within- and across-age performances are far from straightforward and not readily interpretable in the absence of a formulation of epistemological categories. This is to say that it is not feasible to make simple mappings of heterogeneity to lack of coherence in the organization of thought, asynchrony to lack of systematicity in development, or inconsistencies to lack of relatedness of thought and action. This is because of the plausibility that individuals develop qualitatively different knowledge systems. If individuals do acquire distinct forms of knowledge, then sources of between-domain heterogeneity and inconsistency would not be reflective of an absence of structure or organization in cognition. Therefore, inquiry into development and behavior requires corresponding analyses of the substance of the knowledge under investigation (Chomsky, 1975; Piaget, 1971).

It is indeed the case that domains of knowledge are typically distinguished through formal disciplines in scientific and philosophical study (e.g., logic, mathematics, physical or natural sciences, psychology, sociology, ethics) and that such distinctions have something to do both with different types of theorizing (the subjective side) and different aspects of the environment investigated (the objective side). Analogously, children's developing knowledge may have components, partially based on their interactions with and structuring of different aspects of the environment. In pointing out the divisions among formal disciplines, we solely mean to suggest an analogy for an approach to cognition in ontogenesis. This approach is, however, consistent with the heuristic developmental principle that the advanced form of cognition serves as a standard frame of reference for the interpretation of earlier forms (cf. Kessen, 1966, pp. 63–64). Thus some investigators have found it useful to think of even young children as budding ethicists (Kohlberg, 1968), budding mathematicians (Gelman & Gallistel, 1978), or budding physical scientists (Kamii & DeVries, 1978) and social scientists (Ross, 1981). If this analytic strategy is useful in identifying knowledge domains, it is not because distinct formal disciplines predetermine the boundaries of children's thought but rather because the distinct domains constructed in children's interactions remain recognizable as some of the distinct formal disciplines that comprise thought.

Moreover, there is a growing body of research, in addition to our own, showing that children's thinking includes separate systems (Carey, 1984; Chomsky, 1975; Damon, 1977, 1983; Feldman, this volume; Flavell, 1982; Keil, this volume; Wellman, 1982). In this light, we believe that research on children's thinking requires, at a minimum, defining and categorizing the types of knowledge investigated as well as providing criteria for domain boundaries which would include specification of how broad or narrow domains are construed. (Notice that even in proposing that thought is not bounded by separate domains, definitional criteria would have to be provided for the global "domain"; indeed, this has been a concern of theorists who seek to clarify broader distinctions such as those between cognition and perception, or between conceptual knowledge and functions such as memory or representation.)

Part and parcel of the research process, then, is the formulation of epistemological categories, which includes taxonomies of horizontal organizations. The taxonomic task has traditionally focused on delineation of stages or levels of development and identification of cognitive tasks or items. The performance of children of differing ages is assessed through the standard of the developmental levels charted. However, a multiple classification system is needed so that the relation of task performance to developmental level is organized by a categorization of domains (with their separate developmental-level taxonomies).

The formulation of systems for classifying domains of knowledge is both a definitional and a research issue in that definitions and data interpretations feed back on each other. Our strategy has been one of moving from (a) initial domain identifications and definitional criteria (this includes analysis of the substance of knowledge within various formal disciplines, as mentioned earlier), to (b) the gathering of data bearing on how subjects use and distinguish the categories, to (c) reformulations of the categories and criteria—and so on. Moreover, the identification of conceptual domains includes specification of those elements that are not unique to a single domain but may occur in various guises in different domains. Such elements take their form from the domain in which they are embedded and, thereby, may be conceptualized in varying ways.

From the perspective of domain specificity, a question of some importance is how to determine the range of domains of knowledge. We do not yet have a clear or unequivocal answer to this question. We view the problem as one requiring a good deal of research based on working hypotheses regarding the more fundamental, well-studied domains of knowledge. However, the proposition that domains are functioning, self-regulating systems (in contrast with global structure or an indeterminate number of components) has led us to identify a limited number of fairly broad categories (see Keil, this volume, for a similar view of the range of domains).

We begin this map of cognitive territories by painting with a broad brush and then fitting in layers of detail. The map includes three major levels of structuration: realms, domains, and concepts. At the first and broadest level, it is necessary to make a dichotomous partition between the social realm and the nonsocial realm. The distinguishing property of social knowledge is that it is constructed through subject–subject interaction and is about those interactions; that is to say, the interaction of a subject with another independent source of intentionality. In this sense, social knowledge is essentially knowledge of subjects, whereas nonsocial knowledge is essentially knowledge of objects. Therefore, it is possible to make an explicit connection between the social–nonsocial dichotomy and a basic epistemological distinction, the subject–object dichotomy. This analysis provides the rationale for the proposal that social and nonsocial realms comprise a fundamental partition of conceptual organization.

A comparison of the major epistemological and developmental landmarks within social and nonsocial knowledge reveals a parallelism between the two realms. Knowledge within the nonsocial realm develops, most typically, in interaction with physical media; subjects organize these interactions according to (at least) three major epistemological concerns: (a) there is a concern with understanding the empirical properties of the media as such, leading to physical and causal knowledge; (b) a concern with understanding the intrinsic forms of organization in which the media are embedded, leading to spatial and temporal knowledge; and (c) a concern with the most general, intrinsic forms of organization of the subject–object interactions themselves, leading to logical and mathematical knowledge. Knowledge that develops in interaction with social media is constructed according to (at least) three corresponding epistemological concerns: (a) a concern with the empirical properties of the media itself, which leads to knowledge of persons or psychological systems (this is the aspect in which social knowledge is most explicitly self-referential and causal); (b) a concern with the forms of organization in which the media are embedded, leading to knowledge of social systems (or of the matrix of sociocultural institutions from families to governments); and (c) a concern with the intrinsic and obligatory forms of organization of the subject–subject interactions themselves, leading to moral knowledge. [From a philosophical perspective, the parallels between logical necessity and moral obligation are captured in Rawls's (1971, p. 4) dictum that as "first virtues of human activity, truth and justice are uncompromising."]

Table 1 illustrates the first step in our strategy of domain identification, which involves both considerations concerning the substance of different epistemological categories and practical considerations concerning different subject–object interactions. The categories listed are not intended as an exhaustive partition of all knowledge domains, but only of the most readily identifiable and well-studied ones. Second, the table indicates that nonsocial domains do not map one-to-one onto social domains. Within the nonsocial realm, we distinguish pairs of domains for each epistemological concern. That is, the concern with empirical properties comprises the separate domains of causal knowledge and physical knowledge; the concern with objective forms comprises the temporal domain and the spatial domain; and the concern with the organization of interactions comprises the logical domain and the numerical domain.

It should also be clear that this analysis makes use of the distinction between the media of interaction and the object of interaction. By media, we refer to the materials a subject interacts with, as they might be recorded by an objective observer (without access to the subject's interpretation). By object, we refer to the epistemological significance of the materials, which is determined by the specific type of knowledge the subject brings to bear on these materials. Thus, when children manipulate physical materials so as to investigate spatial relations, the physical media are construed as spatial objects and used in the service of spatial knowledge (rather than, for instance, being used in the service of numerical or causal knowledge). As a further illustration, consider a subject shaking hands with another person. Lacking further information about how the subject construes the media, it is not possible to determine the object of the interaction. The object may be the other

Table 1. Taxonomy of Social and Nonsocial Knowledge Domains.

Realm	Domain	Object Characteristic	Knowledge Characteristic[a]
Social	Person knowledge	Empirical content of subjects	contingent
	Societal knowledge	Sociocultural matrix in which media are embedded	contingent
	Moral knowledge	Abstracted form of subject–subject interactions	obligatory
Nonsocial	Physical knowledge Causal knowledge	Empirical content of objects	contingent
	Spatial knowledge Temporal knowledge	Objective matrix in which media are embedded	contingent
	Logical knowledge Mathematical knowledge	Abstracted form of object–object interactions	necessary

[a]Specification of knowledge characteristics refers to definitional criteria for the substance of domains. *Contingent* refers to inferences about the object of the interaction; *necessary* and *obligatory* refer to inferences about the organization of the interaction itself.

person *qua* person; or the other person *qua* representative of a social institution (the conventional handshaking activity suggests this as at least a reasonable guess); or the other person *qua* agent/recipient of moral conduct. Moreover, the object of this interaction may not be a social object at all. It could conceivably be the person *qua* thing with certain physical properties (warmth, skin texture); with spatial properties (height, length, area, volume); with mathematical properties (four fingers and a thumb make five digits). The domain-specific model proposes that media are construed in a variety of ways, but not arbitrarily. It is proposed that a particular domain is most closely connected with a particular type of interaction (Turiel, 1983a). In certain interactions, two or more domains may be equally applicable; such situations entail issues of domain coordination, discussed in some detail in a subsequent section.

As we have noted previously, the principal strategy for domain identification involves distinguishing types of subject–object interactions (for examples of empirical observations of different types of interactions related to the moral and social organizational domains see Much & Shweder, 1978; Nucci & Nucci, 1982a, 1982b; Nucci & Turiel, 1978). The assumption is that there are different types of theorizing (the subjective side) appropriate to different objects of thought (the objective side). As shown in the preceding example, the object of thought is not given directly by the medium of the interaction; it is partially determined by the medium but also partially determined by the type of theorizing applied. Accordingly, it is possible to describe this strategy of domain identification as one of distinguishing the fundamentally different ways that media can be construed, interpreted, or treated as objects of thought.

Finally, in this model the term *concept* is reserved for the functioning substructures of a given domain. The major distinction between concepts and domains is their relative autonomy. Whereas domains are autonomous, concepts within a domain are highly interdependent and do not function independent of the epistemological concerns of the domain. For instance, a few of the concepts found within the mathematical domain are those of set, correspondence, whole number, fraction, and the operations, addition, subtraction, multiplication, and division. These are fundamentally interdependent since whole numbers can be defined as correspondences between sets or as results of operations; fractions can be defined as ratios of whole numbers or through the notion of division; and so forth. In a context of mathematical problem solving or experimentation, the application of certain concepts rather than others is determined by the organizational properties and concerns of the knowledge domain as a whole rather than by these conceptual substructures independently. A similar dependency exists for concepts of social rules and authority in the context of the moral and social–organizational domains (Damon, 1977; Tisak, 1983; Turiel, 1983a).

Domains of Social Judgment

In Table 1, the social domains are listed before the nonsocial domains. This ordering was chosen for the table simply because it represents the order in which definitional criteria and related research are discussed here. It may appear that considering the social domains before the nonsocial ones is counterintuitive, since it is often thought that logical–mathematical and physical concepts are structurally and developmentally prior to social concepts. However, the domain-specificity model does not assume that the development of social concepts is dependent on prior constructions in the nonsocial realms (see Turiel, 1983b, for discussion of this issue). In fact, domain-specific criteria, and corresponding research, have been more explicitly elaborated in the social than in the nonsocial realms. In the nonsocial realm, domains in the present sense have been identified at least implicitly, but research has typically focused on development of conceptual reasoning rather than on the criteria by which reasoning may be said to belong to one domain rather than another. We begin with analyses of how seemingly heterogeneous and inconsistent social judgments may, in fact, reflect structural coherence.

Initial efforts in the research program on social reasoning were directed at determining whether and in which ways children form judgments about fundamental social domains (e.g., Davidson et al., 1983; Nucci, 1981; Pool, Shweder, & Much, 1983; Smetana, 1981; Turiel, 1978; Weston & Turiel, 1980). Most of those studies were on morality and convention, though some work has examined distinctions between moral judgments and psychological concepts that fall under the domain of knowledge of persons (Jancaterino, 1982; Nucci, 1977). Although we are still elaborating upon judgments about these fundamental domains and pursuing hypotheses regarding criteria by which they are distinguished, much recent effort has turned to investigation of domain coordinations and thought–action relations (Smetana, 1982; Stoddart & Turiel, 1984; Turiel, 1983a; Turiel & Smetana, 1984). The methods and constructs used in the study of relations between domains were available as a consequence of the prior isolation and description of domain-specific knowledge.

Perhaps more than any other discipline, philosophy has provided a basis for definitional criteria for the domains. The working definitions guiding our research on concepts of convention and moral judgments, which borrowed liberally from some of those philosophical treatments, have been presented in detail elsewhere (see especially, Turiel, 1983a). However, a brief review, with a focus on the central distinguishing features, provides the background for discussion of those empirical findings illustrating the coexistence of different types of judgments.

In line with philosophical perspectives on convention (Lewis, 1969) and the constitutive systems in which they are embedded (Searle, 1969), *conventions* are defined as behavioral uniformities reflective of shared knowledge that coordinates interactions within social systems. In Searle's (1969, pp. 33–42) use of the concept, the meaning of certain forms of behaviors (rules, uniformities) is defined by the constituted system in which they are embedded. For instance, the rules of games are constitutive in the sense that they define the game and derive their meaning from the (constitutive) system. Correspondingly, there are events (e.g., the marriage ceremony) that derive their meaning from constituted institutions (e.g., the "institutional fact" of marriage). In this sense, constitutive systems take the form of "X counts as Y in context C" (Searle, 1969, p. 35).

The concept of constitutive systems is complemented by Lewis's (1969) analysis of conventions as shared knowledge revolving around arbitrary acts that serves the function of coordinating social interactions. That is, conventions are symbolic elements serving coordinating and efficiency functions in social organization:

> The meanings and functions of conventions, then, are contextual; they are closely related to the social system in which they occur. From the viewpoint of a constitutive system, conventions are relative to the context in which they exist. In a differently constituted system, a particular convention (or set of conventions) may not exist or may exist in another form. (Turiel, 1983a, p. 38)

In contrast, our use of the term morality is not defined by coordinations of social functions, nor is its meaning contingent on institutional arrangements. Moral judgments, as indicated in Table 1, are abstracted from social interactions and pertain to obligatory forms of organizing social interactions. Moral prescriptions are based on concepts of welfare (harm), justice (comparative treatment and distribution), and rights. The analytic distinction between moral imperatives and the constitutive standards, conventions, or practices of social organizations and institutions has been stressed in contemporary philosophical treatises (e.g., Dworkin, 1978; Gewirth, 1978; Rawls, 1971). In those perspectives, moral concepts are characterized by obligatoriness, generalizability, and impersonality. As concepts of welfare, justice and rights, morality is not defined by existing social arrangements but does constitute principles applicable to the actual arrangements of society. Therefore, morality provides a basis for one type of evaluation of existing social arrangements.

It is our contention that this type of horizontal taxonomy of social domains is a necessary step in research on the organization and development of judgments in the social realm. However, global characterizations of social judgments that include moral and social organizational components are not uncommon. Insofar as there are domains in children's thinking to be distinguished, such global characterizations are likely to yield evidence of heterogeneity in reasoning and continuity in development (since diverse types of judgments would be treated as one type).

An example can be seen in recent interpretations of research findings from studies using taxonomies of vertical organization in moral judgments (Kohlberg, 1969, 1971; Piaget, 1932) that encompass what we have delineated as morality (welfare, justice, and rights) and convention (social coordination in constitutive systems). On the basis of research with these global characterizations of the domain, it is claimed by some—most notably by Rest (1983)—that the evidence does not support the proposition that children's moral judgments develop through a sequence of qualitatively distinct stages. The evidence cited for rejecting the stage concept in moral judgments includes findings of heterogeneity and inconsistency across situations, tasks, and types of measures (i.e., measures of production vs. recognition and preference), as well as findings indicating that stage changes are gradual rather than abrupt over time.

In this type of analysis, the starting point is a global characterization of reasoning, which produces findings of heterogeneity and inconsistencies. Then with reference to the presumed structural criteria of homogeneity, synchrony and abruptness, the evidence is evaluated as disconfirming the stage concept. Earlier we maintained that the conjunction of a global characterization of reasoning with the criteria of homogeneity and consistency in evaluating empirical evidence for structural coherence is inadequate since heterogeneity and inconsistency can stem from different domains (we also maintained that abruptness is a totally inappropriate criterion). It is to be expected that a global characterization of the social realms would yield inconsistencies and anomalies as a consequence of the failure to isolate domains. Before considering the bearing of the empirical evidence on this issue, however, further commentary on the epistemological categories is needed, because in addition to arguing against the stage concept on the basis of those findings of heterogeneity and inconsistency, Rest also rejects the validity of the domain distinctions and argues against a narrowing of the globally defined sequence of moral judgment development. More specifically, on mainly two a priori definitional grounds, Rest dismisses research efforts at drawing domain boundaries between morality and convention. The stated grounds for rejecting the distinction amount to pointing out ways in which the domains can be combined.

One of the grounds is that concepts of justice are used in evaluating social structure. Rest claims the distinction is invalid, because "concepts of justice provide standards by which to judge social organization." We are not in disagreement regarding the idea that morality has a bearing on the evaluation of social organization (Turiel, 1983a, chap. 3). But this certainly does not mean that morality is defined by social organization; nor does it have much bearing on whether they can be analytically separated. Conventions are defined by the constituted social system (in Searle's sense of X counting as Y in context C) in which they serve the function of coordinating interactions (shared knowledge) for a designated goal (e.g., efficiency, stability, winning the game). It is in this sense that morality is not defined by social organization and is analytically independent of the existing constitutive

---

[1] That morality, which pertains to social interactions at the social organizational level, can be defined and analyzed by criteria separate from social organization is eloquently stated by Rawls (1971) at the very start of his treatise on justice and its applications to the structure of society:

> Justice is the first virtue of social institutions, as truth is of systems of thought. A theory however elegant and economical must be rejected or revised if it is untrue; likewise laws and institutions no matter how efficient and well-arranged must be reformed or abolished if they are unjust. Each person possesses an inviolability founded on justice that even the welfare of society as a whole cannot override. For this reason justice denies that the loss of freedom for some is made right by a greater good shared by others. It does not allow that the sacrifices imposed on a few are outweighed by the larger sum of advantages enjoyed by many. Therefore in a just society the liberties of equal citizenship are taken as settled; the rights secured by justice are not subject to political bargaining or to the calculus of social interests. The only thing that permits us to acquiesce in an erroneous theory is the lack of a better one; analogously, an injustice is tolerable only when it is necessary to avoid an even greater injustice. Being first virtues of human activities, truth and justice are uncompromising. (pp. 3–4)

Rawls (1971) goes on to say that "Some measure of agreement in conceptions of justice is, however, not the only prerequisite for a viable human community. There are other fundamental social problems, in particular those of coordination, efficiency, and stability" (p. 6).

system (cf. Gewirth, 1978; Rawls, 1971).[1] The criteria for morality (see Table 1) of obligation and noncontingency imply that it is applicable to different types of interactions among people, including those at the interpersonal and social organizational levels (these criteria also imply that it is not defined by existing social arrangements). With regard to the organizational level of interactions, research has shown that individuals' concepts of some—but not all—societal laws, prohibitions, and authority directives involve the application of moral concepts (Damon, 1977; Laupa & Turiel, 1984; Lockhart, 1980; Tisak, 1983; Turiel, 1983a, chap. 5). Consequently, the observation that welfare and justice considerations apply to and are affected by social institutions and social structure does not constitute an argument against the epistemological distinction between morality and convention.[2]

From the perspective of domain specificity, the issue of how morality and constitutive systems are linked is one that we have addressed through investigation of how individuals coordinate domains. The coordination of domains is also relevant to the other a priori definitional grounds offered by Rest for questioning the distinctions between morality and convention. Examples are given that appear to be aimed at showing (a) that violations of conventions can have harmful (moral) consequences (the example is "wearing a bikini bathing suit to the funeral of one's father"), and (b) that actions entailing harm can be conventionally sanctioned (the example is "the ritualistic circumcision rites in some societies").

As these examples were cited in the form of refutation of the distinction, it must first be noted that examples do not an argument make; nor do examples substitute

---

[2] It was also argued by Rest that conventions are merely social practices for which individuals do not perceive any moral rationale or social utility. Implied in this assertion is the idea that any social practice perceived as important or useful would be judged with a moral rationale. Moreover, it was implied that insofar as it has been found that children distinguish between morality and convention, it is because they were comparing important social practices (thus moral ones) and unimportant social practices (thus conventional ones). Defining convention through the lack of moral rationale or social utility is highly discrepant with philosophical analyses, such as those by Lewis (1969), of convention as serving important functions in coordinating social interactions. It is also contradicted by findings of research with children and adolescents. Two aspects of the body of research findings are relevant in this regard. One is that judgments about conventional practices, even when on a quantitative dimension they are rated as less important or severe than moral practices (which is normally the case), are linked to more general conceptions of social organization. In distinguishing morality and convention, subjects do not solely discriminate between items on a quantitative basis. Rather, their concepts of convention are based on a social organizational rationale that includes coordination, regulation, and authority seen as serving functions with social utility. The second aspect is the finding that certain issues considered important, nevertheless, were not perceived to have a moral rationale and were judged with nonmoral criteria. Examples of such issues include sex-role regularities (Carter & Patterson, 1982; Stoddart & Turiel, 1984), public nudity (Weston & Turiel, 1980), and abortion (Smetana, 1982). One study (reported in Turiel, 1983a, chap. 4) explicitly examined the relation between qualitative dimensions of judgment and the quantitative attributes of importance by assessing children's judgments regarding certain conventional transgressions they rated as more serious than a moral transgression. In brief, children conceptualized the more seriously rated transgressions by conventional criteria while conceptualizing the less seriously rated transgression by moral criteria. More generally, it should be noted that a complex relation holds between (a) the qualitative domain distinction of morality and convention, and (b) quantitative differences that have overlapping distributions (its appears that morality is generally, but not exclusively, perceived to be of more importance).

for empirical analysis of the cognitive processes by which judgments are made regarding such situations. Citing examples of this sort is logically akin (and as logically inadequate) to stating that the existence of hermaphrodites renders invalid the distinctiveness of the biological categories of male and female. The point has been previously made in a cogent way by Chomsky (1980) with reference to distinctions between language and cognition:

> Suppose I were to say that something in the room is green, and suppose somebody were to respond, "Well, that is not so because there is something that is white"—that wouldn't convince me that I was wrong in saying that something in the room was green. Correspondingly, if I say that some properties of language use and structure are determined in the initial state by language-specific principles, it does not convince me that I am wrong if I am told that some aspects of language use and structure are related to other aspects of cognitive development—that is a simple point of logic. (pp. 137–138)

To paraphrase Chomsky, it is not convincing to be told the distinction is invalid because some aspects of conventional use and structure are related to other aspects of moral judgment. The examples cited include components from two domains and are of theoretical interest from the viewpoint of domain mixture and coordination. The first example represents what has been shown in previous research to involve, with regard to moral judgments, a second-order phenomenon (Turiel, 1983a, chap. 6); that is, an issue primarily judged as conventional can have moral implications under certain circumstances (this is consistent with our earlier proposition that media of interaction can be interpreted in different ways). The second example represents a mixed domain situation of the type previously investigated (Smetana, 1982; Turiel, 1983a; Turiel & Smetana, 1984) and of the type evident in the social–psychological experiments cited at the outset (Haney et al., 1973; Milgram, 1974).

As discussed earlier, prior to considering between-domain coordinations it is necessary to isolate domains and accurately describe judgments within domains. We cannot stress too strongly the importance of task and situation analyses that specify the types of social variables implicated, in conjunction with horizontal taxonomies, prior to analyses of judgments about mixed-domain situations.[3] Studies guided by this strategy, which examined within-domain concepts and domain distinctions, illustrate how heterogeneous and inconsistent or opposing judgments can reflect coherence within domains.

The proposition of domain specificity, in the way we have presented it, entails

---

[3] Much of the research that has led to a global characterization of moral judgment development is based on stimuli of a compounded, mixed nature. Subjects' judgments are assessed through situations embodying a combination of variables (e.g., stories that include conflicting moral issues, legal issues, personal and conventional considerations). Thus far, no serious efforts have been made to separate variables in those particular stimuli. However, research from a domain perspective shows that children make distinctively moral judgments at younger ages (and in different ways) than was found when using the compounded stimuli. This suggests the need to revise taxonomies of vertical organization (including that presented by Rest), in which children's competencies are underestimated.

assessments of different aspects of subjects' cognition. In addition to analysis of interdomain coordinations and levels of within-domain development (vertical organization), assessments are needed of judgments reflecting domains and their boundaries (horizontal organization). Along with the formulation of epistemological categories and specification of the types of domains listed in Table 1, data on individuals' task performance should bear on whether and how their thinking maps onto those categories. Toward this end, two dimensions of social judgments, one pertaining to vertical organization and the other to horizontal organization, have been assessed. The vertical dimension (labeled justification categories), which entails analyses of forms of reasoning that undergo developmental transformations, is considered in a later section. The horizontal dimension, which we have referred to as criterion judgments (Davidson et al., 1983), includes a set of formal characteristics for the identification and definition of domains. Analyses of criterion judgments basically are designed to ascertain the ways in which subjects delineate identifying features of a domain of knowledge. In turn, through comparisons of criterion judgments, characteristics of domain boundaries can be described.

In studies of moral and conventional criterion judgments, the typical procedure has been to present subjects with sets of stimuli in the form of issues or events corresponding to each domain, as determined by the definitional criteria. In a given study, subjects have been posed with a series of standard questions designed to tap their ways of applying criteria to the stimuli. In line with the strategy of shifting back and forth between domain definitions and related data gathering, the questions posed to subjects have dealt with dimensions like intrinsic–arbitrary, context-dependent–independent (including rule-contingent and authority–determined), generalizable–relative, impersonal–personal.

Using a variety of methodological contexts, procedures, and stimuli, it has been found that the criterion judgments applied to morality differ from those applied to convention. These differences occur at an early age (by 3 or 4 years in studies by Smetana, 1981, in press) and are maintained across a wide age range (to 19 years of age, as studied by Nucci, 1981). The early onset of distinctively moral concepts is also evident in the prekindergarten characteristic-to-defining shift for moral act terms reported by Keil (this volume). It also appears that young children's interactions with events have a bearing on their judgments. In the Davidson et al. (1983) study 6-year-olds distinguished previously familiar (probably reflecting an experiential factor) moral and conventional stimuli more clearly than unfamiliar stimuli. Older children were able to distinguish between domains with regard to both previously familiar and unfamiliar events. By the ages of 8 to 10 years there is an increased ability to abstract elements from unfamiliar events and apply existing domain-specific schemes.

Stated in general terms, the research on criterion judgments shows that moral prescriptions are judged to be obligatory, noncontingent on specific social rules or authority dictates, nonalterable on an arbitrary basis, impersonal and applicable across situations (corresponding to the criteria of obligatory, context-independent, impersonal, and generalizable). In contrast, conventions are judged as contingent on aspects of social organization (rules, authority, and existing practices), relative to the context, and alterable by consensus (corresponding to the criteria of contingency, contextualism, and relativism).

This body of research, details of which are presented elsewhere (e.g., Nucci, 1981; Nucci & Turiel, 1978; Pool et al., 1983; Smetana, 1981), provides substantial evidence of the coexistence within individuals of different kinds of judgments; that is, of heterogeneity. Children are, for instance, relativistic with regard to some events and universalistic with regard to other events. In some situations they uphold authority, in others they do not; similarly, rules are accepted in some instances and rejected in others.

These heterogeneous (and seemingly inconsistent) judgments can be approached from two perspectives that would lead to different conclusions about coherence in social thought. If the research findings were viewed solely from the perspective of the thought of the individual (i.e., a general social–cognition structure), then it would appear that heterogeneity and inconsistency is the rule. Correspondingly, a lack of systematicity with regard to age-related changes would seem to be the rule from the perspective of a global characterization of vertical organization that orders moral judgments by the dimensions of (a) personal and pragmatic, (b) authority and hierarchy, and (c) justice and equality. If, however, the findings are considered from the perspective of domains of knowledge, then it becomes apparent that along with the between-domain heterogeneity, there is coherence within domains. The criterion judgments group together to reflect a domain of thinking about conventions in constitutive systems, on the one hand, and a domain of moral judgments, on the other. In these cases, therefore, heterogeneity is not to be taken as an indication of lack of structure in the development of children's social knowledge.

One source of heterogeneity, therefore, is judgments about events that differ by their domain. However, the specific focus of a social cognition task presented to subjects also has bearing on the domain of judgment applied to it. As noted earlier, an event constitutes material potentially interpretable from the perspective of varying domains. Accordingly, the parameters of the task (e.g., the foci of the questions posed to subjects) regarding a particular event can determine the domain of judgment elicited and, thereby, constitutes another source of heterogeneity. For instance, with regard to the same action, questions can be posed to inquire into the subject's moral evaluations or to inquire into the subject's conceptions of the psychological factors that produce those actions. A moral evaluation may be elicited through questions such as whether the actor should have committed the act or whether the act is right or wrong. Psychological knowledge may be elicited through questions about how persons will act under given circumstances or about what leads persons to act in certain ways.

A recent study (Jancaterino, 1982) explored this issue by assessing 6- to 10-year-old children's moral evaluations and personal or psychological knowledge regarding the act of inflicting physical harm. One group of subjects was posed with evaluative questions ("Is it all right or not all right for X to hit Y?"). Consistent with findings mentioned earlier, almost all children negatively evaluated the act with noncontingent judgments (e.g., it would be wrong to hit, even if a school had a rule permitting it). A second group of subjects was presented with descriptions of varying situational contexts (such as schools and families that prohibited or permitted hitting). These subjects were posed with questions of psychological prediction: whether or not a child would (not should) commit the act. Most subjects predicted that behavior would be influenced by the situational context (hitting in a school that permitted it and refraining from hitting in a school prohibiting it). Insofar as there was a conflict between the family and school, most subjects predicted that behavior would be influenced by the family rule. In this study, too, judgments might appear inconsistent, were they not approached from a domain perspective. More generally, the study demonstrates that (a) children have explanations of the psychology of persons' actions which are not the same as their moral evaluations, and (b) different stimulus tasks can elicit different domains of judgment about similar actions.

Coordinations of Social Judgments

The Jancaterino (1982) study provides a good example of how an event can be approached in a coherent way from the perspective of more than one domain. It also provides an empirical demonstration of how, with regard to the same event, children's concepts about the moral-domain side of the coin may be distinct from their concepts about the psychological-domain side of the coin. While in this study the questions posed to subjects probed for these two types of concepts, it is not difficult to imagine a similar process occurring in naturalistic situations. When observing a person inflicting harm on another, one may make a moral evaluation of the act and at the same time (perhaps as a means toward preventing repetition of it) attempt to determine the psychological factors causing the act.

The same kind of differential approach to domains occurs with regard to actions from the conventional domain that under certain circumstances may have moral implications. An example is the violation of a convention that causes others offense and psychological harm ("wearing a bikini to a funeral" fits this type). It is important, in this regard, to realize that people can become offended or indignant on other than moral grounds; those who subscribe to a particular convention may feel offended when it is violated, and this need not have a moral connotation. However, in certain instances a conventional violation may be intended as a means of insulting or humiliating a specific person or persons. In such cases as well as others, transgressions of conventions can have moral implications. This is a second-order phenomenon (Turiel, 1983a), insofar as there is a reformulation of events primarily conceptualized as conventional into their symbolic moral signifcance. The primary and secondary accessing of domains has been demonstrated by an examination of adolescents' judgments regarding an instance of public nudity that had the effect of offending others (Turiel, 1983a). Some of the subjects judged the event solely according to criteria for conventionality. Others, however, approached it primarily through the conventional domain to make judgments about the class of event, but secondarily accessed the moral domain to account for particularistic reactions of others.

Another type of domain mixture occurs in those multifaceted situations that include components from more than one domain. Many social situations are complex, so that they implicate different and sometimes conflicting social goals and require individuals to coordinate domains of judgment. Examples like "ritualistic circumcision rites in some societies" loosely correspond to a situation entailing domain mixture.[4] Empirically based examples are available from the social–psychological work mentioned earlier. The unusual and glaring behaviors (e.g., administering electric shocks to another in adherence to instructions from an experimenter) evident in these situations may be due to experimental manipulations that heighten conflicts between domains.

Two interrelated aspects of the interpretation of developmental data emerge from a consideration of domain coordinations. One is that mixed-domain situations can be a source of conflict and contradiction. Recent research (Smetana, 1983; Turiel, 1983a) designed to examine reasoning about events with a mixture of moral and conventional components documents that, across ages, individuals do attend to each of the components. In attending to more than one component of the situation, subjects' modes of relating domains include both interdomain conflict and associated inconsistencies as well as interdomain coordination (e.g., the two are balanced in a solution to the problem).

Situations combining components from more than one domain can also produce nonlinear developmental patterns. An example comes from a study (Stoddart & Turiel, 1984) of children's concepts of sex roles, which include psychological (gender identity) and conventional (role stereotypes) components. Assessments of subjects' evaluations of deviations from stereotyped sex-role activities yielded U-shaped behavioral growth patterns of the type extensively studied by Strauss (1983). The youngest (kindergarteners) and oldest (eighth graders) subjects judged the sex-role deviations to be more wrong than did subjects in the middle groups (third and fifth graders). Measures of criterion judgments and justifications showed that the nonlinear age trends stemmed from subjects' efforts at coordinating the conventional and psychological components. In particular, it appears that the adolescents were unable to adequately coordinate their psychological attributions with their conventional judgments.

These considerations also lead us to speculate that models of moral judgment development (Rest, 1983) based on global definitions of morality are actually, through a methodology that does not separate variables, tapping cognitive processes used in relating components of different domains. This would account for discrepancies in the age of attainment of differentiated moral judgments, observed during

---

[4] Isaiah Berlin (1981) presents a nice example of a situation that illustrates how the principle of equality can be subordinated to other ends: "Inequality in the organization of an orchestra there patently is; the reason for it is the purpose of orchestral playing—the production of certain sounds in certain ways which cannot, in fact, be achieved without a measure of discipline which itself entails some degree of inequality in the distribution of authority" (p. 91). For further discussion, see Turiel and Smetana (1984).

adolescence in research guided by the global approach and in early childhood by the domain approach.

As noted earlier, mixed-domain situations bear on another potential source of inconsistency: the behavioral expression and application of social judgments. The relation of moral judgments to behavior, of course, has traditionally been of great concern in social and developmental psychology. This concern is usually translated into a research question in the form of whether there is consistency or inconsistency between judgment and behavior. In our view, however, an understanding of judgment–action relations is not adequately informed by determining whether or not there is consistency (in fact, a review by Blasi, 1980, shows that levels of consistency ranging from low to high have been obtained in correlational studies of moral judgment and behavior). Greater attention needs to be given to the fact that when behavioral choices are made in situations of a multifaceted nature that involve domain mixture, simultaneous patterns of consistency and inconsistency would be expected to occur. In such cases, interrelations of horizontal organization intersect with behavioral outcomes. When this intersection of domains with behavioral choices is taken into account, the evidence indicates that judgment is related to action.

We can make these ideas more concrete through the components and behaviors in the experimental situations of Milgram's (1974) studies. Recall that in some experimental conditions the majority of subjects adhered to the experimenter's instructions to administer shock to another person, whereas in other conditions the majority refused to do so. As analyzed by Turiel and Smetana (1984), the situational context in these experiments includes domain-associated components of a moral and social–organizational nature. The experimental situation includes inflicting physical harm (the moral component) as well as a hierarchical organization with an authority ostensibly implementing activities serving scientific ends. The two components are in a relation of conflict: To avoid inflicting harm, the authority must be contradicted and the scientific aims thwarted, whereas to maintain them the moral prescription must be violated. Furthermore, in some experimental conditions the salience of the harm component was increased, while in other conditions the salience of the organizational component was decreased (for more detailed "task" analyses of the experimental situations, see Turiel & Smetana, 1984).

The mixture of domains in the experimental situations is reflected in subjects' behaviors. In some conditions the behavior of a majority of subjects was more in line with the moral component and in other conditions with the social organizational component. However, a close analysis of the sequence of behavior throughout the experiment indicates that that there was a fair amount of conflict in the process (Milgram, 1974; Turiel, 1983a). Most significant, for the present purposes, is that either behavioral outcome actually can be seen to represent both consistency and inconsistency between judgments and action. Although judgments were not assessed in the studies, the following interpretations are plausible (Turiel & Smetana, 1984). For subjects who continued with the experiment the behavior would be consistent with their conventional–organizational judgments and inconsistent with their moral judgments. The behavior of subjects refusing to continue administering the shocks would be consistent with their moral judgments, but not with their organizational judgments. Moreover, that the subject's domain accessing interacts with the parameters of the situational context is evidenced by the finding that subjects were much more likely to refuse to continue the experiment when either the salience of the harm component was increased or the salience of the organizational component was decreased.

Recent research assessing judgments in relation to behavioral decisions (women's decisions to continue or abort a pregnancy) supports the hypotheses stemming from the interpretation of the Milgram experiments (Smetana, 1982). Other social–psychological studies can also be seen to induce domain conflicts. One example is the Haney et al. (1973) study, in which there was a simulation of prison life (24 hours a day over a one-week period) with subjects assigned to roles of guards and prisoners. To state it very briefly, participants' roles and authority status in the situation produced real and significant contradictions with their moral orientation to others. In this situation, as in the Milgram experiment, behavior that may appear inconsistent with moral reasoning needs to be interpreted with reference to subjects' attempts at coordinating moral and social organizational concerns.

It is apparent, then, that mixed-domain situations are also a source of empirical heterogeneity and inconsistency. Individuals display heterogeneous behavior in different contexts and inconsistencies between judgment and behavior are evident. Both consistencies and inconsistencies between judgment and action reflect the interaction of subjects' domain orientations with multifaceted situational contexts. Taking this interaction into account can contribute to an understanding of the systematic relationship between thought and action as it pertains to development.

Domains of Nonsocial Judgment

In listing domains in the nonsocial realm (Table 1), with their parallels in the social realm, it was noted that criteria have been identified more explicitly for the social than the nonsocial domains. However, with the taxonomy in Table 1 as a starting point we can consider criteria for domains of nonsocial judgments. In particular, research on the development of logical and numerical reasoning provides evidence for such domain differentiations and shows that some of the previously observed heterogeneity stems from tasks requiring interdomain coordinations. These analyses are somewhat different from those of the previous sections. We focus on a comparison of two types of studies: those using within-domain tasks and those using mixed-domain tasks requiring coordinations. In our view, much of the research in this area has yielded apparent findings of heterogeneity and asynchrony because of the use of mixed domain tasks.

First, let us note briefly the formal distinction between the noncontingent domains (logical and mathematical) and the other nonsocial domains. As Frege (1884/1974) eloquently points out, the major distinction is between knowledge generalized directly from empirical data and that which is imposed by the subject for the sake of organizing the data:

> While I am not in a position, simply by thinking of it differently, to alter the colour or hardness of a thing in the slightest, I am able to think of the Iliad either as one poem, or as 24 Books, or as some large Number of verses. . . . It marks, therefore, an important difference between colour and Number, that a colour such as blue belongs to a surface independently of any choice of ours. . . . The Number 2, on the other hand, or 100 or any other Number, cannot be said to belong to the pile of playing cards in its own right, but at most belong to it in view of the way in which we have chosen to regard it. (pp. 28–29)

These forms of organizing the data are characterized by necessary inferences, whereas empirical knowledge is not. Thus, 4 is necessarily greater than 2, and a whole is necessarily greater than any one of its parts; whereas the sun does not necessarily have a greater diameter than the earth (in another possible universe it might not have), and space itself is not necessarily Euclidian (indeed, it is now empirically probable that it is non-Euclidian). Epistemologists have traditionally assumed that only contingent knowledge can arise through experience and have therefore been at pains to explain the existence of necessary knowledge. One major explanatory trend has been to assert the existence of innate ideas (Plato, *The Meno*) or innate forms of structuring experience (Kant, *Critique of Pure Reason*). Another has been to deny the distinctiveness of necessary knowledge, interpreting it as a generalized sense of certainty based on observed empirical regularities (J. S. Mill, *System of Logic*). Both these tendencies remain evident in contemporary theories of cognitive development.

However, we do not assume that only contingent knowledge can arise through experience (earlier we also proposed that moral knowledge is noncontingent and arises through experience). In addition to interaction with contingent properties of objects, experience also includes interaction with organizational properties of action itself. For instance, infants interact with quasi-continuous media such as Play-Doh by biting, pulling apart, pressing together, and so on. The formal aspects of such actions include operations of composing and decomposing, which intrinsically reflect relations of part to whole. On some occasions infants focus on empirical information (hardness, taste, etc.) and thereby construct knowledge of physical objects. On other occasions, they focus on the composing and decomposing properties of action *for their own sake* and thereby construct precursory knowledge of part–whole relations (Langer, 1980). As is indicated by this example from research with infants, knowledge about the organizational properties of action is possible from the earliest stages of development. Such knowledge contains the prerequisite elements and regulations for the eventual acquisition of necessary inference (class inclusions, etc.), upon completion or stabilization of the knowledge structures (in the sense described earlier in this chapter).

There are two fundamentally distinct ways in which subjects' actions impose order on the media and therefore two ways of theorizing about the necessary organizational properties of these interactions. On the one hand, actions may be directed at intensive aspects of the media; in such cases, objects are construed as exemplars of certain qualities (e.g., color) or in terms of relations between qualities (e.g., red is a subclass of color). On the other hand, actions may be directed at extensive aspects of the media, in which case objects are construed as individual elements or constants such that intensive qualities are irrelevant (e.g., in counting a set of red, blue, and green objects as a single series, color is irrelevant).

As a further illustration, children impose intensive classification schemes directed at the similarities of objects with respect to predicate properties such as color, texture, or functional utility ("edibles," "throwables," etc.). Subjects also impose relational schemes based on binary predicates such as "lighter than," "smoother than," and degree of variation from some established prototype (e.g., given that a small ball is the prototypical throwable, then other objects can be antisymmetrically ordered by degree of suitability for throwing). The domain of logical knowledge is derived from interactions in which children focus on these intensive forms of organization. The distinctive aspect of the logical domain is its lack of concern with quantities or magnitudes that concern individual elements, because individual elements as such play no role in intensive reasoning, except for their role as exemplars of qualitative predicates. Consequently, quantitative judgments in the logical domain are limited to part–whole comparisons. Thus, if a class P (e.g., red) is properly included in a class Q (e.g., colors), then by intensive reasoning alone, the extent of Q is greater than that of P (by some indeterminate amount). Similarly, intensive binary predicates such as "less than" specify a class of ordered pairs which express a qualitative difference. In a series A, B, C conforming to such a relation, one may conclude that the difference between A, C is greater than, or includes, the difference between A, B; but the absolute amount of these differences is indeterminate or irrelevant from the intensive point of view.

For the interested reader Table 2 is included to illustrate the major formal characteristics that distinguish the numerical from the logical domain. The most central of these is that numerical knowledge begins by considering objects as individual elements rather than as exemplars of intensive predicates. It follows that the basic part–whole relation in numerical structures is that of *membership* of individual elements in a set rather than that of *inclusion* of subclasses in a class, as is the case in the logical domain. The fundamental distinction between sets and classes is that sets can be constructed by assignment of arbitrary elements not possessing any common intension. For instance, "the Rock of Gibraltar and the exhibits at the Smithsonian" is a valid set whose elements have no obvious common intension (prior to the ad hoc intension created by assigning them to the same set). In general, then, a set is defined by listing its extension; or, if convenient, by giving an algorithm that generates such a list (e.g., "let X be the set of integers that divide 1,000 evenly"). The relation of set inclusion may then also be defined extensively,

Table 2. Comparison of Logical and Numerical Domains.

	Logical Domain	Numerical Domain
Basic elements	*n*-place predicates (P, Q, R . . .)	individual constants, or terms (a, b, c . . .)
Basic aggregates	classes corresponding to predicates (P, Q, R . . .)	sets of individual elements (A, B, C . . .)
Part–whole relations	intensive inclusion (e.g., $P \subset Q$)	1. membership (e.g., $a \in A$) 2. set inclusion (e.g., $A \subset B$) defined in terms of membership
Basic unary operations	complementation (A′)	inverse ($-a$) reciprocal ($^1/a$)
Identity elements	empty class (Λ)	null set ($\emptyset$) additive identity (0) mult. identity (1)
Basic binary operations	union ($P \cup P' = Q$) intersection ($P \cap P' = \Lambda$)	addition ($a + -a = 0$) multipl. ($a \cdot {}^1/a = 1$)
Operational properties	resorption ($P \cup Q = Q$) ($P \cap Q = P$) tautology ($P \cup P = P$) ($P \cap P = P$)	iteration: (e.g., $2 + 4 \neq 4$) ($2 \cdot 4 \neq 2$) ($2 + 2 \neq 2$) ($2 \cdot 2 \neq 2$)

in terms of the membership relation: if all members of X are members of Y, and not all members of Y are members of X, then X is properly included in Y.[5]

Developmentally, the separation of logical and numerical cognition may be attributed to the distinct organizational properties of actions that can be described as tautological and iterative, respectively. An example of tautological actions is when children sort objects according to predicate properties (large vs. small; blue vs. nonblue; etc.) their repetitions reflect a series of tautologous assertions ("another blue, another blue, another blue, . . .", or in symbols, $B + B + B \ldots + B = B$). On the other hand, when children accumulate objects into a set, they proceed by suppressing intensive predicates, and thus their repetitious actions reflect a series of iterative compositions ("another one, another one, another one . . .", or in symbols, $1 + 1 + \ldots + 1 \neq 1$). That is, iterative actions are those that produce distinct results with each repetition. Recent research has shown that children attend to both these forms of organization-in-action as early as the sensorimotor period when playing with objects such as colored block, rings, dolls, and Play-Doh (Lan-

[5] Clearly, the distinction made here between logical and numerical reasoning lies closer to classical than to modern interpretations of the subject matter of logic. Classically, the definition of necessary logical inference involved an idea of the intension of the conclusion being included in the intension of the premises; for example, "all men are (included in) mortal; all bachelors are (included in) men; therefore all bachelors are (included in) mortal." In contrast, modern (e.g., first order) logic was invented with a view to formalizing mathematics; thus extensive aspects (individual constants, iterative operations, etc.) receive heavy emphasis. Consequently, in our view modern mathematical logic is a branch of mathematics rather than of logic.

ger, 1980; Forman, 1982; Sugarman, 1983). On some occasions infants and young children direct their activities by focusing on intensive properties including color, shape, and proximity. On other occasions they construct sets of elements extensively, mixing various intensive properties in a single construction, and go on to construct elementary relations and operations upon individual elements.[6] Robert, Cellerier, and Sinclair (1972) present an interesting microanalysis of a lengthy period of interaction by a 4-year-old with a collection of toy swans, ducks, and fish. During the course of the interaction, the subject clearly alternates between intensive forms of organization (constructing arrays of like elements without regard to numerical correspondence) and extensive forms (constructing arrays in numerical correspondence without regard to class predicates).

Further information about the possible separate origins of logical and numerical reasoning comes from a recent study (Davidson, 1983), which was conducted to extend the Piaget, Grize, Szeminska, and Vinh-Bang (1968/1977) analysis of the logic of functions. According to the Piaget et al. model, sensorimotor action schemes give rise during the period of 2 to 4 years to cognition of functional relations, i.e., $y = f(x)$, whose causal and operatory properties are poorly differentiated. By age 4 to 6 years, this cognition of functions is well consolidated into a structure with certain limitations (lack of reversibility, most notably); this structure provides the necessary preconditions for the construction of concrete operational logic found at about age 7 or 8 years. Based on findings such as those reviewed earlier, the Davidson study hypothesized that acquisition of operational numerical knowledge should be better accounted for by specifically prenumerical forms of function cognition than by conceptual abilities falling under the logical domain. Accordingly, 5- to 7-year-old children were tested on a battery of numerical, logical, and function-cognition tasks. One variety, the "intensive function" tasks, involved functions defined on intensive predicates (e.g., functions represented by the exchange of one color for another). A second variety, the "extensive function" tasks, involved functions defined on discrete, individual elements (e.g., functions represented by matching specific objects to specific positions). These measures did not entail any explicit knowledge of the numerical concepts (arithmetic operations and number conservation) nor of the logical concepts (class inclusion, vicariance, and seriation) tested.

Performance on extensive function tasks was a stronger predictor of numerical competence than were the logical class and order concepts tested, explaining 62%

[6] Since even infants construct sensorimotor precursors of logical and numerical knowledge, it is worth pointing out that findings of precocious competencies in young children (say, 2-year-olds) is not necessarily evidence of innate foundations for domain-specific cognitive development; nor, of course, are findings of late-blooming competencies necessarily evidence that logical and numerical development occurs by cumulative empirical generalizations. The explanation for either type of finding needs to consider (a) the actions that children of the given age, or younger, employ to structure their environment; (b) evidence of whether children focus on their organizational properties, e.g., perform such actions for their own sake; and (c) the association between these forms of logical or numerical experimentation and the specific (precocious or late-blooming) competency in question.

of the variance in a composite measure of arithmetic and number conservation abilities. Thus, the hypothesis that numerical concept development derives from specific prenumerical function abilities received support. Second, there were distinctions among the various function notions tested: Performance on extensive function tasks predicted the numerical concepts but none of the logical concepts; performance on intensive function tasks predicted all logical concepts but none of the numerical concepts. These results support the basic contention of the Piaget et al. (1968/1977) model—that the cognition of functions lays the groundwork for concrete operational thought—while also suggesting that the model should be amplified to take account of distinctions between specifically prenumerical and specifically prelogical forms of function cognition.

Coordinations of Nonsocial Judgments

As we have noted previously, the media of interaction are interpretable from a variety of different conceptual frameworks. In spite of this, as the preceding examples show, children's interactions with physical media are not random but rather are structured, with each step of the interaction directed by the particular concerns of some delimited epistemological domain. This observation contrasts with a view in cognitive development according to which young children's spatial, logical, and numerical knowledge are seen as relatively undifferentiated. Thus, for example, children's failure to conserve number has been attributed to lack of differentiation between spatial and numerical reasoning (Piaget, 1941/1965) and early performance on classification tasks to lack of differentiation between spatial and logical reasoning (Inhelder & Piaget, 1959/1964). An especially intriguing example that appears to show the absence of differentiation between intensive and extensive concepts is given by Greco, Inhelder, Matalon, and Piaget (1963). In this study, the experimenter and child removed counters one by one from a large set and a small set, respectively. The young subjects stated that the subset removed from the large set would be more numerous than that removed from the smaller set, evidently thinking that subsets inherit numerosity from their sets of origin, as they might inherit intensive properties such as color.

However, a reinterpretation of current evidence leads us to hypothesize that young children's performances on this task and on traditional number-conservation tasks are not due to lack of differentiation between knowledge domains, but rather to (a) incomplete elaboration of knowledge within, and (b) problems of coordination between, domains that are themselves differentiated. This interpretation implies that even young children should apply domain-specific reasoning in appropriate ways when given tasks that are *both* within reach of their domain-specific competencies *and* within reach of their domain-coordination competencies. Data on the development of the numerical domain bear out this interpretation.

For example, Starkey (1983) found that children as young as 2 years understand numerical addition and subtraction when the extent of numerical knowledge required is strictly limited to set sizes less than 4 elements and the need to coordinate numerical with other knowledge is eliminated (by using an apparatus that screens all spatial cues, etc.). Mehler and Bever (1967) found that very young children recognize that adding elements to a small set, while also crowding up the spatial extent of the set, increases its numerosity. In this task, the degree of elaboration of numerical knowledge required for a solution is again limited to very small set sizes. Although the task does involve coordination of numerical knowledge with spatial knowledge, here the two types of judgment are *not in conflict.* As Piaget (1968) has pointed out, 2-year-olds' numerical judgment of "more elements" is compatible with their judgments of spatial extent, which reflect a concern with topological notions of proximity and crowding. Finally, number conservation is, by definition, a problem in coordinating distinct conceptual domains, when these distinct judgments *are in conflict.* That is, the classical conservation task is set up so as to require a coordination of quantitative numerical judgments (e.g., "nothing added or subtracted") with quantitative spatial judgments (e.g., "longer equals more"). The Greco et al. variant, mentioned earlier, sets up a conflict between intensive and extensive judgments.

The data indicate that children as young as 2 years solve numerical problems by applying domain-appropriate reasoning when the problems are simple enough and when domain coordinations are not required. Young children also apply domain-appropriate reasoning on tasks requiring domain coordination—but only when the judgments to be coordinated are not in conflict. At age 6 or 7 years children can apply domain-appropriate numerical reasoning while also coordinating judgments from separate domains that are explicitly in conflict.

Therefore, reports of children's inconsistent numerical competencies are seen to reflect orderly developmental patterns when interpreted from the point of view of domain coordination. As another example, consider the findings alluded to in the introduction of this chapter. Children who at age 8 are almost certainly able to solve standard seriation problems, nevertheless do not when the items to be seriated differ imperceptibly by .2 mm; they resort to the type of categorical reasoning typically found in much younger children (A, B, C, D, are "small"; E, F, G, are "big") (Piaget, 1980). Notice that in standard seriation-of-size tasks, the subject applies a logical order scheme to spatial media (lengths, heights, diameters, etc.). This is a case, then, of coordinating logical and spatial judgments that are not in conflict and are in fact mutually supportive. In the task with imperceptible .2 mm differences, subjects can easily perceive a marked difference in size between the first and last elements, A and G; this should be sufficient evidence to infer that the same order scheme is indeed applicable. However, this logical inference conflicts with the very salient spatial (perceptual) judgment that every element is equal in size to its immediate neighbor. Correct resolutions under these conditions of conflict are not found until about age 11. Inconsistencies result, if performance failures on such tasks are interpreted in terms of domain differentiation rather than domain coordination. Based on the conflict task, one might conclude that spatial and logical concepts are not adequately differentiated earlier than 11 years, whereas using hierarchical classification tasks, one gets an estimate of 7 or 8 years. Such findings do not constitute inconsistencies from the viewpoint that different problem contexts present varying degrees of difficulty in the coordination of differentiated domains.

A somewhat similar age discrepancy is found between the acquisition of the notion of numerical unit as assessed by number conservation tasks, and that as assessed by measurement tasks. The discrepancy results from the relative difficulty of coordinating numerical knowledge with spatial judgments about discrete, discontinuous objects, on the one hand, and with spatial judgments about continuous objects (e.g., lengths or areas), on the other hand. In the former case, the objects are perceived as qualitative units (i.e., as wholes), and this spatial judgment is compatible with the relevant numerical interpretation of the objects as quantitative units. In the latter case, the objects are also perceived as qualitative units (i.e., the length or area is an integral whole): however, this spatial judgment is in conflict with the relevant numerical concept, which involves constructing an arbitrary yet systematic decomposition of the whole object into equal fractions. Consequently, the notion of measurement entails an even higher degree of coordination between numerical and spatial cognition than does that of number conservation and is achieved a year or two later.[7]

Numerical and logical reasoning constitute two distinct ways of theorizing about the organization of interactions. "Criterion judgments," in the sense discussed in the preceding section on domains of social knowledge, have not been investigated explicitly in these domains. However, there is good indirect evidence that even young children apply such criterion judgments, for example, in the sequential application of intensive and extensive considerations to guide their interactions. This interpretation would account for the observed overall coherence in children's spontaneous activities with objects, as reported by Forman (1982), Langer (1980), Robert et al. (1972), Sugarman (1983), and others. The media in these activities permit assimilation to various domains (physical, spatial, logical, numerical); if the interactions nevertheless appear systematic and meaningful rather than random, it is because they are directed at each step by the specific concerns of a single, bounded domain.

There is evidence, then, that does not support the view that logical, numerical, and other nonsocial domains are themselves undifferentiated early in development, only to become differentiated with the advent of operational cognition. Evidence that has been advanced for nondifferentiation can be interpreted in terms of domain coordination: Subjects may experience difficulty with certain interdomain coordinations even when the domains are well differentiated. When domain coordination is not a problematic task variable, even young children apply differentiated domain-

---

[7] These observations suggest that the issue of horizontal décalage in development can be approached through analyses of progressive interdomain coordinations. The décalage notion has often been viewed as an ad hoc adjustment to structural developmental theory, which substantially compromises its claims about qualitatively distinct stages. In contrast, by the present interpretation, décalages are a necessary feature of developmental stages inasmuch as they reflect qualitative differences in interdomain coordination.

appropriate reasoning. Apparent inconsistencies in development, such as asynchronies in the application of numerical knowledge or seriation concepts under varying task conditions, are seen to reflect the systematic development of increasingly adequate levels of interdomain coordination.

Developmental Processes and Representations of Cognitive Growth

We began this chapter with examples of heterogeneity in judgments on different types of tasks and inconsistencies between judgment and behavior. We have placed such empirically derived examples into a theoretical context of domain specificity in development and the associated proposition of the coexistence of heterogeneity, homogeneity, inconsistency, and consistency. Within the context of a domain taxonomy of horizontal organization, findings reflecting the coexistence of these polarities are neither anomalous nor self-contradictory. In concluding these analyses, however, it still remains for us to draw the implications of the proposition of domain separations for the issue of continuities and discontinuities in the process of development.

One clear implication is that explanations of development require consideration of multiple dimensions at the intersection of horizontal and vertical organization (i.e., the domain distinctions, shifts within domains, relations between domains, shifts in domain coordinations, and separation of within-domain and interdomain tasks). From this perspective, it is possible to pursue hypotheses, at varying poles, pertaining to general qualitative developmental changes in the coordination of domains or to specific qualitative changes within the domains.

It could be proposed that in the context of domain distinctions the locus of development is in the interaction of domains. In other words, it may be that general sequential changes occur in children's abilities to coordinate knowledge from different domains. As noted earlier, such a theory is lacking at the present time. However, an example of the possible parameters of development in situations requiring domain coordinations might be that there is a progression from the fusion of domains, to their differentiation, to their differentiation and integration. Our own hypothesis, though, is in the direction of viewing the locus of qualitative changes specifically within domains. One reason for this hypothesis is our expectation that competence in coordinating domains is at least partially determined by relations between the child's level on each of the domains in question and the complexity of the coordination task. To state it briefly, the expectation is that young children will be able to coordinate domains when their level of competence on each domain matches the task requirements and the task embodies a straightforward relation of the domain components (e.g., they are not in conflict). In contrast, even older children may not be able to coordinate domains well when their level on one or both domains does not match the task requirements or the task embodies a complex relation of domain components.

If the coordination of domains is task-related in the above sense, then straightforward general age-related stages of interdomain coordination are not likely to be found. Our proposition, instead, is that the locus of developmental transformations lies within domains and that analyses of within-domain development should account for three factors: structural discontinuity (qualitative differences on the vertical dimension; e.g., developmental changes in justification categories), functional continuity (the criteria by which the qualitatively different judgments are in the same domain; e.g., commonalities across ages in criterion judgments), and the processes of change from one level to the next (involving deformation of one form of reasoning in concurrence with the formation of a subsequent form).

These factors have a bearing on interpretations of data on task performance at different ages and representations of development. First, the proposition that change encompasses both structural discontinuity and functional continuity implies that representations of change must also capture the within-domain relations of earlier and later structures. Accordingly, when interpreting idealized developmental curves for within-domain age changes, a method is needed to represent the functional continuities.

As an example, consider the type of curves in Fig. 1, which are sometimes taken as good representations of stage development. Figure 1 may be regarded as the hypothetical development of a single domain, which appears in (for example) three qualitatively distinct forms during the relevant period (e.g., from birth to 20 years

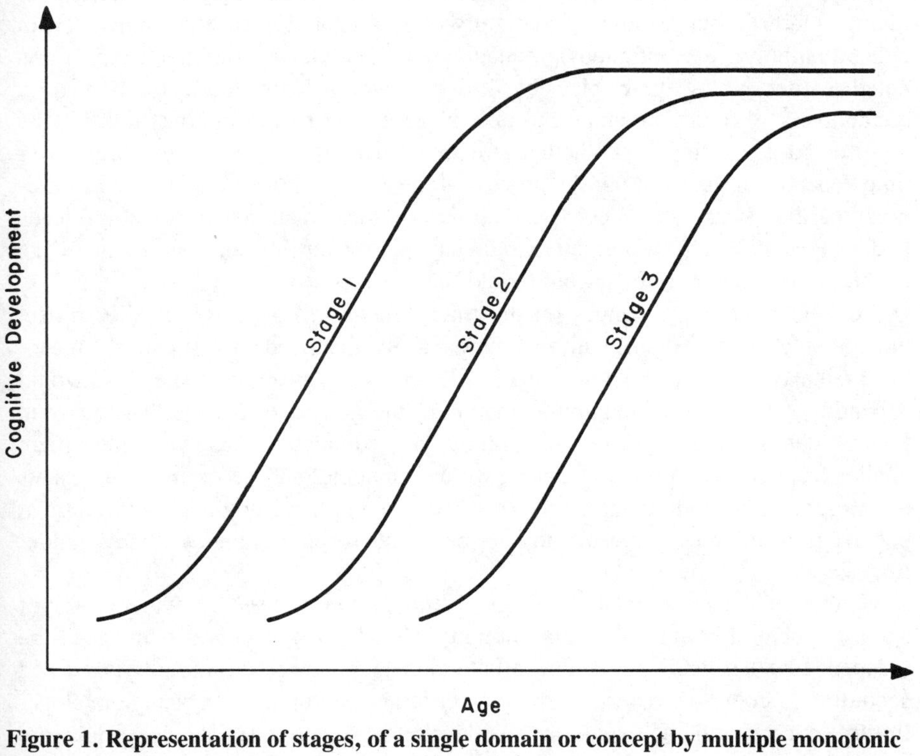

Figure 1. Representation of stages, of a single domain or concept by multiple monotonic curves.

of age). Although this representation is meant to reflect discontinuities between stages, it fails to represent the domain connections and continuities in a sequence because it suggests that successive stages have distinct and independent origins. With respect to structural discontinuity, the graph indicates the coexistence of different levels of a concept or domain. From the viewpoint of the domain-specific model, the acquisition of each successive form of cognitive regulation entails a deformation of the prior form. It is not that prior cognitions disappear but that they have the subordinate role of elements in the subsequent structure. Therefore, earlier structures are the basis for, and part of, later structures.

We expect that the coexistence of structures would be found between domains. Developmental sequences with separate origins, such as those depicted in Fig. 1, can be regarded as representing distinct domains. A change in the notations in Fig. 1 from stages to domains (e.g., A, B, C) as well as depiction of the curves as nonconvergent would render the graphic representation a little closer to the position we have put forth. Such changes would not be sufficient, however, because the representation would still fail to capture qualitative discontinuities in vertical organization.

It is important that we be explicit as to which aspects of such graphic representations are insufficient. It is not that the curves are insufficient because of the *temporal* continuity represented in age changes. On the temporal dimension continuity (gradual change) and discontinuity (abrupt change) can be associated with either qualitative (discontinuous) or quantitative (continuous) cognitive changes. As noted earlier, it is quite plausible that transitions from one structure to a qualitatively different structure can take place over a period of time and thus be represented as continuous. Whether structural changes are gradual or abrupt over time depends on the nature of the process of the change. For example, one proposition regarding structural change (i.e., a process of deformation and formation) leads to the expectations that it normally would take place over time and, with time or age as the independent variable, would yield the type of curve in Fig. 1.

Conversely, abrupt changes on the time dimension do not necessarily reflect qualitative changes. Again, with time or age as the independent variable, the meaning of the type of curve shown in Fig. 2, with its plateaus and inclines, would depend on what is measured and the nature of the process of change. The curve in Fig. 2 could readily portray quantitative or cumulative changes rather than qualitative changes. As one example, we can imagine that the ordinate represents the weight of an individual who experienced a not uncommon growth spurt at puberty (the first rate increase) and later began intensive training as a body builder (the second rate increase).

Neither of the representations (Figs. 1 and 2) can be used as evidence for or against structural or qualitative discontinuity. Developmental curves with age as the independent variable do not denote structural change, since discontinuity is defined according to cognitive criteria rather than criteria of abruptness or change in slope. In turn, curves with such constructs as development as the dependent variable are unable to represent transformations such as changes of regulations to elements.

Figure 2. Representation of stages by a single curve with plateaus and inclines.

These graphic representations are insufficient, therefore, because they are incomplete. A more comprehensive example of developmental curves representing structural change is provided by Keil (this volume). In the studies of conceptual structure for word meanings, Keil describes a shift from characteristic to defining features. The graphs depicting this shift (see Keil's chapter for details of the research and the graphic representations) contain intersecting curves, which show that the emphasis on defining features increases with age while the emphasis on characteristic features decreases. Those curves more adequately represent developmental patterns because of the inclusion of dual processes. Keil's findings can be interpreted to show that change involves an interaction of the process of deformation in concurrence with formation. Stage transition is defined by this shift, rather than temporal abruptness; the intersecting curves illustrate discontinuity in development but are themselves continuous, with fairly constant slopes.

The general or idealized form of such curves comes close to representing the essential properties of developmental stages and transitions. A given level of structural stability is characterized by the predominance of a type of concept. Thus, in Keil's examples, early reasoning is defined by a concern with characteristic features. The subsequent stage is characterized both by the ascendancy of defining features as well as the transformation of the status of the prior concept in the new structure. That is, in this example, it should not be assumed that subjects cease paying attention to characteristic features; these remain important components of judgment, but they become subordinated to the defining criteria.

In a previous section we mentioned another example of this formation–deformation process with regard to shifts in moral judgments during childhood. More extensive cross-sectional and longitudinal analyses have been done of age-related shifts within the conventional domain (Turiel, 1978, 1983a). A sequence was described in which there is a successive acceptance and rejection (or what may be termed affirmation and negation) of convention. Transitions include a phase of negation, characterized by a critical orientation toward the existing way of thinking, followed by the construction and affirmation of a more advanced set of concepts (such periods would be represented by the intersection points in curves, like those of Keil). For example, at around 6 years of age children conceive of conventions as empirical social uniformities. By age 10, they conceive of conventions as concrete rules or authority expectations. This shift in the regulatory component of societal reasoning involves a period of deformation in which empirical uniformities are increasingly rejected as a basis for evaluating conventions, and a period of formation in which uniformities begin to be treated as elements that fall under, or instantiate, concrete rules and authority expectations. By age 15, individuals evaluate conventions according to yet another regulation, a concern with social organization. This shift involves an intervening period in which rules and authority are rejected as regulatory components and begin to be viewed as elements subordinated to the more general and adaptive concern with social organization.

In the model of formation and deformation in the conventional domain, transitions are gradual, involving conflicts, negation and nonlinearity. Transitions are identifiable phases that do not appear to occur in sudden, abrupt fashion. Similar examples are found in children's performance in balancing tasks (Karmiloff-Smith & Inhelder, 1974/1975) and in children's reasoning about intensive quantity (Strauss & Stavy, 1982). In previous sections we have suggested how the coexistence of contrary orientations actually reflects coherence of separate domains in the underlying cognitive organization. These observations concerning transition periods indicate that coexistence of inconsistencies within a domain also may reflect the process of within-domain development. That is, the concurrence of less adaptive and more adaptive cognitive regulations and judgments is an essential feature of the formation–deformation process described, and as such is an important component in the "vertical organization" of cognition.

References

Asch, S. E. (1956). Studies of independence and conformity. A minority of one against a unanimous majority. *Psychological Monographs, 70,* (No. 9).

Beilin, H. (1968). Cognitive capacities of young children: A replication. *Science, 162,* 910–921.

Berlin, I. (1981). *Concepts and categories: Philosophical essays.* New York: Penguin Books.

Blasi, A. (1980). Bridging moral cognition and moral action: A critical review of the literature. *Psychological Bulletin, 88,* 1–45.

Carey, S. (1984). Are children fundamentally different kinds of thinkers and learners than adults? In S. Chipman, J. Segal, & R. Glaser (Eds.), *Thinking and learning skills* (Vol. 2). Hillsdale, NJ: Erlbaum.

Carter, D. B., & Patterson, C. J. (1982). Sex-roles as social conventions: The development of children's conceptions of sex-role stereotypes. *Developmental Psychology, 18,* 812–824.

Chomsky, N. (1975). *Reflections on language.* New York: Pantheon.

Chomsky, N. (1980). On cognitive structures and their development. A reply to Piaget. In M. Piatelli-Palmarini (Ed.), *Language and learning: The debate between Jean Piaget and Noam Chomsky.* Cambridge, MA: Harvard University Press.

Damon, W. (1977). *The social world of the child.* San Francisco, CA: Jossey-Bass.

Damon, W. (1983). The nature of social-cognitive changes in the developing child. In W. Overton (Ed.), *The relationship between social and cognitive development.* Hillsdale, NJ: Erlbaum.

Davidson, P. (1983). *The development of numerical reasoning in children: The role of classes, relations, and functions.* Unpublished doctoral dissertation, University of California, Berkeley.

Davidson, P., Turiel, E., & Black, A. (1983). The effect of stimulus familiarity on the use of criteria and justifications in children's social reasoning. *British Journal of Developmental Psychology, 1,* 49–65.

Dworkin, R. (1978). *Taking rights seriously.* Cambridge, MA: Harvard University Press.

Flavell, J. H. (1971). Stage-related properties of cognitive development. *Cognitive Psychology, 2,* 421–453.

Flavell, J. H. (1981). Structure, stages and sequences in cognitive development. In A. Collins (Ed.), *Minnesota Symposia on Child Psychology* (Vol. 15). Hillsdale, NJ: Erlbaum.

Flavell, J. H. (1982). On cognitive development. *Child Development, 53,* 1–10.

Flavell, J. H., & Wohlwill, J. F. (1969). Formal and functional aspects of cognitive development. In D. Elkind & J. H. Flavell (Eds.), *Studies in cognitive growth: Essays in honor of Jean Piaget.* New York: Oxford University Press.

Forman, G. E. (1982). A search for the origins of equivalence concepts. In G. E. Forman (Ed.), *Action and thought.* New York: Academic Press.

Frege, G. (1974). *The foundations of arithmetic.* Oxford: Blackwell. (Original work published 1884).

Gelman, R., & Gallistel, C. R. (1978). *The child's understanding of number.* Cambridge, MA: Harvard University Press.

Gewirth, A. (1978). *Reason and morality.* Chicago, IL: University of Chicago Press.

Greco, P., Inhelder, B., Matalon, B., & Piaget, J. (1963). *La formation des raisonnements recurrentiels.* Paris: Presses Universitaires de France.

Haney, C., Banks, C., & Zimbardo, P. (1973). Interpersonal dynamics in a simulated prison. *International Journal of Criminology and Penology, 1,* 69–97.

Inhelder, B., & Piaget, J. (1964). *The early growth of logic in the child.* London: Routledge & Kegan Paul. (Original work published 1959).

Jancaterino, W. (1982). *The relationship between children's understanding of social influence and their moral evaluations of harm.* Unpublished doctoral dissertation, University of California, Santa Cruz.

Karmiloff-Smith, A., & Inhelder, B. (1974/1975). If you want to get ahead, get a theory. *Cognition, 3,* 195–212.

Kamii, C., & Devries, R. (1978). *Physical knowledge in preschool education.* Englewood Cliffs, NJ: Prentice-Hall.

Kessen, W. (1966). Questions for a theory of cognitive development. *Monographs of the Society for Research in Child Development, 31,*(5), 55–70.

Kohlberg, L. (1968). The child as moral philosopher. *Psychology Today, 2*(4), 24–30.

Kohlberg, L. (1969). Stage and sequence: The cognitive-developmental approach to socialization. In D. A. Goslin (Ed.), *Handbook of socialization theory and research.* Chicago, IL: Rand McNally.

Kohlberg, L. (1971). From is to ought: How to commit the naturalistic fallacy and get away with it in the study of moral development. In T. Mischel (Ed.), *Psychology and genetic epistemology.* New York: Academic Press.

Langer, J. (1980). *The origins of logic.* New York: Academic Press.

Langer, J. (1982). From prerepresentational to representational cognition. In G. E. Forman (Ed.), *Action and thought.* New York: Academic Press.

Langer, J. (1983). Concept and symbol formation by infants. In S. Wapner & B. Kaplan (Eds.), *Toward a holistic developmental psychology*. Hillsdale, NJ: Erlbaum.

Laupa, M., & Turiel, E. (1984). *Children's conceptions of adult and peer authority*. Unpublished manuscript, University of California, Berkeley.

Lewis, D. (1969). *Convention: A philosophical study*. Cambridge, MA: Harvard University Press.

Lockhart, K. L. (1980). *The development of knowledge about uniformities in the environment: A comparative analysis of the child's understanding of social, moral, and physical rules*. Unpublished doctoral dissertation, University of Pennsylvania, Philadelphia.

Mehler, J., & Bever, T. G. (1967). Cognitive capacity of very young children. *Science, 158,* 141–142.

Milgram, S. (1974). *Obedience to authority*. New York: Harper & Row.

Much, N., & Shweder, R. A. (1978). Speaking of rules: The analysis of culture in breach. In W. Damon (Ed.), *New Directions for Child Development. Vol. 2: Moral Development*. San Francisco, CA: Jossey–Bass.

Nucci, L. (1977). *Social development: Personal, conventional, and moral concepts*. Unpublished doctoral dissertation, University of California, Santa Cruz.

Nucci, L. (1981). The development of personal concepts: A domain distinct from moral or societal concepts. *Child Development, 52,*114–121.

Nucci, L., & Nucci, M. (1982a). Children's social interactions in the context of moral and conventional transgressions. *Child Development, 52,* 403–412.

Nucci, L., & Nucci, M. (1982b). Children's responses to moral and social conventional transgressions in free-play settings. *Child Development, 53,* 1337–1342.

Nucci, L., & Turiel, E. (1978). Social interactions and the development of social concepts in preschool children. *Child Development, 49,* 400–407.

Piaget, J. (1932). *The moral judgment of the child*. London: Routledge & Kegan Paul.

Piaget, J. (1965). *The child's conception of number*. New York: Norton. (Original work published 1941).

Piaget, J. (1968). Quantification, conservation, and nativism. *Science, 162,* 976–979.

Piaget, J. (1970). *Structuralism*. New York: Basic Books.

Piaget, J. (1971). *Psychology and epistemology*. New York: Viking.

Piaget, J. (1980). *Experiments in contradiction*. Chicago, IL: University of Chicago Press.

Piaget, J., Grize, J-B, Szeminska, A., & Vinh-Bang. (1977). *Epistemology and psychology of functions*. Dordrecht, Holland: D. Reidel. (Original work published 1968).

Piaget, J., Inhelder, B., & Szeminska, A. (1960). *The child's conception of geometry*. London: Routledge & Kegan Paul.

Pool, D. L., Shweder, R. A., & Much, N. C. (1983). Culture as a cognitive system. Differentiated rule understandings in children and other savages. In E. T. Higgins, D. N. Ruble, & W. W. Hartup (Eds.), *Social cognition and social development: A socio-cultural perspective*. Cambridge, England: Cambridge University Press.

Rawls, J. (1971). *A theory of justice*. Cambridge, MA: Harvard University Press.

Rest, J. R. (1983). Morality. In P. H. Mussen (Ed.), *Handbook of child psychology* (*4th ed.*). Vol. 3: J. H. Flavell & E. Markman (Eds.), *Cognitive development*. New York: Wiley.

Robert, M., Cellerier, G., & Sinclair, H. (1972). Une observation de la genese du nombre. *Archives de psychologie, 41,* 289–301.

Ross, L. (1981). The "intuitive scientist" formulation and its developmental implications. In J. Flavell & L. Ross (Eds.), *Social cognitive development: Frontiers and possible futures*. Cambridge, England: Cambridge University Press.

Searle, J. (1969). *Speech acts*. London: Cambridge University Press.

Silverman, I., & Briga, J. (1981). By what process do young children solve small number conservation problems? *Journal of Experimental Child Psychology, 32,* 115–126.

Smetana, J. (1981). Preschool children's conceptions of moral and social rules. *Child Development, 52,* 1333–1336.

Smetana, J. (1982). *Concepts of self and morality: Women's reasoning about abortion*. New York: Praeger.

Smetana, J. (1983). Social–cognitive development: Domain distinctions and coordinations. *Developmental Review, 3,* 131–147.

Smetana, J. (in press). Preschool children's conceptions of transgressions: The effects of varying moral and conventional domain-related attributes. *Developmental Psychology.*

Starkey, P. (1983). *Some precursors of early arithmetic competencies.* Paper presented at the Bienniel Meeting of the Society for Research in Child Development. Detroit, MI.

Stoddart, P., & Turiel, E. (1984). *Children's concepts of deviations from gender-associated uniformities.* Unpublished manuscript, University of California, Berkeley.

Strauss, S. (1983). Appearances, disappearances, nonappearances, and reappearances of various behaviors: Methodological strategies of experimentation and their implications for models of development. In T. G. R. Bever (Ed.), *Regressions in mental development: Basic phenomena and theories.* Hillsdale, NJ: Erlbaum.

Strauss, S., & Stavy, R. (1982). U-shaped behavioral growth: Implications for theories of development. In W. W. Hartup (Ed.), *Review of Child Development Research* (Vol. 6). Chicago, IL: University of Chicago Press.

Sugarman, S. (1983). *Children's early thought.* Cambridge, England: Cambridge University Press.

Tisak, M. (1983). *Children's conceptions of authority within different systems of social rules.* Unpublished doctoral dissertation, Stanford University.

Turiel, E. (1978). The development of concepts of social structure. In J. Glick & A. Clarke-Stewart (Eds.), *The development of social understanding.* New York: Gardner Press.

Turiel, E. (1983a). *The development of social knowledge: Morality and convention.* Cambridge, England: Cambridge University Press.

Turiel, E. (1983b). Domains and categories in social cognitive development. In W. Overton (Ed.), *The relationship between social and cognitive development.* Hillsdale, NJ: Erlbaum.

Turiel, E., & Smetana, J. (1984). Social knowledge and action: The coordination of domains. In W. M. Kurtines & J. L. Gewirtz (Eds.), *Morality, moral behavior, and moral development: Basic issues in theory and research.* New York: Wiley.

Wellman, H. M. (1982). Metamemory revisited. In M. T. H. Chi (Ed.), *Trends in memory development: Vol. 2. Contributions to human development.* Basel, Switzerland: Karger.

Weston, D., & Turiel, E. (1980). Act-rule relations: Children's concepts of social rules. *Developmental Psychology, 26,* 417–424.

6

On the Structure-Dependent Nature of Stages of Cognitive Development*

Frank C. Keil
Department of Psychology
Cornell University

In the last several years there has been a strong trend in cognitive development research raising serious questions about the reality of stages in cognitive development. In increasing numbers, investigators are challenging stage theories such as those of Piaget, Bruner, and Vygotsky. The pattern has been to question whether young children really lack a certain type of competence that older children have; and the studies have generally argued that, if careful enough tasks are used, young children are not qualitatively different from older ones. Of course, if they are not qualitatively different, then there can be no stages of cognitive development for that type of knowledge.

One of the most influential critiques along these lines was Fodor's (1972) retrospective review of Vygotsky's (1934/1962) *Thought and Language* which argued that there has been little evidence for general representational shifts in the course of development. Instead, Fodor suggested that roughly the same computational and representational systems are shared by individuals of all ages and that what develops is an increasing ability to use these systems in a wider and wider range of tasks.

Rozin (1976) echoed this notion when he argued for increasing access to underlying competencies as one of the main mechanisms of developmental change. In his view, children may not differ from adults in terms of basic underlying abilities or competencies but more in how easily they can apply those abilities to a wide range of potentially relevant tasks.

These theoretical proposals have been supported by a number of different research programs that seem to argue against qualitative changes in either mode of representation or of processing. For example, in contrast to the Piagetian accounts of number conservation, Gelman and her associates (e.g., Gelman & Gallistel, 1978) have found that preschoolers are able to conserve number, if certain task factors irrelevant to the conservation paradigms are controlled. Thus, if the size of the numerical display is made small enough and if the two displays to be compared

* Preparation of this paper and much of the research described herein was supported by National Science Foundation grants BNS-81-02655 and BNS-78-06200. Many thanks to Sydney Strauss, Iris Levin, Michael Kelly, Tammy Globerson, Ruth Berman, and Barbara Bauer for making very helpful comments on this manuscript.

are presented in an appropriate manner, 3-year-olds are able to show the hallmarks of conservers. Gelman argues that these children normally fail on the traditional Piagetian tasks, not because they are unable to engage in concrete operational thought, as the Piagetian account would hold, but because memory limitations and other irrelevant factors interefere with the application of their basic competencies.

A different example is research on classification, where researchers have again argued that very young children can engage in types of cognition that meet all the criteria of concrete operational thought. Rosch, Mervis, Gray, Johnson, and Boyes-Braem (1976), for example, showed that the same child who seemed to classify in a completely preoperational way for some classes of objects could behave in a concrete operational manner with other objects that used the more salient differences between members of basic level categories. Similarly, Markman (Markman & Siebert, 1976) has shown that the ability to apprehend subset/superset relations is present in preschoolers, if the classses are united by certain concepts that make them be viewed as collections rather than classes. A group of blue blocks is judged correctly to be less than all of the blocks present, if all of the blocks are referred to as "the pile of blocks" and not if referred to as "blocks."

The same sort of account has been repeated in several other domains as well. One final example is causal knowledge, where a number of investigators such as Bullock, Gelman, and Baillargeon (1982) argue that young preschoolers, if assessed carefully enough, show almost all the aspects of mature causal knowledge. Bullock et al. describe a series of studies concerned with three separate principles that adults use to understand causal relations: determinism (physical events must have causes), priority (causal events must precede effects), and mechanism (causes bring about effects by a kind of transfer of "causal impetus"). In contrast to earlier work suggesting dramatic qualitative shifts in the nature of causal thinking, they describe a series of studies showing that children as young as 3 years adhere to each of these principles. They conclude that

> This means that the development of causal understanding is more a process of learning where, when, and how to apply the rules of reasoning than figuring out what those rules might be. . . . The differences that exist [between preschoolers' causal thinking and that of adults, F. K.] arise not because the child and adult think about things in fundamentally different ways, but because the child's thought is more constrained by context, complexity and verbal demands, limiting the scope and flexibility with which the child can apply his or her knowledge. (Bullock et al., 1982, p. 251)

What develop, according to these accounts (e.g., Gelman & Gallistel, 1978; Rozin, 1976; Fodor, 1972), are a variety of factors, none of which strongly suggests qualitative change. Thus, there is increasing access to the same competence, enabling its use in an ever broadening range of tasks. There is also the adoption of new strategies, which represent not so much a change in competence as of choice. Finally, there is increasing differentiation and complexity of knowledge, but not necessarily any change in its fundamental nature.

Elsewhere (Keil, 1981) I have interpreted research along these lines as suggestive of a constraints-oriented view of cognitive development of the sort that

Chomsky (1980) has proposed in linguistic theory. Under this view, acquiring complex knowledge would be impossible, if all of us did not share certain a priori constraints that restrict the range of hypotheses to be made about the structure of what is learned. Throughout development, the child is guided by sets of constraints that represent a kind of competence that is always present, namely, the set of possible, or natural, knowledge structures in a domain. In this account the main phenomena of development are increasing differentiation and increasing access. If our knowledge structures and procedures in various domains are always guided by constraints, then little qualitative change would appear possible.

To summarize, one line of recent research in cognitive development argues that, whereas children may differ from adults, they do not differ in any qualitative way that suggests general stages of mental representation or processing. This perspective might nonetheless be regarded as structuralist in that much of the work is devoted to uncovering structural descriptions of the basic competencies and constraints in various domains of cognition. One can be interested in the structural principles of mind without being a stage theorist. The converse, being a stage theorist without being a structuralist, does not seem possible. The very notion of a cognitive stage would seem to require structural descriptions of knowledge at various points in development. Without such descriptions it is difficult to know how to evaluate whether the developmental differences are qualitative versus quantitative.

I still embrace the constraints perspective; it represents a fundamental component of any account of knowledge acquisition. It is incorrect, however, to assume that this perspective and the above trends in cognitive development research imply that there cannot be clear qualitative, stagelike changes in development as well. Demonstrations of core competencies in surprisingly young children are not the entire story; nor are increasing access and gradual differentiation along the strict guidelines of a priori constraints the only types of change. While each of these undoubtedly exists, and while the previously described studies do suggest that there is very little evidence for across the board developmental shifts in manner of representation or processing, there are other types of change that do not fit into the patterns just described.

Cases Where Qualitative Shifts Do Occur

Some types of conceptual change have traditionally been used as support for the idea of global representational shifts in development. Consider, for example, a pattern of change in conceptual structure that has led developmental theorists over the years to propose general shifts from the wholistic to the analytic, or from the concrete to the abstract, or from the perceptual to the conceptual (e.g., Werner, 1948; Bruner, Olver, and Greenfield, 1966; Anglin, 1984).

Each of these theories has been grounded in robust observations about the sorts of errors that children seem to make in using and talking about concepts; however, each of them has also failed to recognize how closely the shifts are tied into the particular concepts involved. My students and I have been investigating a shift in conceptual structure for word meanings that appears to reflect the same phenomenon that motivated these theories. We have described it as a shift from representations based on features that are merely characteristic of an entity to representations that emphasize defining features. There are numerous instances from past literature that illustrate the idea. For example, many investigators have noted that young children seem to represent the meanings of kinship terms solely in terms of the properties of typical examples or symptoms (e.g., Vygotsky, 1962; Piaget, 1928; Haviland & Clark, 1974; Landau, 1982). Thus a grandmother might be initially represented as an elderly woman with a certain set of behavioral dispositions. Only later does the the child begin to realize that these characteristic features are only characteristic and that what really counts for someone to be a grandmother are the various kinship relations that define the concept.

We (Keil & Batterman, 1984) have documented repeatedly the occurrence of such a shift by using stories that have either all the defining features of a concept but highly uncharacteristic ones, or vice versa. Thus a child will first state that the story with highly characteristic features but incorrect defining ones describes a valid instance, while rejecting the one with correct defining features but highly uncharacteristic ones. An older child will do the opposite. The +characteristic/−defining (+c/−d) story for ''uncle'' described a fellow ''roughly the age of your father who brought you presents on your birthday, visited a lot on weekends and watched football with your dad, but who wasn't related to anyone in the family.'' The −c/+d story described ''a 2-year-old infant who was your father's brother.'' For almost all terms studied, one sees crossovers from an emphasis on defining features to an emphasis on characteristic ones, as shown in Fig. 1.

These results might at first be interpreted as evidence for a general stagelike transition in cognitive development from characteristically based concepts to more definitionally based ones. There is an additional finding, however, that challenges such an interpretation: the shift can occur at radically different times for different concepts. Thus, most preschoolers may have already made the shift for some concepts, while 12-year-olds still have not made it for others. It is difficult to argue that such a shift represents merely a horizontal décalage for a general stage, when the stage seems to cover such a large age range. This is especially true, given that comparable shifts occur in most adults, as will be described later. If one does wish to call it a décalage, then the décalage becomes the entire developmental phenomenon and the general stage becomes irrelevant.

This claim that the occurrence of the shift is dependent on local knowledge structures gains further support from two additional lines of research: work with terms from different semantic fields and demonstrations that the shift occurs between novice and expert adults. In a follow-up study, we have looked at how and when such shifts occur for closely related sets of terms that form what are normally known as semantic fields (cf. Lehrer, 1978). These would include kinship terms (e.g., aunt, uncle, grandmother), cooking terms (e.g., boil, bake, fry), and moral act terms (e.g., lie, steal, cheat). In all cases we had two stories, one that described an entity or activity that had many of the most salient characteristic features of a word, but incorrect defining ones, and one that had the opposite.

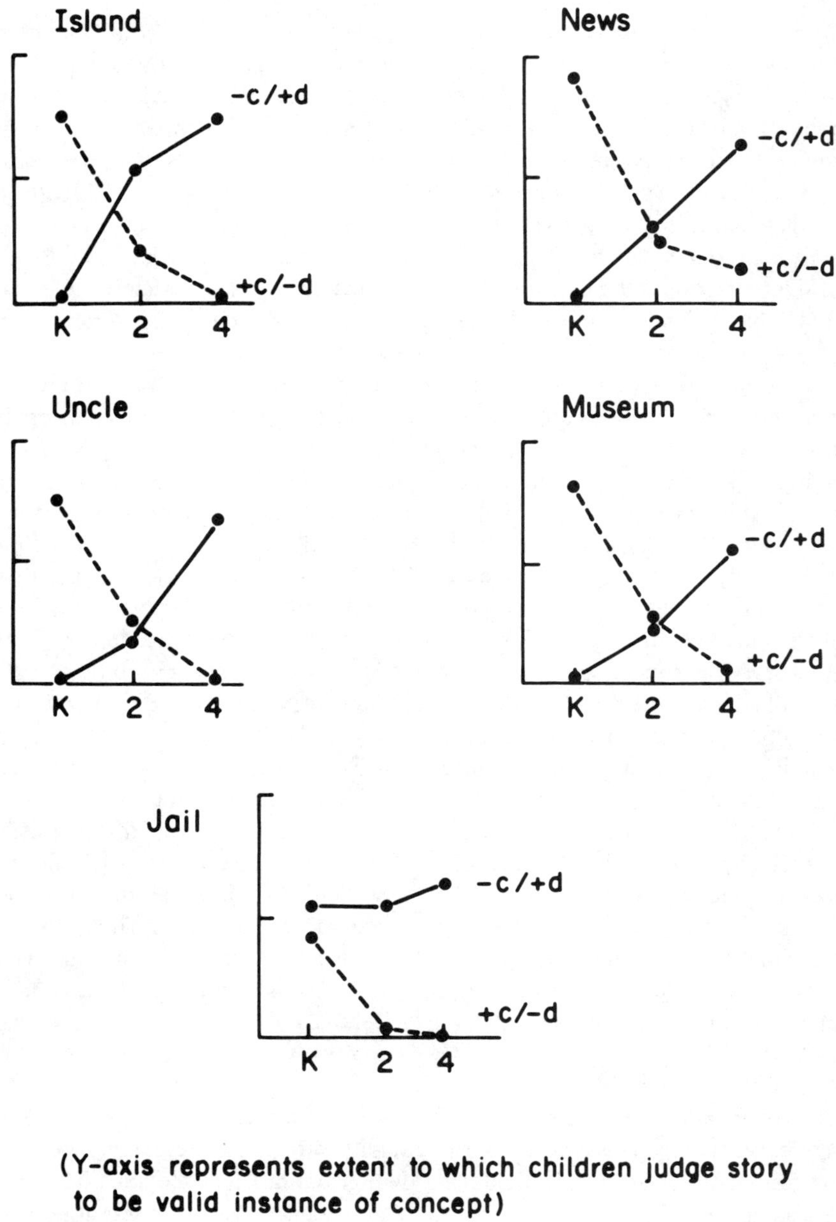

Figure 1. Examples of the characteristic-to-defining shift for five terms; *Y*-axis represents extent to which children judge story to be valid instance of concept.

The results support the claim that the shift is structure dependent, because it occurs at roughly the same time for terms within each field but at widely differing times for the sets of terms across fields. Thus, children generally shifted at a much earlier age for moral act terms than they did for cooking terms. Figure 2 illustrates how the shift occurred at different ages depending on the concepts involved. The crossover points on each graph can be construed as roughly the ages at which the shift occurred. This pattern of results does not fit well with the proposal that there is a general-content independent stagelike change.

A second example comes from Chi, Feltovich, and Glaser (1981), who examined differences between adult experts and novices in knowledge of classical physics. While their method was quite different and simply involved having the subjects sort out various physics problems into equivalence classes, the results showed a shift from novices to experts that is very much like the characteristic-to-defining shift that we have uncovered. Novices tended to classify problems based on highly characteristic features that had little to do with underlying principles of physics. Thus they would sort all the ramp problems in one pile and all the pendulum ones in another, and so on. The experts sorted on a completely different basis that involved reference to basic laws of physics such as Newton's Force Laws, Conservation of Energy, etc. The occurrence of such shifts in adults again argues against an across-the-board developmental shift. The expert/novice contrast is not in itself a complete description of child/adult differences, but it does illustrate the limitations of a general stage theory explanation for the characteristic to defining shift.

A final example of the characteristic-to-defining shift comes from a training study that we are just now transcribing and analyzing. It is particularly relevant because of the major role that training studies have played in discussions of general stages of cognitive development. In essence, we attempted to teach children the meanings of novel words that come from various semantic fields. For example, we taught children the meanings of new cooking terms (e.g., sauté, baste) and new agricultural terms (e.g., mulch, irrigate) by giving them explanations that included clear definitions as well as several typical examples. The following two passages illustrate the explanations we gave, including the special care we took to present the defining properties twice, once at the beginning and once at the end of the story:

> *Baste:* To baste is to wet a food with something while it's cooking. For example, at Thanksgiving, your mom probably basted your turkey by brushing on melted butter and meat juices every half hour so the meat would stay moist.
>
> So to baste is to wet something every so often with a liquid while you're cooking it.
>
> *Mulch:* To mulch is to put something over the land so that things will grow better on it. I'm sure you've seen people mulch the land in Ithaca. People often put lots and lots of wood chips around rose bushes or young plants here in the winter to keep them from freezing and to keep the water in the soil.
>
> So mulching is when you cover the ground to help things you've planted there grow better.

After making sure that the children understood the explanations for these novel words, we followed the procedure used in prior studies and asked them whether

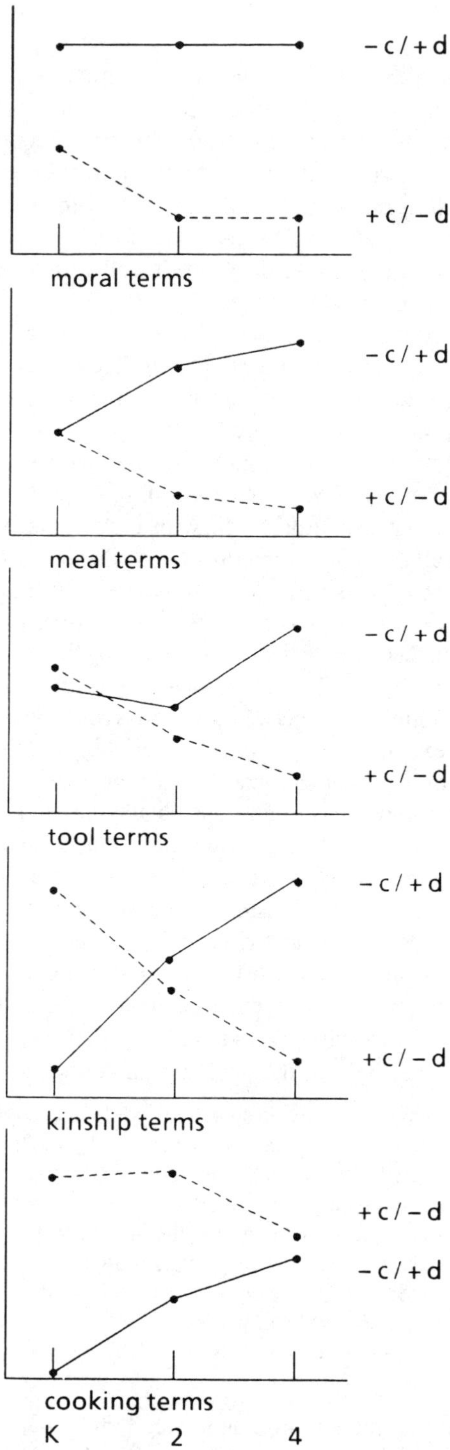

Figure 2. The characteristic-to-defining shift in terms from five different semantic fields.

+characteristic/−defining and −characteristic/+defining stories described valid instances. We had two hypotheses, and it looks as if both will be supported. The first was that, even when given an explanation with a clear definition, younger children may nonetheless tend to represent the meaning in terms of characteristic features. Based on preliminary scoring, there is a significant main effect of age with an increasing reliance on defining features in older children. A second expected pattern was that the shift would not occur at the same time for all concepts; and this, too, appears to be happening. Children tended to shift at an earlier age for the cooking terms than they did for the agricultural terms, although the effect is not as strong as in the semantic field study described previously.

This study illustrates that the effectiveness of training may depend not so much on a general cognitive awareness afforded by a particular stage of cognitive development as it does on the degree of knowledge in the domain in which the novel concept is being taught. Only when a child is beginning to master several more familiar cooking terms does s/he begin to develop an understanding of what the underlying dimensions are that are relevant to the various cooking verbs and what sorts of relations are likely to be defining when learning new ones. More generally, success and failure in training studies should not necessarily be regarded as evidence for a general stage of cognitive development when a more local knowledge-specific explanation will suffice.

The characteristic-to-defining shift is only one case of several where a general stagelike phenomenon has been shown to be more dependent on specific knowledge structures than was previously thought. Much of the classsical developmental literature assumed that younger children had qualitatively different memories and memory procedures that reflected a general change in representation and/or processing. Recent studies, however, suggest an alternative account. For example, Chi (1978) and others have used the expert/novice distinction to show that expert kindergartners can show adultlike performance on tasks involving their own expertise and childlike performance with less familiar domains, whereas adults who are novices in the children's area of expertise might show the opposite pattern. Consequently, differences in memory performance reflect not so much changes in underlying competence as they do the increasing enrichment of knowledge in different domains.

The discussion up to this point leaves us with several loose ends. One concerns the relations between the two views that have been used to criticize the stage theories, the early competency view and the knowledge enrichment view. How do these two views fit with each other? Can they be put together into one theory or must one flourish at the expense of the other?

A second issue concerns interactions between various domains of cognition. What happens when a qualitative shift occurs in one particular domain that is used in a wide range of cognitive tasks? Do such cases exist, and if so do they not in effect result in general stages of cognitive development? Second, to what extent are there interdependencies between domains such that knowledge must reach a certain form in one before a shift can occur in the other?

A third and related issue asks what else there is beyond expert/novice differences to qualitative changes in development. It is unsatisfying to assume that children simply acquire knowledge in the same manner that adults become experts. Common sense suggests that children often go about acquiring knowledge in fundamantally different ways than adults and that one cannot simply map the adult model in its entirety onto children.

Reconciling the Early-Competency and Knowledge-Enrichment Views

I have argued that two lines of research raise questions about the validity of general stages of cognitive development; but each does so in very different ways. The early competence view argues that the basic properties of knowledge are the same throughout development. Thus the formal structural properties of knowledge in each domain show very little qualitative change. The constraints version of this view argues that knowledge structures at all points in development honor fundamentally the same sets of constraints. The mains types of change are increasing access to the same underlying structures and increasing differentiation of present structures along the guidelines of constraints. The increasing access may result in what look like qualitative changes in representation but which are not; and the increasing differentiation presumably does not even appear to be qualitative. This view on its own, therefore, not only argues against general stages of cognitive development but also against stages specific to local domains. Other early competence views might not put as explicit an emphasis on constraints, but the general critique of stage theories remains the same.

The knowledge enrichment view makes few commitments either way about stagelike changes, but the described examples suggest that, whenever stagelike changes do occur, they are functions of local knowledge systems and the particular properties of what is being learned, not general, more endogenous changes in cognition.

What these two views have in common, therefore, is a questioning of developmental patterns that purportedly show changes in cognitive structures that are global in nature. Where they differ is that an absolute constraints view disallows any sort of qualitative change while the knowledge view is more noncommittal.

Not surprisingly, I will suggest that both views are needed for an adequate account of cognitive development. It is difficult to doubt that there are universal constraints on knowledge that seem to guide how that knowledge is acquired and the structure it takes even in very young children. It is equally difficult to doubt that local expertise can sometimes result in dramatic changes in the nature of specific knowledge structures.

The system I envision consists of a conceptual core or core competency that is defined by sets of rigorous constaints and which is supplemented by periphery that is governed much more loosely by more domain-general (i.e., content-independent) constraints. Obviously, the relative proportions of core and periphery vary greatly depending on the type of knowledge involved, but virtually all natural learning situations involve elements of both.

The conceptual core lays down a skeletal framework within which the periphery differentiates. Thus, the constraints are rarely if ever overridden by the periphery; nonetheless dramatic shifts can occur in the periphery, provided they honor the general boundaries of the skeleton. Because this core is only a skeletal framework, it does not completely constrain the structures within it and therefore allows changes that might be called qualitative. Initially, the patterns of learning in the periphery and the resultant representations might be surprisingly homogeneous across domains and may use many common types of learning procedures. But as the learner becomes more and more knowledgeable in each domain, the particular properties of each domain begin to dominate and the knowledge representations begin to diverge, so that there is an increasing plurality of representational types with increasing knowledge. Put differently, at the level of the periphery, development consists of a gradual divergence of knowledge structures, as the particular properties of what is learned in each domain begin to have dominance over the earlier influences of general learning procedures.

This model can be illustrated more concretely with examples from the characteristic-to-defining shifts. At the periphery virtually all word meanings start off being represented in roughly the same manner: namely, sets of characteristic features, perhaps arising out of familiar exemplars. Repeatedly, for a wide range of terms we have found that the earliest stage of knowledge seems to be represented in terms of some sort of typicality structure. This is true for many different types of nouns as well as verbs and adjectives. Nouns as different as artifact and natural-kind terms both start out being represented by lists of their most typical properties. Similarly verbs ranging from "lie" to "boil" to "irrigate" are initially construed as applying to events that have all the characteristic features of those words. Finally, adjectives such as "tall" are initally understood in terms of characteristically tall things (Keil & Carroll, 1980).

As children learn more about each of these concepts, their knowledge structures show a shift away from characteristic features. For some terms the shift is toward what seem to be necessary and sufficient features. These are words that have been discussed so far and include kinship terms, terms referring to social conventions, and the like. They also include artifacts, although here the shift is to a definition that corresponds to the function intended by the creator of the artifact.

Even with these words with clear definitions, the final adult-knowledge structures may differ substantially from domain to domain, because the sorts of relations necessary to describe each can be intrinsically different. Thus, the kinds of relations describing various cooking terms are likely to be very procedural in nature while, in the case of kinship terms, the relevant procedures (i.e., those of reproduction) are referred to in only the most fleeting manner. Moreover, the differences between terms across the various syntactic categories may also cause different sorts of representations such as the noun/verb differences described by Gentner (1981).

The differences in final knowledge structures are more dramatic, if one considers the natural-kind terms. As philosphers of language have now argued for some time, these terms do not seem to have simple definitions or anything close to a set of necessary and sufficient features (cf. Putnam, 1975; Kripke, 1972). In more recent work we have demonstrated that children also start off with instance-bound representations for natural-kind terms and shift away from them; but they shift to a very different sort of knowledge structure than for artifacts. One research paradigm that we have used extensively consists of describing to children certain transformations on objects that change many of the most salient characteristic features and nothing else. Thus a lion might have various parts of its body shaved and then be dyed with stripes. The child is asked whether the end product is a lion or a tiger. (Photographs of the before-and-after states are shown to the children and are, in fact, photographs of a real tiger and a real lion.)

We find that young children will often assume that changes in characteristic features cause changes in identity, whereas older children do not; and this is true, not just for natural kinds that are animals, but also for plants and minerals. The older children do not shift to a belief in simple definitions, however, but rather to a belief in complex theories of a more causal nature. They move away from representing the meaning merely in terms of typical properties to an underlying causal theory and often a theory that explains the origins of the natural kind. For example, for animals, older children and adults often refer to the animal's parents as the basis for its identity.

We have compared the same children's responses to stories about artifacts and find that, while in the early stages both sorts of terms are represented in roughly the same manner, they diverge dramatically in how they are represented in older children and adults.

More generally, characteristic features or other sorts of representations based on typicality information seem to be very pervasive in the early stages of knowledge acquisition, not just for the meanings of words but also for syntactic categories (Maratsos, 1983) and perceptual patterns (Kemler, 1983). There are probably several other general representational strategies and types that are employed in the early stages of knowledge acquisition and which give way to more local structural influences with increasing knowledge. The important point here is that when qualitative change occurs, it is of a certain sort—from representations that are initially set up with relatively little attention to the peculiarities of what is learned to representations where these peculiarities dominate. This few to many mapping of representational types may be the primary way in which qualitative changes occur. Thus, early in concept acquisition when general strategies predominate, concepts from a wide range of domains may show many representational properties that are derivative from those strategies. Later, however, the concepts tend to diverge more and more from each other in terms of their particular structural properties.

Above and beyond these qualitative shifts, however, there is a conceptual core that exerts constraints on the knowledge at any point in its development. Again, the case of word meaning acquisition provides a useful illustration. While young children will frequently assume that merely changing the characteristic features of natural kinds can change them into a different natural kind, they do not allow just any changes to do so. They are perfectly willing to let one animal be changed into another animal, or a plant to be changed into another plant, or an artifact into another artifact; but they are unwilling, for example, to say that a toy mouse that acquires all the characteristic features of a real mouse has been changed into a real mouse. This is the case, even though the photographs of the toy mouse and the real mouse might be more perceptually similar than other within-category picture pairs such as the tiger and the lion. The children appear to have some basic principled categorical distinctions such as those between artifacts, animals, and plants that overrule characteristic features. Within each of these categories, children frequently judge changes of characteristic features to change identity; but they are very reluctant to have changes in characteristic features move an object into a different category. These broad categories are the same ones that I have focused on at length in other work and have been called *ontological categories* or *basic categories of existence* (Keil, 1979). One of the most important aspects of that knowledge is that it seems to obey certain structural constraints from even the earliest ages.

One can find evidence for constrained cores in other natural domains such as natural language syntax (Chomsky, 1980), visual pattern perception (Marr, 1982), and causal reasoning (Bullock et al., 1982). With other domains that are more artificial—such as knowledge of chess—it is clear that there will be no constraints specific to the domain. Our knowledge of chess may be indirectly constrained, however, by constraints on cognitive faculties used in chess such as spatial maps and causal thinking.

Domains and Constraints: Some Essential Distinctions

There are a number of different senses that the terms *domain* and *constraint* can take, and it is essential, before proceeding any further, to draw some distinctions between these various senses. As with so many other terms in psychology such as schema, stage, and competence, these terms have come to mean different things to different people. Each of these senses has value, but it is all the more important to know what they are.

There are at least three different ways in which knowledge can be constrained, and these different types of constraints lead us naturally into making distinctions among different senses of *domain*. One sense of constraint, the sense that I have emphasized here, is the domain-specific a priori. These are the sort of constraints that set up the core competencies discussed above. They demarcate broad areas of cognition such as moral reasoning, language, knowledge of spatial layout, causal reasoning, visual pattern perception, and the like. They are specifically applicable to one, and only one, of these broad areas and are often instrumental in defining the scope of these areas. They are called domain-specific with a sense of domain that refers to these broad areas of cognitive competence. A priori domain-specific constraints are either present throughout development or emerge at certain points in a maturational fashion relatively independent of experience, just as beards emerge in men. Some of the universal constraints on syntax may emerge in this fashion (cf. Gleitman, 1981).

Domain-general, a priori constraints apply across many different sorts of knowledge, if not all types, and because they apply to such a wide range of knowledge types, they tend to be less restrictive. The more general the range of application, the more flexible and diluted a constraint is likely to be. The sorts of constraints that arise here appear to be more closely linked to learning procedures or patterns of processing and might include the general tendency to form prototypes or similar representations based on typicality information, the limits of short-term memory (if that is a unified concept), patterns of memory interference between related types of knowledge, and properties of general associative models. It is suggested that these constraints are almost invariably present from the earliest stages of concept acquisition and that they play an important role in structuring the "periphery" in initial periods of acquisition of natural concepts and in structuring the entire concept in the initial stages for more artificial sorts of knowledge. If these domain-general constraints were more restrictive and showed dramatic change at various points in development, they might be responsible for across-the-board stagelike changes. It is suggested, however, that this does not occur.

Expertise-specific, post priori constraints are perhaps the most obvious. What you learn influences or exerts constraints upon what you will learn next, not only in terms of the process of learning itself but in terms of how it comes to be represented. If one has a rich knowledge of cooking terms, for example, then one will have different representations of novel cooking terms than a novice, differences that may be striking and qualitative in nature. These constraints are products of the structure of what is learned, and their restrictiveness and generality depend on the knowledge involved. It is assumed here that the various areas of expertise that exist in our world have many unique structural properties that become all the more pronounced with increasing expertise. Thus novice-to-expert transitions may frequently represent qualitative shifts in manner of processing and representation.

These different types of constraints illustrate at least two important senses of domain. The domains that are demarcated by a priori, domain-specific constraints are relatively small in number and represent broad areas of competency for which humans have evolved special structural predispositions. The expertise domains are potentially unlimited in number because they are dependent on the varieties and intensities of human experiences. They tend to be much more local and can subdivide into finer and finer distinctions as demands on more and more local areas of expertise increase. There are countless examples of these sorts of differentiating areas of expertise in the history of technology. Computer science and its various subdisciplines are an excellent example.

All these sorts of constraints and both senses of domain are involved in virtually any learning situation, and it thus becomes very easy for theorists from different perspectives to end up talking at cross-purposes, because they may be talking about different things. In addition, depending on the type of knowledge to be acquired and depending on the degree to which the learner has mastered that knowledge, there will be dramatically different weights put on each of those components.

I interpret Turiel and Davidson's recent work (see this volume) as particularly relevant to the domains issue. They argue that there are seemingly paradoxical trends of not only homogeneity and consistency but also heterogeneity and inconsistency in the development of social concepts, and they suggest that the solution to this contradiction is the recognition of different domains of social judgment. The heterogeneity comes from the differences between such domains as social conventions, morality, and psychological concepts. The homogeneity arises from the internal coherence of each domain resulting from shared organizing principles and the like. This heterogeneity/homogeneity compromise results in a level of domain organization that may be comparable with the domains demarcated by the domain-specific, a priori constraints that I have discussed here and elsewhere (e.g., Keil, 1981). As Turiel and Davidson state

> The domain-specific proposition is that there is neither a general structure of mind as a whole to be identified nor that there are so many domains of knowledge that we are left with a series of elements with no systems of organization. (Chap. 5, this volume)

Broad domains such as social conventions, morality, and psychological concepts might have each been of sufficient importance in evolution and sufficiently different from one another in basic structure that humans have developed specific predispositions to process and represent each in qualitatively different ways. The optimal or most efficient representations and methods of learning for each type of concept might differ considerably and thus, as demands were placed on skills in these areas, these manners of learning and representation may have become increasingly divergent.

An additional claim made here is that clear qualitative shifts are only likely to be observed in the more local expertise domains within each of these broader social domains. It is not clear if Turiel and Davidson would endorse this claim as well, but it at least seems consistent with their account. Thus, there may not be stagelike changes for the broad field of social conventions, but within more local areas of expertise such as table manners and traffic rules there might be much more stagelike changes (cf. Lockhart, 1980).

Interactions Between Domains

If my account is correct, then qualitative changes will be observed only in certain well-specified situations, namely, where a child is moving from a general representational format to more detailed knowledge structures that reflect the particular properties of a given domain. A necessary corollary is that the qualitative shift is always restricted to specific knowledge domains and that there are no general stages of cognitive development constituting across-the-board shifts in representation or manner of thinking. It is now necessary to show how such knowledge specific shifts might nonetheless occasionally give the appearance of much more general shifts by means of interactions between domains.

Consider the situation where a child gains a new level of mastery of some particular domain of knowledge that is crucially involved in a wide range of cognitive tasks. As soon as the child gains mastery in that one domain, it might result in qualitative shifts across all the various domains that use that knowledge. So, at least in principle, it is possible for a qualitative shift in a particular type of knowledge to result in dramatic changes for a much broader group of cognitive domains. The problem is trying to find clear cut examples of such a phenomenon. If one believes that a certain set of logical operations are crucially involved in successful performance of a wide range of cognitive tasks as Piaget did, one might argue that shifts in the nature of such logical operations can have far-reaching effects on virtually all cognitive tasks. This particular example, however, has been quite controversial (see Osherson, 1974; Brainerd, 1978), and it is not clear that one universal logical competence is responsible for success on as wide a range of tasks as conservation, classification, and seriation. Moreover, even if such a competence is implicated in all these tasks, it is even less clear whether it changes in fundamental qualitative ways in the course of development (Osherson, 1974; Brainerd, 1978).

There are probably no cases where a domain that itself undergoes qualitative shifts is an indispensable component in all other cognitive tasks. On a smaller scale, however, there are many cases where local knowledge structures can cause changes in performance in at least a few other domains. Thus, the child's emerging arithmetic abilities will obviously enhance any task requiring such competence; but it is not clear if dramatic qualitative shifts are observed in the development of arithmetic abilities. Moreover, even if there were qualitative changes in arithmetic knowledge, it does not necessarily follow that these changes would cause qualitative shifts in performance on tasks that required such knowledge.

One clearer case involves the ability to comprehend metaphors. In some recent work (Keil, 1985) we have shown that the ability to comprehend metaphors is again largely a function of the sophistication of knowledge one has in various cognitive domains. Thus, if two domains are used in which young children are knowledgeable, they will readily comprehend metaphors between those two domains such as metaphors about animal functions being applied to automobiles (e.g., the car is hungry, the car is thirsty, the car is tired, etc.). If their knowledge is not well articulated in one or both of the domains in a metaphor, then the children will often only understand the literal meaning.

On occasion, increasing maturity of knowledge in one domain might have more far-reaching effects, because that domain interacts with several others. One example is human personality traits which have been used metaphorically with cars, books, plants, computers, and weather conditions, among others. If the child's knowledge in the personality domain undergoes a dramatic shift, then metaphorical competence in general might show a dramatic change, because it is now possible to map relations from personality traits onto several other domains that have already been well articulated in their own right. In our own work, we saw some indications of such an effect for terms referring to human feelings and personality.

A different example of how a local shift can have broader effects is seen in children's emerging awareness of the nature of nonphysical entities, in particular, mental entities. I have argued elsewhere (Keil, 1983) that early on, children have little or no awareness that there are things that exist in the world that are not in themselves physical objects such as events, mental entities, and various other abstractions. Their knowledge of the ontological category of nonphysical objects appears to emerge rather suddenly and dramatically (see Keil, 1983, for details). This initial awareness is confined to ontological knowledge; but once the insight occurs, it can influence a wide range of tasks that benefit from an awareness of categories other than the physical. Thus, any of the mental abilities known as metacognitive skills (e.g., metamemorial, metalinguistic, metaattentional) cannot develop until the child is aware that there is such a thing as memory, or language, or attention. More important, the various metacognitive strategies (such as memory heuristics) that arise out of these abilities and greatly enhance task performance may also be dependent on such an awareness. In this manner, the child's very specific awareness that there are mental entities can have a dramatic impact on a wide range of tasks.

I suspect that such cases where a local insight or change has broader effects are not uncommon. I doubt that the influence is ever as sweeping as the proposed role of logical operations, but it could nonetheless give an illusion of a much broader cognitive change than actually occurred. Even as adults, we are often surprised with how some new local expertise helps us better understand many other more familiar phenomena. One example is knowledge of the basic principles of operation of the computer. People who acquire such expertise often use it to make conjectures about the nature of human cognition, social organizations, biological models, and the like. Metaphorical extensions are particularly relevant here; for if a new domain has within in it relationships that give new insight into the nature of other more familiar domains, then metaphorical extensions from that new domain into the old ones have a strong impact on thinking in those domains.

Domain interactions can also mask developmental patterns. Thus, Turiel and Davidson (this volume) describe how different subjects can perceive the same task as drawing on different domains of knowledge, such as moral versus conventional knowledge. Depending on which domain they choose to access to make a judgment or decision, they may show very different patterns of responses. Similarly, in development children may shift in which domain they predominantly access to solve a problem and give the resulting appearance of a stagelike shift in competence, while in actual fact, they possess both competencies throughout. This then is a modified version of Rozin's (1976) proposal about increasing access and development, one that can mimic stages.

In sum, if one grants the existence of cognitive domains with unique organizing principles, patterns of representation, and modes of processing, those domains can closely approximate behaviors that have been taken as evidence for broad stagelike changes. Moreover, beyond the mere plausibility of such accounts, this may be the favored alternative given the empirical difficulty of finding support for general stages.

Beyond Expertise: Some Differences Between Experts and Novices and Adults and Children

It seems strongly counterintuitive to suggest that the major qualitative changes in childhood are no different from changes that occur between adult novices and experts. We are all uncomfortable with the notion that development and learning are simply the same thing. Moreover, a little reflection reveals that there are important differences between the two. One of the most important is that the child is much closer to being a universal novice, something that no adult ever could be; and because of this, to the extent that there are well-articulated domains and rich interactions between them, there will be strong differences between children and adult novices.

An adult who is a novice in one domain has considerable expertise in others and, if any of these are relevant to performance or learning in the novel domain, the adult is likely to exhibit a very different pattern of learning than the young child. To use the metaphor example again, an adult who is a complete novice about cognitive psychology might nonetheless find it considerably easier to learn about than a young child because of all the other knowledge domains that help one understand cognition better, such as knowledge of how computers work, for example. It is clear that much of the formal instruction to adults relies on comparisons to related concepts from other domains of greater expertise.

In general, adults have a much richer cognitive context within which they can learn about new concepts. It is this context that often enables them to learn more efficiently than young children. Their acquisition patterns should more closely approximate those of children in situations where what they are learning is both novel and relatively autonomous (that is, not connected to others sorts of knowledge) and cannot benefit from analogies to other domains. Knowledge of chess and dinosaurs might be two such examples, and this may be why Chi chose them in her work with expert children and novice adults.

This broader cognitive context created by rich knowledge in other domains is clearly an important difference between expert and novice adults and children and adults and is most relevant to the concerns of this chapter. There are other differences as well, however, that at least deserve mention. Adults may develop new general learning strategies that help them acquire knowledge in a wide variety of domains such as general studying or reasoning skills. Many of these may be a reflection of an emerging metacognitive awareness that enables them to develop and use strategies. Strategies may also become more robust with increasing age, and the older child or adult may become more able to apply them across a wider range of situations, that is, they become less task- and content-bound (cf. Brown, Bransford,

Ferrara, & Campione, 1983). Thus, while many strategies emerge and continue to operate in experts in a fully content-dependent manner, some do become increasingly detached from the original domain of acquisition. Adults may also learn differently from children because certain basic processes that they share with children become more efficient (cf. Brown et al., 1983). While not the same as a capacity difference, this could have essentially the same effect, as Case et al. (Chap. 1, this volume) have argued.

It is not clear how independent either of these alternatives (strategies and efficiency of basic processes) can be from expertise in local domains, and Brown et al. (1983) argue that, to a certain extent, they must be intertwined. Nonetheless, at least in principle, they represent two other reasons to distinguish between expert/novice differences and developmental ones.

Conclusions

I have tried to sketch out some ways in which qualitative changes are and are not likely to occur in cognitive development. The qualitative/quantitative issue can be overemphasized, because there are many indeterminate cases. Nonetheless, people who spend a lot of time with children frequently have strong intuitions that dramatic restructurings of knowledge are occurring, and these intuitions need to be explained. I have argued that two recent research trends in development, reflecting the early competency view and the knowledge-enrichment view, both raise serious challenges to the notion of across-the-board stages of cognitive development. I have interpreted them as also raising challenges to changes in representation that are purely endogenous, as if the child simply goes through certain stages that then have dramatic impacts on all aspects of their interactions with the world. Instead, qualitative shifts mostly occur as the child goes beyond general representational schemes such as collections of characteristic features to learn more about the particular properties of each domain. At the same time, however, the early competency view is involved in that there are constraints on cognitive domains that restrict the nature of the resulting knowledge structures in a more skeletal fashion.

Perhaps the most important theme that these two approaches share is an emphasis on cognitive content and structure rather than on content-independent procedures and heuristics that apply indiscriminately to all types of knowledge. Such processes may only be seen at the earliest stages of acquisition of knowledge and, even then, they are likely to be heavily influenced by constraints if the domain is a cognitively natural one. For more artificial domains such as chess, the primary shift is from general representational formats to those reflecting the properties of chess, with little reference to universal constraints. Attempts to explain stagelike transitions in terms of general formal structural or procedural properties that are imposed on all types of knowledge are not likely to succeed. The most powerful constraints are almost always domain-specific and inapplicable to other domains; and expert knowledge in different domains reflects much more the unique properties of each domain than it does general principles of representation or processing. Because the structures of each of our mental faculties tend to be so different from each other, and because the structures of the types of things there are in the world vary so greatly, developmental change is extremely sensitive to these differences and cannot be modeled in terms of vast homogeneous modes of thought that ignore this rich diversity of structure in both the mind and the world.

References

Anglin, J. M. (1984). The child's expressible knowledge of word concepts: what preschoolers can say about the meanings of some nouns and verbs. In K. E. Nelson (Ed.), *Children's language* (Vol. 5). Hillsdale, NJ: Erlbaum.

Brainerd, C. J. (1978). The stage question in cognitive developmental theory. *Behavioral and Brain Sciences, 1,* 173–182.

Brown, A. L., Bransford, J. D., Ferrara, R. A., & Campione, J. C. (1983). In J. H. Flavell & E. M. Markman, (Eds.), *Handbook of child psychology: Vol. 3. Cognitive development.* New York: Wiley.

Bruner, J. S., Olver, R. R., & Greenfield, P. M. (1966). *Studies in cognitive growth.* New York: John Wiley.

Bullock, M., Gelman, R., & Baillargeon, R. (1982). The development of causal reasoning. In W. F. Friedman (Ed.), *The developmental psychology of time.* New York: Academic.

Chomsky, N. (1980). *Rules and representations.* New York: Columbia University Press.

Chi, M. T. H. (1978). Knowledge structures and memory development. In R. S. Siegler (Ed.), Children's thinking: What develops? Hillsdale, NJ: Erlbaum.

Chi, M. T. H., Feltovich, P. J., & Glaser, R. (1981). Categorization and representations of physics problems by experts and novices. *Cognitive Science, 5,* 121–152.

Fodor, J. (1972). Some reflections on L. S. Vygotsky's *Thought and language. Cognition, 1,* 83–95.

Gelman, R., & Gallistel, C. R. (1978). *The child's understanding of number.* Cambridge, MA: Harvard University Press.

Gentner, D. (1981). Verb semantic structures in memory for sentences: Evidence for componential representation. *Cognitive Psychology, 13,*(1), 56–83.

Gleitman, L. (1981). Maturational determinants and language growth. *Cognition, 10,* 105–113.

Haviland, S., & Clark, E. (1974). This man's father is my father's son: A study of the acquisition of kin terms. *Journal of Child Language, 1,* 23–47.

Keil, F. C. (1979). *Semantic and conceptual development: An ontological perspective.* Cambridge, MA: Harvard University Press.

Keil, F. C. (1981). Constraints on knowledge and cognitive development. *Psychological Review, 88,*(3), 197–227.

Keil, F. C. (1983). On the emergence of semantic and conceptual distinctions. *Journal of Experimental Psychology: General, 112*(3), 357–385.

Keil, F. C. (1985). *Semantic fields and the acquisition of metaphor.* Unpublished manuscript.

Keil, F. C., & Batterman, N. (1984). A characteristic-to-defining shift in the development of word meaning. *Journal of Verbal Learning and Verbal Behavior, 23,* 221–236.

Keil, F. C., & Carroll, J. J. (1980). The child's conception of "tall": Implications for an alternative view of semantic development. *Papers and Reports on Child Language Development, 19,* 21–28.

Kemler, D. G. (1983). Exploring and reexploring issues of integrality, perceptual sensitivity, and dimensional salience. *Journal of Experimental Child Psychology, 36,* 365–379.

Kripke, S. (1972). Naming and necessity. In D. Davidson & G. Harman (Eds.), *Semantics of natural language.* Dordrecht, Holland: Reidel.

Landau, B. (1982). Will the real grandmother please stand up? The psychological reality of dual meaning representations. *Journal of Psycholinguistic Research, 11,*(1), 47–62.

Lehrer, A. (1978). Structure of the lexicon and transfer of meaning. *Lingua, 45,* 95–123.

Lockhart, K. L. (1980). *The development of knowledge about uniformities in the environment: A comparative analysis of the child's understanding of social, moral, and physical rules.* Unpublished doctoral dissertation, University of Pennsylvania.

Maratsos, M. (1983). Some current issues in the study of the acquisition of grammar. In J. H. Flavell & E. M. Markman (Eds.), *Handbook of child psychology: Vol. 3. Cognitive development.* New York: Wiley.

Markman, E. M., & Siebert, J. (1976). Classes and collections: Internal organization and resulting holistic properties. *Cognitive Psychology, 8,* 561–577.

Marr, D. (1982). *Vision.* San Francisco: Freeman.

Osherson, D. N. (1974). *Logical abilities in children,* Vol. 1. Hillsdale, NJ: Erlbaum.

Piaget, J. (1928). *Judgment and reasoning in the child.* London: Routledge & Kegan Paul.

Putnam, H. (1975). *Mind, language and reality.* New York: Cambridge University Press.

Rosch, E., Mervis, C. B., Gray, W. D., Johnson, D., & Boyes-Braem, P. (1976). Basic objects in natural categories. *Cognitive Psychology, 8,* 382–439.

Rozin, P. (1976). The evolution of intelligence and access to the cognitive unconscious. In J. M. Sprague & A. A. Epstein (Eds.), *Progress in psychobiology and physiological psychology.* New York: Academic Press.

Vygotsky, L. S. (1962). *Thought and language* (E. Hanfmann & G. Vakar, Trans.). Cambridge, MA: M.I.T. Press. (Original work published 1934)

Werner, H. (1948). *Comparative psychology of mental development* (2nd ed.). New York: International Universities Press.

7

*Stage/Structure Versus Phase/Process in Modelling Linguistic and Cognitive Development**

Annette Karmiloff-Smith
Medical Research Council Cognitive Development Unit
London, England

Introduction

The introductory section of this chapter deals with some terminological spring-cleaning, so that my use of particular terms is understood in the way I intend, together with a brief discussion of how notions such as stage and structure apply to different types of theory. The following three sections discuss three different ways in which linguistic and cognitive development can be compared: structural comparisons, conceptual comparisons, process comparisons. The third of these, process comparisons, will be developed at some length in an attempt to show that a process-oriented approach can be a most fruitful way of reconsidering the relationship between linguistic and cognitive development. In the concluding section, I shall—inter alia—raise some questions regarding cross-species comparisons, arguing that it is not language that differentiates human from nonhuman cognition but rather a specific feature of the *processes* underlying both human cognitive and linguistic development.

In recent papers (Karmiloff-Smith, 1983a, 1984) I have stressed a distinction which I shall again use here between stage, phase, and level, because these terms are frequently used interchangeably (and thus atheoretically) across the developmental literature and even within single papers.

A *stage* designates a stretch of time that is characterized by a qualitative change (i.e., a *new* internal *organization* and not merely an additional component) with respect to both the preceding and the following stages. It is clearly not sufficient to state that children pass through three different stages simply because they fail a task at age 4, partially solve it at age 6, and completely solve it at age 8. This is a total misuse of the term *stage* which focuses on the task rather than on the child's

* This paper was prepared for the Tel Aviv Workshop on Human Development entitled ''Stage and Structure,'' October 1–17, 1983. A few paragraphs reproduce with permission parts of my Keynote Address to the Stanford Forum on Research in Child Language entitled ''Language Development as a problem-solving process'' which appeared in Papers and Reports on Child Language Development, Vol. 22, 1983, printed by the Board of Trustees of the Leland Stanford Junior University.

changing cognition. For the term stage to have any theoretical depth, it must in my view be *explicitly specified* with respect to its underlying organization. Further, the stage approach usually implies that once a child has reached, say, stage 3, s/he does not return to stage 1 in normal interactions with the environment. Although the structures underlying some behavior may be symptomatic of the immediately preceding stage, one would not expect a stage theory (e.g. Piaget's) to allow for a child to be at stage 3 in a certain number of problems and stage 1 in others. Rather, the child would be argued to be predominantly at stage 3 with respect to most problems and at stage 2 with respect to some others. The latter is usually explained away by the inherent complexities of the task.

There is another theoretical approach which could invoke stagelike developmental patterns, but where the underlying structure of particular stages might differ radically from one domain to another. Thus, in contrast to Piaget's all-encompassing notion of a stage, in a modular view of the human cognitive system (e.g., Keil, 1981) the notion of stage might still be used, even though, within any particular age range, a child's structures would vary in his or her approach to different domains. Stage 2, say, in one domain might look very different to stage 2 in another domain. Moreover, a modular theory could allow for stage transitions to take place at very different ages according to the domain. But modular theories do not necessarily imply stagelike development, of course.

In nonmodular theories such as Piaget's, the stage notion necessarily focuses on structures. By contrast, in the case of modular theories, a systems approach rather than a structuralist one may turn out to be the most appropriate. Indeed, from a modular viewpoint it would be theoretically crucial to focus on the systemic *interactions* between outputs of modules and between the modules' possible stagelike developments.

However, if one does wish to argue for analogy *across* domains, stagelike structures, tied to a sequence of age ranges, are not the only way to approach the issue. Another theoretical approach is to invoke changes which, unlike stages, are considered to be *recurrent.* For this type of change, which is the basis of the model which will be presented later in this chapter, I shall use the term *phase,* with the hypothesis that children (and adults, too, for that matter) attack new problems by going through the same three phases, both within the various parts of particular domains and across different domains. Within nonmodular theories such as Piaget's, the stage concept refers to equivalence of *structure* across domains, whereas the phase concept focuses on cross-domain equivalence of *process.* Moreover, unlike stages, the recurrent-phase approach means that at any age across the developmental span, a child will be simultaneously at phase 1 for a certain number of problems, at phase 2 for others, and at phase 3 for yet others, and, according to the hypothesis, faced with a new acquisition problem, the child will again pass through the same three phases. Unlike the hypothetical use of stages discussed earlier for modular theories where stage 1 in one domain could differ significantly from stage 1 in another domain, the recurrent-phase concept argues for *equivalence* of each particular phase across domains.

An additional advantage of the recurrent-phase concept is that it can be extended to the analysis of microdevelopmental change, whereas the stage concept cannot. If children are asked to solve a problem which is within their cognitive capacity but for which a novel solution needs to be worked out on the spot, they may be seen to pass through two or more of the phases within the space of a single experimental session. While stages are loosely age-related, phases clearly are not.

The term *level* will be used to describe changes *within,* rather than across, domains (although *sequence* or *step* might have been better terms). However, having used the term *level* in recent papers, I shall retain that term for consistency's sake. Like stages, but unlike phases, levels are not recurrent. Once the child is at, say, level 3 in a specific domain, s/he does not return to level 1. *Phases* refer to *general processes* across domains. By contrast, *levels* account for *specific* changes within a particular domain, but they can be lifted to a greater degree of generality by accounting for the levels in terms of phases. Thus my account of *level* 2 in a linguistic task will obviously differ in content from my account of *level* 2 in, say, a physics task. Analyzed in terms of levels, the two domains may appear to have no developmental links. But the two levels can be shown to have deep equivalence of process when reanalyzed in terms of *phase* 2.

It is essential with a recurrent-phase approach that one does not do injustice to the richness of content within specific domains. Pitching the level of generality so that the sequence of phases is applicable *across* domains and yet still captures essential details of processes underlying behavior *within* particular domains has been my constant concern in developing the process-oriented model which will be discussed in the section on "Process Comparisons Between Linguistic and Cognitive Development."

When discussing possible analogies between linguistic and cognitive development, which is the purpose of this chapter, it seemed essential first to differentiate between terms such as stage, phase, and level and their theoretical implications. The following three sections deal with different approaches to the relationship between linguistic and cognitive development.

Structural Comparisons Between Linguistic and Cognitive Development

As discussed earlier, the majority of stage theories are structuralist. Can underlying logical structures be used to yield interesting relationships between linguistic and cognitive development? Most structural comparisons between linguistic and cognitive development have been explicitly or implicitly based on Piagetian theory. For Piaget, structures are a coherent set of mental operations which can be applied to different aspects of an individual's psychological interaction with the environment. Those in favor of structural comparisons have thus tended to account for the child's linguistic level by invoking elements of Piaget's structural account of sensorimotor, preoperational, and concrete-operational stages in the nonlinguistic cognitive domain.

Within this framework, researchers argue that children need to have acquired a certain stage of cognitive structure *before* being able to acquire the hypothesized corresponding linguistic structures. It is Sinclair, the major psycholinguistic interpreter of Piagetian theory, who first took Piagetian structural descriptions as the framework for explaining language acquisition. In her 1967 book Sinclair states: "To carry our hypotheses further, the syntactic structures will show isomorphisms with the logical structures described by Piaget" (p. 149). Later, Sinclair used what she called the structural "isomorphism" (although in her writings the isomorphism has remained no more than an unspecified "analogy") to argue that sensorimotor action schemes were the prerequisite for and the explanation of the way in which the child initially acquired language. Taking the grouplike structure of sensorimotor action schemes invoked by Piaget, Sinclair argues that: "This 'group' of action patterns also serves as a heuristic model for the acquisition of syntax and it is precisely what the child (every child) has learnt during the first eighteen months or so that provides him with the *necessary and sufficient* presumptions to start this acquisition" (Sinclair, 1975, p. 232, italics added). As this is stated, it is one of the most extreme structuralist positions, which totally negates language-specific constraints. Sinclair's arguments have largely concerned very early child language and have remained at a very general level of comparison, although at times she has suggested somewhat more specific links (Sinclair, 1971, p. 126). In her writings, Sinclair has consistently maintained that cognitive development determines the course of linguistic development. Nothing language-specific is invoked about linguistic constraints on structure. The mere application of previously existing nonlinguistic cognitive structures to linguistic input is used to explain language acquisition. No attempt has been made by the Genevans to specify *how* the application of the cognitive structures to language could actually take place. All that is offered are some suggestions about plausible analogies, together with the unquestioned assumption that cognitive structures antedate corresponding linguistic structures through all respects of the developing linguistic system.

In a School where so much structural underpinning is given to children's use and understanding of language, it is surprising that Piaget and his disciples should have neglected to notice that he was imputing structures underlying adult linguistic knowledge to very early child language. Thus in his well-known discussion of "the slug," he implied that his daughter, Jacqueline, had the capacity to use definite articles and pronouns intralinguistically, i.e., as discursively referential. However, it can be shown that Piaget's cognitive explanation of this linguistic interaction may be misguided. The example was as follows:

> J. at 2.6: *"That's not a bee, it's a bumble bee. Is it an animal?"* But also about 2,6 she used the terms "the slug" for the slugs we went to see every morning along a certain road. At 2,7 she cried: *"there it is!"* on seeing one, and when we saw another ten yards further on she cried: *"there's the slug again";* I answered: "But isn't it another?" J. went back to see the first one. "Is it the same?" *"Yes."* "Another slug?" *"Yes."* "Another or the same?" The question obviously had no meaning for J. (Piaget, 1951, p. 225)

Analyzing J's use of the expression "the slug," Piaget suggests that the child is looking upon the two slugs, not as different members of the *class* of slugs (which would require structures well beyond his daughter's stage of cognitive development) but as successive reappearances of a single member, a sort of semi-individual, semigeneric prototype (Piaget, 1951). However, basing the hypothesis on an adult interpretation of the child's use of determiners is tenuous. A series of experiments (Karmiloff-Smith, 1979a) showed that determiners such as *a, other, the,* etc., have very different functions in early child language compared to adult usage. For small children, the definite article and the pronoun function as deictics. They point to the extralinguistic context; they do not refer anaphorically to the intralinguistic context. In Piaget's earlier-cited example (in French, *encore la limace*), J was not saying, "There's the (anaphoric reference to previous discourse) slug again," but rather something like "another the slug." The use of the definite article points deictically to a slug being focused on extralinguistically, and *encore* can mean both "another" or "again" in French. Moreover, *same* can mean *same one* (identity) or *same kind* (analogy) in French, and young children prefer the analogy meaning (Karmiloff-Smith, 1977). Thus, when J walked back to look at the first slug before answering Piaget's question, it is conceivable that she was not checking whether it was the same *one* (anaphoric interpretation) but rather whether it was the same *kind.* Piaget's question may thus have been very meaningful to J in that for small children "another slug" (another one) can indeed be simultaneously "the same" (kind).

Thus, with respect to language/cognition comparisons in the very young child, not only is there a need for greater structural specification, but it is also essential to exercise extreme caution in making use of the child's early language functions —as if they coincided with adult functions—in attributing properties of constrained cognitive structures (e.g., the semiparticular/semigeneric prototype) to the child.

Beyond the very young child, the Genevan school has also attempted to make structural comparisons between linguistic and cognitive development for older children at the preoperational and concrete operational levels. One such example is to be found in Sinclair's (1967) study of class inclusion structures and children's linguistic understanding of quantifiers. Briefly, Sinclair argued that for children to be able to understand quantifiers such as the French plural definite article *les* versus the French plural partitive *des* ("the" vs. "some of the"), it was necessary for them first to have acquired the underlying structures of class inclusion in their nonlinguistic cognitive development. However, my work (Karmiloff-Smith, 1979a) was an attempt to demonstrate that many *language-specific* issues were involved for the child in acquiring the meaning and uses of quantifier terms such as *a, the, some,* etc. One important linguistic fact is that these terms are plurifunctional. They can act as deictics, anaphors, quantifiers—and have many other functions. Thus, for the child it is not merely a matter of finding the correct linguistic term to express quantifier notions that have already been acquired cognitively. Rather, the child has also to come to grips with the fact that plurifunctionality means that there is not a simple one-to-one form/function relationship between linguistic terms and their meanings. Depending on the function, depending on the linguistic emphasizers that are added (e.g., "all *the*" vs. "*some*," rather than just "*the*" vs. "*some*"), children understand and use quantifiers in different ways at different ages. Thus, irrespective of whether one accepts the plausible structural relationship between cognitive class inclusion and linguistic quantifiers, it is still essential to understand exactly what the *linguistic* problem is for the child.

Another area of structural comparisons concerned the passive construction. Sinclair and Ferreiro (1970) argued that children's understanding of passive sentences was dependent upon the logical operation of reversibility. Taking a Chomskyan position, these authors argued that passive sentences were the result of a transformation on the deep structure of the corresponding active sentences. This linguistic operation was then equated, without specification, to the logical transformation involving reversibility, which is imputed to children as part of the structured wholes operant at the concrete operational level. The Genevan argument was that the logical structures of concrete operations were a precondition for children's linguistic understanding of the passive construction.

However, the direct meaning and/or transformational relationship between active and passive sentences has come under serious attack, and, in the developmental psycholinguistic literature, Dewart (1975) showed that even a small amount of discourse context will enable children to understand passives well before the concrete operational level. This demonstrates not only that the structural link may be a tenuous one but also that the linguistic analysis of passives as isolated sentences, rather than as part of discourse, is to be questioned. Further, it is not clear, even when invoking the criterion of reversiblity for the understanding of passives, how reversibility is connected with the other necessary components of the structure underlying concrete operations. Although the Genevan school frequently stresses the idea of *structures d'ensemble,* rarely are the set of mental operations specified as *a whole* when extending the structural model to language.

Metalinguistic development has also been a candidate for structural comparisons (e.g., Papandropoulou & Sinclair, 1974; Van Kleeck, 1982). To take an example this time from outside Geneva, Van Kleeck has attempted to link the child's growing capacity in metalinguistic awareness to preoperational and concrete operational stages in cognitive development. Tasks were chosen in metalinguistic awareness that ostensively required the cognitive operations of reversibility and decentration. Van Kleeck found that children who failed such tasks were under 6, and she concluded that they must therefore be preoperational (although they were not tested in cognitive tasks). She also found that children who succeeded in the metalinguistic tasks were over 6, and they were thus classified by Van Kleeck as concrete operational. However, there is no evidence here of any causal link between the metalinguistic and presumed operational levels of the child. All we have again is an age relationship, together with the underspecified argument that the metalinguistic tasks require reversibility and decentration.

In general, the examples discussed here and many others like them show that the use of stages, and their underlying structures, has thus far not been very fruitful in comparing linguistic and cognitive development. Stages have been used at too general a level and in an ad hoc fashion. There is need for far more detailed specification in both the cognitive and linguistic aspects being compared, if our understanding of possible structural relationships is to be enhanced.

My own view is that structural comparisons alone will not enlighten the language/cognition issue. However, some general aspects of stagelike structures may turn out to be a necessary complement to other approaches such as the process-oriented one discussed later. Moreover, structural comparisons have considered the cognition/language relationship unidirectionally and have neglected the bidirectional interactions which may actually take place. This is also true of conceptual content comparisons to be discussed in the next section.

Conceptual Comparisons Between Linguistic and Cognitive Development

The search for structural analogies discussed in the last section placed explanation at a broad level, independent of particular content. Thus, for example, reversibility, or the broader notion of concrete operations as a whole, were used to explain development in various areas of language. Another more specific approach to the relationship between linguistic and cognitive development, but also within a stage/structural framework, has been to invoke *specific* conceptual knowledge as a prerequisite for the acquisition of corresponding linguistic terms. The focus here is less on the underpinning structural organization than on the specific *content* of the concepts built via such structures.

Typical examples of conceptual comparisons can be seen in work on temporal and spatial relations. For instance, Ferreiro (1971) made an extensive psycholinguistic study within the Piagetian framework of the development of temporal relations in children's language. Her psycholinguistic results were then compared with children's cognitive development. For older children, Ferreiro compared their *temporal* judgments in cognitive tasks with their use and understanding of *temporal* expressions in linguistic tasks. However, as far as the younger children were concerned, the author was unable to find a suitable cognitive task for testing temporal relationships. She therefore used the conservation-of-liquids task as a comparison with her results on the linguistic tasks. Ferreiro thus made a conceptual comparison for older subjects and a structural comparison for younger subjects. Such comparisons within the same study, slipping from the structural to the conceptual, are rather unsatisfactory in pursuing the language/cognition issue. On the one hand, comparisons are made at a *very general* level, arguing that underlying the problems experienced by younger children in understanding or using temporal terms is the lack of a cognitive structure. On the other hand, for older children a *very specific* link is suggested between the *temporal* notions that the child has acquired and the corresponding *temporal* expressions. Thus, the comparison is either too general or too specific to be usable in extrapolating the language/cognition comparison to other aspects of the developing linguistic system.

A further problem concerns the very assumption that *temporal* concepts are directly or solely linked with *temporal* expressions. The argument underlying this sort of approach is that the acquisition of the *specific* conceptual knowledge is the prerequisite for the acquisition of the corresponding linguistic terms. This implies that language is constrained by cognitive developments also at a level more specific than the general structural one. Thus, when studying *temporal* terms, comparisons have been made with the corresponding *temporal* concepts. Likewise, when studying *spatial* terms, comparisons have been made with the corresponding *spatial* concepts. Yet when children discover something relevant about contrasting spatial terms, for instance, they may well apply that knowledge also to temporal terms. In other words, it is arguable that children could deduce from the newly acquired contrastive spatial terminology the fact that language encodes, say, polar contrasts. They could then scan the input model to discover other linguistic contrasts, e g., apply the general notion of encoding contrasts linguistically, discovered via spatial concepts, to try to understand expressions about, say, time, without necessarily awaiting the full development of the corresponding temporal concepts. This allows for a two-way arrow, where cognition clearly does affect language development but where also the *processes* of linguistic encoding operations (and not narrowly solely their specific content) can affect cognition.

Another area where invoking cognitive constraints for linguistic development may, in fact, also have a linguistic processing explanation is in the area of time and aspect. Bronckart and Sinclair (1973) showed that very young children first use verb inflections to indicate aspect rather than temporal relations. Thus, children used the distinctions between past tense, present tense, and so forth, to indicate whether an action had a long or short duration, whether it gave rise to a result or not, etc. They did not use tense inflections initially to express the temporal relation between the event described and the time the utterance about it was made. In Bronckart and Sinclair's study, these results were given an interesting cognitive explanation. However, the child's use of tense to express aspectual features rather than temporal features may also be partially explained by the fact that the input frequently contains redundant markers, whereas small children expect each marker to have a separate function (Karmiloff-Smith, 1979a, p. 232). In saying, for example, "Yesterday I went . . . ./Tomorrow I will go . . . ," the temporal indication is expressed twice, on the adverb and in the verb morphology. Naturalistic data show that for a period of development *yesterday* means for the child *any time in the past* and *tomorrow* means *any time in the future*. If these words are interpreted by the child to encode the temporal relations, it is not surprising that the child uses the verb inflections to indicate something else, i.e., the aspectual features. Very young children initially seem to expect one-to-one mapping between form/function pairs.

In general, language/cognition comparisons as well as cross-linguistic comparisons are too content-restricted (Karmiloff-Smith, 1978). Furthermore, many of the problems discussed in the previous section with respect to structural comparisons clearly also apply to conceptual comparisons.

Despite my criticisms of the structural and conceptual approaches, I do not deny that *some* structural analogies and *some* conceptual analogies may have a role to play in explaining the relationship between linguistic and cognitive development. Slobin (1973), for example, has argued that when the a priori conceptual knowledge maps more or less directly onto some linguistic meanings, then this will pace the order of language acquisition. However, when such conceptual mapping is not obvious, children use principles which are specifically linguistic. Others, too, have reviewed and discussed the general structural/conceptual underpinnings of language acquisition (e.g., Bowerman, 1976; Campbell, 1979; Cromer, 1976). But we are still far from having established the *precise* nature either of the contributions of cognitive structure and concept or, more importantly, of their interaction with explanations based on process-oriented approaches. In the rest of this chapter, I wish to argue for a much more prominent place to be given to comparisons of *processes* than has hitherto been the case.

Process Comparisons Between Linguistic and Cognitive Development

Although process-oriented approaches to developmental phenomena have been undertaken within the information-processing framework (e.g., Siegler, 1981), these have come under heavy attack from various sources (see, for instance, Strauss & Levin, 1981). Such process-oriented analyses are often very task-specific, and, consequently, domain-specific. A further criticism can be made that these approaches are "asemantic." If we take, for example, Siegler's Production System (PS) description of the way in which children solve the balance-scale problem, it can be noted that children develop different rules with age. Initially the child makes a decision regarding balance by taking only weight differences into consideration; later in development, the child first uses a weight comparison, but if this yields no clear-cut decision, only then does the child subsequently consider distance from fulcrum; finally, at the end of the development sequence, the child takes weight and distance simultaneously into consideration by comparing the relevant weight/distance ratios. My argument that PS accounts are "asemantic" is based on the fact that from the first rule through to the acquisition of the last rule, the concept of weight for the child is assumed not to change. However, it is implausible that weight at the initial level is for the child identical to her/his concept of weight at the final level. Children's representations of concepts such as weight change and enrichen with development. Thus, as formulated in PS theory, the rules operate on empty (i.e., asemantic) values from the point of view of the child's developing concepts. A PS account of this sort is thus more of a surface description of behavioral changes than of underlying representational changes. Nonetheless, the level of detailed specification of process which PS accounts offer is an important aspect which structural approaches eschew.

The rest of this chapter will present some of my own attempts to accomplish a process-oriented comparison between linguistic and cognitive development. I shall take four areas of comparison, two from language acquisition and two from nonlinguistic cognitive acquisitions. Three of these will concern macrodevelopmental changes, i.e., across age, and one covers microdevelopmental changes, i.e., changes within the space of a single experimental session. I shall submit a three-phase, process-oriented model to account for the development to be seen in these four areas. The next subsection outlines the model; the following ones illustrate it from four areas of research. The model is based on my argument that, across many age levels and across many domains, children pass through three recurrent phases in the way in which progress takes place both macrodevelopmentally and microdevelopmentally. Some may consider this an overly generalized statement, but I would contend that much depends on correctly pitching the level of the explanatory account and on clearly distinguishing between behavioral and representational change.

A Three-Phase Model of the Processes Underlying Linguistic and Cognitive Development

Phase 1

The first phase, which I have termed the *procedural phase,* is characterized in the following way. The behavioral output of the child is generated by predominantly data-driven processes. Data-driven covers those processes which are generated by adaptation to external stimuli. In such cases, it is predominantly the environment which is in control of the behavior (see Lindsay & Norman, 1972, for discussion of such processes in adult behavior). The procedures children generate are success-oriented, and children use as a process method a simple evaluator of match–mismatch between present state and goal state. In the case of language, this would mean the evaluation of the child's output with his/her representation of adult input, and the progressive narrowing of the distance between present state and goal state. Once the distance is repeatedly zero, the child will have attained what I term *procedural success*. Procedural success is achieved by the child taking carefully into account the environmental stimuli, using both positive and negative feedback in adapting to the particular problem s/he is solving. Each behavioral unit is rich and elaborate, successfully achieving its goal. However, each adaption is merely juxtaposed to the previous behavioral unit. During this first phase of procedural success, there are no attempts by the child at overall organization which would link behavioral units one to another into a consistent whole. One might say that, so far, there is no "domain" or "subdomain" from the viewpoint of the child. The adult observer may interpret the child's behavior as if it were generated from a single, richly integrated representation, but for the child the behavioral units consist of a series of isolated, yet well-functioning, procedures. For the phase-1 child, the contents of procedures cannot be accessed for representational reorganization. What we witness at phase 1 is behavioral change, the narrowing of the distance between present state and goal state, and isolated representational adjunctions. Procedures at this phase have the status of "tools" for successfully reaching a goal. But it is essential to note that by the end of phase 1, the child manifests the stable, consolidated use of well-functioning, automatized procedures. Now, if the child were merely success-oriented, development in any particular area should stop at phase 1. This is not, however, the case.

Phase 2

Phase 2 provides striking evidence that the human child goes "beyond success." Phase 2 is characterized by the fact that the implicit knowledge embedded in phase 1 procedures is rewritten into what I have called Explicit$_1$, i.e., the component parts are defined explicitly in the internal representation, although they are not accessible to consciousness (for full discussion of the relationships between conscious access versus implicit and explicit representations, see Karmiloff-Smith, in press). Thus, at phase 2 children work on their earlier (successful) procedural representations as problem spaces in their own right. During this phase the behavioral output is generated predominantly by a top–down control mechanism. External stimuli are very secondary to an internal representation which is imposed on the environment and contrains the particular behavioral manifestations.

Contrasted with phase 1, the behavioral output of phase 2 often appears less elaborate in detail, limited in scope, and lacking the flexible use of negative feedback. In fact, the child may totally ignore or violate environmental information, but this is as a consequence of an overall organization-oriented procedure dominating the behavioral output. The child now treats the previous series of isolated problems as a single one. To unify the whole, the child generates a simplified procedure which allows him/her to have control over and link the previously isolated procedures into a unified representational framework. There is thus a loss of the richness of the phase-1 adaption to negative feedback, but a gain in that the simplified single approach to all problem parts affords a unifying of the isolated procedures of phase 1.

Thus, unlike the earlier success-oriented procedures, phase-2 children go beyond successful attainment of the goal and generate organization-oriented procedures which now have the status of "cognitive objects of attention" rather than merely tools. The child predominantly focuses on his/her internal representations rather than on the external enviroment. The organization-oriented procedures ignore environmental data, give rise to externalized marking of what previously remained implicit as well as generate self-repairs to behavior which appeared smooth in the earlier phase. This is a result of the child's endeavor to simplify, unify, and have control over the internal representations. In contrast to phase 1 this may temporarily increase the distance between present state and goal state and result in less success from a behavioral point of view, but not of course from a representational point of view.

It is important to stress that in phase 1, the output is *predominantly* data-driven and in phase 2 *predominantly* top–down, because functioning would clearly be impossible if it were exclusively one or the other.

Phase 3

The third phase, which I call the *conceptual phase,* is governed by a subtle control mechanism which modulates the interaction between data-driven and top–down processes. Neither the environmental feedback nor the top–down process predominates. The child is now in control both of external stimuli and of the internal representations which constrain and guide his/her behavior. Yet the behavioral output at phase 3 is identical or closer to that of phase 1 in that the child successfully reaches his/her goal at both phases. However, this is only true from a *behavioral* viewpoint and not from the viewpoint of the ($Explicit_1$) *representations* underlying the phase 3 child's output and which are richer and more organized than the phase 1 child's isolated procedural representations. Phase 3 is the result of the reorganizational process at work in phase 2 which, once consolidated, can take environmental feedback into account without jeopardizing the overall organization.

Thus, the developmental picture which I paint is not the traditional view of initial failure, subsequent partial, but fragile, success, followed ultimately by robust success. Others have recently concentrated their developmental theorizing on U-shaped behavioral trends, invoking various explanatory frameworks (Strauss, 1982). However, it is essential to keep in mind that this concerns U-shaped *behavioral* trends, thereby restricting the generality of the notion. I shall include in my model processes underlying both U-shaped behavior and non–U-shaped behavior, because the focus here is on *representational* change. U-shaped representational curves, if they exist at all, could only apply to forms of pathological development.

In focusing on the way in which children go beyond success, my argument has been that the human organism (both linguistic and cognitive) incorporates a drive to have control not only over the external environment (the input stimuli) but also, and importantly, over its own internal representations and finally over the intricate interaction between the two. In my analyses of the data to be presented in the next subsection, I will situate the initial level 1 at the time when children can perform successfully on each task, i.e., not as the final level, but as merely the beginning of a cycle of representational change. The *success* of level 1, based on implicitly-linked representations, is, I have submitted, an essential prerequisite of fundamental representational changes which are initiated on the basis of repeated *positive* feedback. My argument has been that failure is rarely the prerequisite for *representational* reorganization (Karmiloff-Smith, 1979b); rather, failure gives rise to representational adjunctions and generates *behavioral* change in phase 1 by the system evaluating and narrowing the distance between the child's goal and the child's present output. But true representational changes are the result of representational *reorganizations,* the prerequisite for which is not negative feedback and failure, but positive feedback and procedural success.

This three-phase model allows, in my view, for fruitful process-oriented comparisons between language and cognition, both from a macrodevelopmental and microdevelopmental perspective. Some selected details of research in both the linguistic and nonlinguistic cognitive domains will now be provided. The data in the next section will be presented in terms of levels within particular domains. The reader is then invited to reread the Section on the three-phase model to evaluate how the account in terms of *levels* can be reformulated within the more general, yet detailed, framework of *phases* which allows for comparisons across domains.

Empirical Data

The first linguistic examples are drawn from a very large number of experiments aimed at analyzing children's gradual construction of the French determiner system. Abstracting from the experiments dealing with gender markers, the study involved 11 experiments covering over 600 3–11-year-olds. Experimental design will not be described here for space reasons (see Karmiloff-Smith, 1979a, for full details). Rather, the pattern of results from two of the many linguistic categories studied will be briefly summarized.

The results fall into three levels. At the first level (up to roughly 5 years of age), the linguistic forms children use are perfectly adequate for the communicative context and reflect correct adult outputs, although I submit that they are not based on the same representations as adults'. But the early successful usage is followed at level 2 (between 5 and 7 years of age) by externally marked distinctions (not in adult grammars) between different functions of the same word, which bear witness to organization-oriented processes. Finally, the correct outputs are again used at level 3 (over 8 years olds), but generated from rich underlying representations of the links between different determiners and their multiple functions. The age ranges given are approximate, because the 3-level sequence differed slightly from one linguistic category to another. Two examples of the three developmental levels with respect to the indefinite article and the postdeterminer *same* are given.

Example A: Indefinite Article, Levels 1 and 3. In French, one form, *un* (or its feminine counterpart, *une*), covers both the meaning of nonspecific reference "a" and the numeral function "one." Unlike English, there is no surface distinction between these two meanings in French. At the first and third levels, children use this one form (e.g., *une voiture*) correctly to convey different functions in different contexts.

Example A: Indefinite Article, Level 2. In contexts requiring nonspecific reference (equivalent to the English "a car"), children of level 2 correctly use the French indefinite article: *une voiture*. However, in contexts where they wish to imply the numeral function (equivalent to the English "one car"), level-2 children use a partitive construction, *une de voiture*, which would normally only be used in colloquial, emphatic language. The tasks did not require such emphasis, and children of levels 1 and 3 would know this colloquial use and yet did not employ it for this task. Thus, level 2 children introduce a surface distinction which is not part of the adult input for that function.

Example B: Postdeterminers, Levels 1 and 3. The postdeterminer *même* ("same") is preposed correctly to imply in different contexts either the meaning "same one" or the meaning "same-kind," e.g., *la même voiture*, meaning the same kind of car or the same one, depending on the context of the use.

Example B: Postdeterminers, Level 2. In contexts requiring the meaning "same one," level 2 children prepose the French postdeterminer correctly, e.g., *la même voiture,* whereas in contexts requiring the meaning "same kind," level-2 children again use a partitive construction, e.g., *la même de voiture*. This distinction is not encoded in this way in adult language.

Many supporting data from other linguistic categories can be found in Karmiloff-Smith (1979a) as well as in the work of Bowerman (1982) on late-occurring errors and Newport (1982) on children's acquisition of American Sign Language.

The examples show that while level-1 children appeared to be using the indefinite article and postdeterminers correctly as adults do, children at level 2 made distinctions which rendered explicit, in the surface forms, the dual functions normally conveyed by a single form. The oldest subjects at level 3 used exactly the same surface forms as the children at level 1, i.e., adultlike forms with no externalized distinctions. I would argue that at level 1, children have acquired the correct forms by success-oriented, implicitly represented procedures, mapping specific forms to specific contexts. In other words, while from the addressee's viewpoint, level-1 children are using a single form to cover several functions, from the *child's* viewpoint the identical surface forms are actually represented as a series of unifunctional homonyms, i.e., a plethora of form/function pairs with no explicit representational link between pairs.

To take a concrete example of the indefinite article, my argument is that the child has one procedure for nonspecific reference which gives rise to the indefinite article as its product, another procedure for the numeral function which also generates the indefinite article, yet another procedure for the naming function which again outputs the indefinite article, and so forth. In other words, level 1 children have a plethora of different procedures which, although they output the same surface forms of behavior, are not organized representationally as such. Had the level 1 child already explicitly represented the cross-procedural analogies, there is no plausible explanation for the surface distinctions introduced later at level 2. Once each separate procedure outputting the indefinite article has become highly automatized, when it is a well-functioning procedural unit, then the child can rewrite in Explicit_1 form the implicit links between procedures. At this metaprocedural level, the child can then work on the analogies and differences across the functions all outputting the same phonological form, the indefinite article. The reorganizational process rewrites the plethora of procedures into a *new* representation whereby ultimately the indefinite article has plurifunctional status. To do this, level-2 children temporarily create externalized handles i.e., new forms for some form/function pairs, for the distinctive functions to which they have become sensitive until, at level 3, they have successfully reorganized their representations such that a single form can carry several meanings. Thus, when level 3 children *behave* as level 1 children, the identical behavior is generated from very different representations.

Once the indefinite article and, say, the definite article have passed from implicit to Explicit_1 representation, then the implicit *relationship* between the two articles can be worked on via the Explicit_1 representations of each separate article. The cycle continues until the whole determiner system is constructed.

It is clear that external pressures from mismatch to model or adult corrections cannot be invoked to explain more than phase 1's *behavioral* changes in the acquisition process. Although negative feedback plays a role at level 1, the outputs of level-2 children show that level 1 ultimate successful output is not sufficient. Children go beyond procedural success and, after repeated positive feedback, they introduce important *representational changes* and not merely representational adjunctions and behavioral changes.

Let us now take a study outside language acquisition, and with an older age range. The study was aimed at analyzing the spontaneous notational systems that children create as external memory devices (Karmiloff-Smith, 1979b). It was also aimed at testing whether the 3-phase model, and specifically, the theoretical concept of metaprocedural processes, could be applied to microdevelopmental change. To test the model, the task had to be solvable by all subjects within the space of the experimental session i.e., they had to be able to reach procedural success, and yet the problem had to present enough of a challenge and novelty to provoke interesting representational and not merely behavioral changes.

Sixty-three subjects between 7 and 11 years participated. The task was as follows. Subjects were shown a 12-m roll of wrapping paper on which was drawn a route from a house to a hospital. At key points there were bifurcations, where one route led to a cul-de-sac while the other permitted the child to continue toward the hospital. There were 20 such bifurcations, some of which were marked with topographical indices such as trees, little men, etc., others with salient zigzags on one of the sides of the bifurcation, and so forth. Children were asked to drive a sick patient in a little ambulance from a house to the hospital. The experimenter unrolled the paper as the child drove the ambulance and rolled up the part already covered (for details, see Karmiloff-Smith, 1979b). The patient was not in the ambulance during the first run, so that the child could make mistakes as s/he drove along, and back up if s/he chose the cul-de-sac. On the second run the patient was placed in the ambulance. During the first run the child was encouraged, after s/he had discovered each time which was the correct route, to mark something on paper (avoiding suggestive terms like "write" or "draw"), so that when s/he came through on the second run with the patient, s/he would avoid making mistakes. In other words, the problem for the child was to create some form of external memory device, i.e., a notational system which could be used on the second run to remember which way to turn at each bifurcation. In taking the notes, the child could copy any of the figural or topographic indices, use a simple left–right system, etc. No suggestions were made by the experimenter as to the form of notation.

Five types of notational system were generated by the subjects. They ranged from more concrete to more abstract systems, examples of which were found at all ages. The analysis focused on modifications that occurred spontaneously, during the one-hour experimental session, to elements within a single notational system. These intrasystemic changes involve modifications where the original forms that children had created already contained *all* the necessary information for the child to succeed in the task, and yet, nevertheless, changes were introduced to them. The selected example, from the numerous ones which occurred, is used to illustrate the fact that modifications to children's representational systems cannot be explained by external constraints of inadequacy (i.e., failure) of the system to attain the goal, but rather are triggered internally via the child's tendency to work metaprocedurally on the representational tool s/he has created.

The example chosen is from a 7-year-old child, typical of children whose external memory code was of the analogical abstraction type, i.e., formed of a series of fork-shaped bifurcations (see Karmiloff-Smith, 1979b, for details of changes to linguistic and other codes). The child initially drew a series of bifurcations that reproduced (i.e., copied, like linguistic matching to adult input) rather faithfully the actual bifurcations on the route. The process was predominantly data-driven and resulted in an adequate drawing for the decoding phase. Later in the same protocol, however, the child suddenly began to render explicit the information that was implicit in her earlier drawings (see Fig. 1). She did this despite the fact that the initial drawings were perfectly adequate for the decision-making part of the task and that she could have continued using this success-oriented procedure throughout. But she now put a cross through the cul-de-sac, implying, "Do not take this direction," and put an arrow on the correct road, indicating the direction to follow. While these additions are obviously redundant with respect to the earlier form generated, they bear witness to the fact that the child was working on her representations, spelling out explicitly the meaning oppositions in the surface drawings, rather than merely generating a success-oriented tool which only represented the meaning oppositions implicitly. Later still in the same protocol, the child dropped these external redundant markings.

The forms drawn during the first and final parts of the session were identical

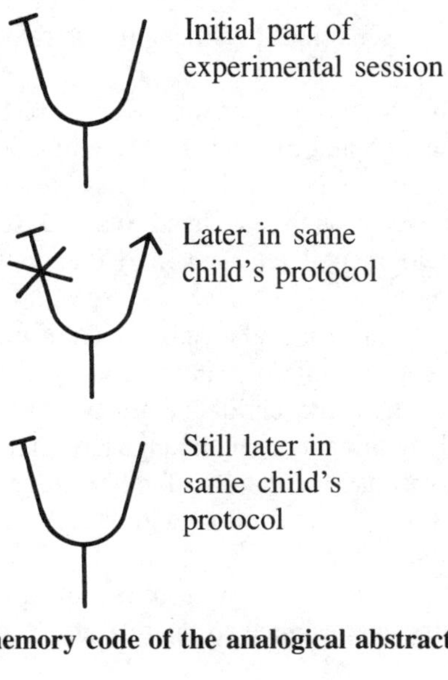

Figure 1. An external memory code of the analogical abstraction type of a 7-year-old.

from the viewpoint of behavioral output. Yet I would argue that their representational status is not the same for the child. By the third part of the protocol, the form was generated from a more powerful underlying representation which now contained explicit information about the meaning oppositions (externalized in part 2). By contrast, in the initial part, the form was a mere mapping between the actual route and the child's drawing, the information being implicitly represented in the success-oriented procedure.

The essential fact to be retained from the brief example presented here, and from the numerous others in the data, is that, when children introduced changes to a well-functioning procedural unit, in no case did they ever go back and alter their initial drawings of earlier decision points, although nothing stopped them doing so and they did go back to correct any errors made. Thus, in each case, children were sensitive to the fact that the initial form carried sufficient information for the decoding part of the task. The changes therefore cannot be explained by the child having anticipated failure of the system at the decoding end of the task. Rather the changes are triggered by the child's drive to rewrite implicit procedural representations into explicit form, thereby gaining control over them. After reaching procedural success, the child steps up to a metaprocedural level, but temporarily, as in the linguistic examples discussed above, needs to mark tangibly and externally the different pieces of information represented implicitly in the procedure generating the form initially used. By this metaprocedural process the child makes explicit the information encoded. Finally, when the child returns to using a form identical to that used initially, the output is, in fact, generated from a new, explicit representation.

This study demonstrates in my view the plausibility of using the 3-phase model not only for accounting for macrodevelopmental change but also for microdevelopmental change. It also lends support to my hypothesis that going beyond success to work on one's internal representations at a metaprocedural level is a pervasive facet of human cognition.

When I originally made this comparison (Karmiloff-Smith, 1979b) between linguistic processes of change and external memory notational changes, I found it satisfactory. However, I have since noted some possible limitations due to the fact that the comparison is between, on the one hand, macrodevelopmental changes and, on the other, microdevelopmental changes, and that in both cases, the studies concerned semiotic codes.

I thus next attempted to push the plausibility of the model further by making comparisons which did not mix micro- and macrodevelopment but involved only macrodevelopment, and where the child's linguistic system was compared to a domain which had clearly nothing to do with language. In this chapter, I will briefly take one of such comparisons (for details of other comparisons, see Karmiloff-Smith, 1984), that of children's story-telling and their use of the physical principle of torques in a block-balancing task.

First, let us look at the block-balancing task (see Karmiloff-Smith & Inhelder, 1974, for full details). It was designed to analyze the development of children's representations and implict theories-in-action, and to ascertain the role for children of negative and positive feedback in attributing the status of counterexample to the exceptions to the theory-in-action underlying their behavior.

Sixty-seven children between 4 and 9 years were given a large number of different blocks and asked to balance them one after another, in any order they pleased, on a narrow metal bar. Examples of three of the block types are illustrated in Fig. 2. Some of the blocks balanced at their geometric center (type A); some balanced slightly off-center due to weights conspicuously stuck to one end of the blocks (type B); in some, lead had been drilled into one end (type C), and the blocks balanced at that end, but looked exactly like type A blocks which, by contrast, balanced at their geometric center. This was designed to lead children into error and to discern the representations which generated their behavior. There were several duplicates of each block type to test rigidity of behavior.

Children's performance can be divided into three levels varying in type of balancing behavior and underlying representations. As the first level, 4- to 6-year-olds were successful at balancing each of the blocks. They simply moved the block back and forth along the bar, via data-driven processes, using the proprioceptive feedback of the direction of falling of the block, until they succeeded in balancing it. However, level-1 children did not make a choice among the different blocks. For example, when they had just succeeded in balancing a type-B block, they did not go on to try another type B block immediately afterward in order to use the previously acquired information about roughly where the block's point of balance might be situated. Rather, level-1 children simply picked up any block as it lay on the table, placed it on the bar anywhere, and corrected the position on the bar until they reached procedural success. They treated each block as an isolated problem, but the elaborate adaption to negative feedback led to the successful balancing of each block. The common information in the procedures was only implicitly defined, so the child could not relate the representations underlying each successful balancing one to another.

By contrast, 6- to 8-year-olds at level 2 were *not* successful in balancing any but the type A blocks, i.e., they only managed to balance those blocks where the

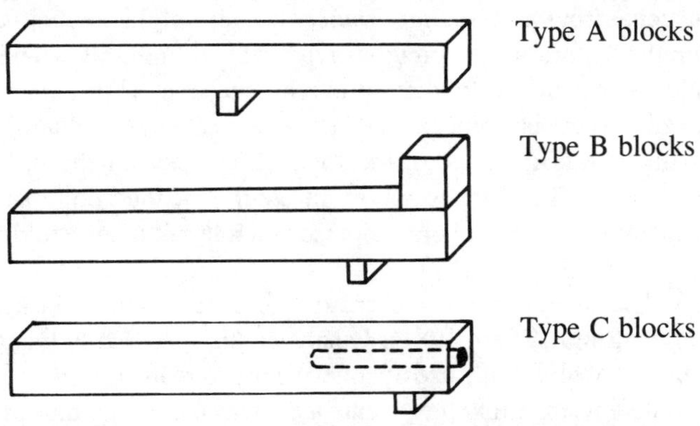

Figure 2. Examples of 3 block types used in the block-balancing task.

geometric center coincided with the center of gravity. For all the other blocks, level 2 children placed them resolutely at their geometric center, and when the blocks fell to one side or the other, the children discarded them as "impossible to balance." They did not make the elaborate corrections that the younger level 1 children made using proprioceptive (i.e., data-driven) feedback. While these level 2 children did not "succeed" on the whole task and ignored the proprioceptive feedback so useful to level-1 subjects, it is important to note that level 2 children used a single unified procedure in attempting to balance all the blocks, i.e., based on a theory-in-action, represented in Explicit$_1$ form internally, that objects balance at their geometric center. Their failure to balance certain blocks was clearly not due to their *inability* to use proprioceptive feedback, because when these children were asked to close their eyes, they balanced types B and C blocks easily. However, with their eyes open, they again resolutely placed each block at its geometric center. These children were using a simplified top–down procedure, generated by an underlying spatial representation of symmetry which guided their approach to all the blocks. This rigid top–down process of imposing a procedure on the environment despite negative feedback, i.e., of internal representations overriding external stimuli, led to children failing to balance types B and C blocks. Level 2 children's procedures were not simply success-oriented, but organization-oriented, trying to link all the blocks into a single representational framework.

At level 3, children tried the geometric center first but rapidly adjusted the position of the block according to the proprioceptive feedback. Behaviorally, this looks rather similar to level-1 children. But representationally, it is different on two accounts. First, level 3 children intitially used a top–down process (based on the geometric center theory-in-action). Second, their attention to proprioceptive feedback and their ultimate success at balancing a particular block were not merely success-oriented but were subsequently used for attempting to balance other similar blocks. Their planning was guided by explicit (although not necessarily conscious) representations. These level 3 children made a clear choice in the order in which they attempted to balance blocks, e.g., after successfully balancing a type-B block, they looked for another B block. For the second type-B block, they did not first try the geometric center, but assessed *in advance,* because of the explicitly represented similarity with the features of the previous type-B block, roughly where the point of balance would be according to their previous success. This contrasts with the behavior at level 1 where each block was treated as a separate problem. At level 3, a control mechanism was modulating the interaction between the data-driven and top–down processes. The child was now in control of the coordination between external information and the explicitly represented internal theory guiding the external behavior.

I shall now discuss developmental changes in a story-telling task. In previous papers (Karmiloff-Smith, 1980, 1981), I analyzed the specifically linguistic aspects of the development of children's use of cohesive markers in narratives. At that time, my aim was to demonstrate that the "sentence" was the wrong unit of analysis for understanding anaphors both in linguistics and in language acquisition. I endeavored to show that rules that may be operative *intra*sententially could not account for the use and nonuse of pronouns, full noun phrases, and zero anaphora *inter*sententially, i.e., in connected spoken text.

My aim here is to take the same linguistic data and reanalyze them from a less linguistic and more cognitive *processing* point of view. The experiment I shall discuss involves asking 170 4- to 9-year-olds to tell stories from picture books with no text. Another 180 children of the same age range described a single picture extracted from the stories. Some of the books had a clear central character with a subsidiary character of the same sex; some had two characters of different sex; some had three characters; and some were merely a collection of six pictures with no obvious connection between the drawings. The design was manipulated so that in some cases pronominal devices would be potentially ambiguous because protagonists of the same sex were involved; in other cases there would be no ambiguity, because natural gender differences would be reflected in pronouns; in some cases the picture contents held the story together, whereas others required the introduction of something extraneous to the picture contents to link them together. The visual stimulus of some pictures purposely led the child into describing a picture in one way but inserted into the story, the identical picture might be described in a different way according to the structure the child imposed on the overall narrative.

Children's performance on these narrative tasks could be divided into three levels. At level 1, the linguistic output is predominantly data-driven with the child generating success-oriented procedures in order to describe the contents of each picture. Each utterance is correct syntactically and has rich lexical variety. A series of correct utterances are juxtaposed locally but lack the organizational property of a single narrative unit. The addressee hears the output as if it were a unit, but I contend that this is not the case with respect to the processes *the child* uses to generate each of the utterances of the narrative.

At level 2, the linguistic output is generated by predominantly top–down control processes by which the child monitors the flow of her connected utterances. Contrasted with level 1, the narrative is poor in lexical variety and weak on story details. In fact, for some stories it violated truth conditions. Yet, as a narrative, it is a single unit, with utterances linked within an overall organization. Unlike the success-oriented procedures of level 1, the level 2 child uses organization-oriented procedures. To organize the whole into a single unit, the child generates a simplified but rigid procedure, preempting initial utterance position for references to the thematic subject of the narrative.

The third level is characterized by the modulated control between data-driven and top–down control processes. Lexical variety is again rich. The story details are much fuller than at level 2 but are coordinated with the top–down control mechanism which organized the narrative into a single unit. Through this process of modulated control the child produces differential linguistic markers according to the dynamics of the progressively unfolding discourse.

A few examples will illustrate these developmental levels of children's narrative production (see Karmiloff-Smith, 1983a, for fuller details). Examples 1, 2 and 3, from a story about a boy and girl fighting over a bucket, are taken from levels 1, 2, and 3, respectively.

Example 1 (4-year-old): "There's a boy and a girl. He's going fishing and she's going to make sand castles. So he takes her bucket, and she tries to grab it back, and he runs off with it, so she sits there crying by the tree. Now he can do his fishing. He got four fish.

Example 2(7-year-old): "There's a boy and a girl. He's going to catch fish, so he takes the girl's bucket and he runs off and catches lots of fish."

Example 3(9-year-old): "There's a girl and a boy. The boy wants to go fishing, so he tries to get the girl's bucket, but the girl won't let him take it, so he grabs it out of her hand and the girl chases after him, but he gets away from the girl and he starts to fish while the girl sits there crying. He goes home smiling with four fish."

In example 1 from level 1, we can note the rich lexical variety. The syntax is locally correct and, given that the two protagonists are of different sexes, all the pronominal references happen to be nonambiguous. The child is successful in that there are no gender errors, no syntactical errors, and no incorrect lexical mappings. But I would argue that while there is some local connection between utterances, the overall output is not generated as a single unit. The utterances are held together by the picture contents. This very fact makes the child's output appear more cohesive than it actually is. Thus, this sounds very much like an organized story for the addressee. However, compared to the narratives from older children of levels 2 and 3, the output is still not yet generated as a single unit by the child. My argument becomes stronger once cases are considered where the natural gender of the two protagonists is the same (see example 4 later). But it is important to stress that the level 1 productions usually appear very successful.

Example 2 is a typical output from level 2. Together with reduced lexical variety, the details of the picture are almost completely lost. The boy in the story is in the subject slot of each utterance and is pronominalized, whereas the girl is hardly referred to at all, which contrasts totally with example 1 from level 1. Despite the loss of detail, the level 2 child is using a top–down process, treating the narrative output as a single unit.

Recall that in example 1 from level 1, the child pronominalized both protagonists throughout. The pronouns were nonambiguous because of the gender distinction. In example 3 from level 3, by contrast, only one of the protagonists, the thematic subject, is pronominalized. Almost all references to the girl are with full noun phrases. The use of these full noun phrases cannot be explained by an attempt to avoid potential ambiguity of reference because, when told with pronouns throughout, the story is completely unambiguous. Clearly, the level 3 child has preempted pronouns to refer to the boy, the thematic subject, and full noun phrases to refer to the girl.

In my view, the pronouns used in example 3 stem from different representations compared to those generated in example 1. In example 1, I would argue that the child has mapped the correct natural gender features onto the word for the extralinguistic referent ("he" refers to the boy in the pictures, and "she" refers to the girl). In example 3, by contrast, the pronoun no longer refers deictically to the extralinguistic referent but rather to the highest node (the thematic subject) in the child's internal representation of the discourse as an organized unit. The level 3 child's pronouns are discursively referential, whereas the level 1 child's pronouns are extralinguistically referential. At level 3, the data-driven processing of details of the story and the top–down processing which treats the discourse as a unit are governed by a modulated control mechanism which generates *differential markers* (pronouns vs. full noun phrases) for reference to the protagonists' differing discourse functions.

Examples 4, 5, and 6, from the three levels, respectively, are taken from a story about a boy and a balloon vendor.

Example 4 (5-year-old): "There's a little boy in red. He's walking along and he sees a balloon man and he gives him a green one, and he walks off home and it flies away into the sky, so he cries."

Example 5 (6-year-old): "There's a boy going along. He gets a green balloon, he lets go of the balloon, and he starts crying."

Example 6 (8-year-old): "A little boy is walking home. He sees a balloon man. The balloon man gives him a green balloon, so he happily goes off home with it, but the balloon suddenly flies out of his hand, and so he starts to cry."

Example 4 demonstrates that when the story content does not by chance facilitate the correct use of referential devices (as in the natural gender distinctions of the bucket story), then the argument that the child is not treating the overall output as a single unit becomes even more convincing. The referential devices used in example 4 are ambiguous for the addressee. They refer to different *extra*linguistic referents, but without establishing *intra*linguistic links between them. However, if the addressee treats each pronoun as a separate unit referring only within each utterance to the person involved in the action it describes, then these deictic pronouns are *not* ambiguous. It is when we, as addressees, treat the pronouns as *discursively* referential that they become ambiguous. My argument is that the level 1 child is not making internal discourse reference but local utterance deictic reference.

Example 5, by contrast, is discursively referential, but suffers from the shortcomings of level 2 outputs, discussed earlier for example 2. Finally, example 6 illustrates the use of differential marking of referential expressions, typical of level 3, when data-driven and top–down processes are modulated by a subtle control mechanism.

These examples and those from other stories not cited here demonstrate that pronouns are not used merely for picking out the distinctive feature (e.g., gender, number) for identifying a referent. Rather, with development, pronouns and full noun phrases come to function differentially as organizational components of a total discourse. Whether we decide to pronominalize or not depends marginally on local sentential constraints and far more on the unfolding state of the total discourse.

The various arguments I have made above with respect to the initial isolation and progressive interaction of data-driven and top–down processes are strengthened by data from children's spontanous repairs during the narrative production, but I refer the reader to other papers for a discusson of these (Karmiloff-Smith, 1980, 1983a; for a very different type of theoretical framework for such comparisons, see Case, this volume).

Conclusions

A central problem in attempting to compare linguistic and cognitive development is that of pitching the level of analysis such that the model has generality and yet captures essential facets of each specific domain. When the level of comparison is pitched too high, we can allow ourselves to do little more than state that cognitive stages pace the course of language acquisition or, more rarely, vice versa. When the level is pitched too low, then comparisons across domains are rendered virtually impossible. For example, an underlying representation of symmetry may explain why children come to create the geometric-center theory. Symmetry may indeed account for a number of procedures children use (Karmiloff-Smith, 1975; Piaget & Karmiloff-Smith, in press). However, invoking the notion of symmetry does not allow for the comparison with, say, children's narrative development. It is the wrong level of analysis for establishing a broad equivalence of process across domains. Likewise, perspective-taking is a concept which might contribute to the analysis of the results of the narrative study. However, once again, although invoking perspective-taking is useful for describing some aspects of the narrative data, it is pitched too close to content to allow us to discern the relationships which might exist between language-production, notational codes and physics. The level at which I have pitched the 3-phase model (see the Section on "A Three Phase Model of the Processes Underlying Linguistic and Cognitive Development") allows, in my view, for detailed process equivalences to be drawn across linguistic and cognitive domains.

A certain number of results require justification of inclusion in the same phase 2. How can the phase 2 child at times show a complete loss of detail compared to phase 1 and at others generate extra information compared to phase 1, i.e., redundant externalized marking? Is this not a basic contradiction? I would argue that these seeming differences are at the behavioral level and are simply the various by-products of the phase 2 children's focus on their internal representations with the consequence of ignoring external stimuli, sometimes rigidly imposing the representation on the stimulus, sometimes explicitly unpacking the components of the representation, creating extra information.

Had I made a comparison between the results of the block-balancing task with, say, children's development of the past-tense marker *-ed,* I might have characterized phase 2 by the notion of overgeneralization. Children first use *went* correctly, then later produce *goed* (or *wented*), only later to use *went* correctly again (Ervin, 1964). But the descriptive term of overgeneralization of a morpho-syntactic rule would not be usable for phase 2 in the narrative study where there is no form of overgeneralization. The same applies to the notion of U-shaped patterns. The descriptive terms of overgeneralization and U-shapes capture behavioral changes, but they do not account for the sort of *representational* changes invoked for metaprocedural phase 2. Moreover, the other terms are far less general in their applicability. Metaprocedural phase 2 focuses on the moment when the implicit content of procedures is rewritten in Explicit_1 form.

The concept of metaprocesses should not be confounded with that of metacognition where *meta* implies conscious verbal explanation on the part of the child. The latter involves a second rewriting in Explicit_2 form (see Karmiloff-Smith, in press, for full discussion of these issues). Within the notion of a metaprocedural phase is not only that of a top–down mechanism but also the idea that children are driven by the need to have control over their internal representations. They *work on* their existing representations, reanalyzing and reorganizing them into new groupings in long-term memory. Thus, metaprocedural processes produce representational change.

Modeling the transition between phases is no easy matter. I have argued that children step up to a metaprocedural level once procedures at any phase become *automatized, consolidated,* and *function* efficiently. These terms are descriptive and require specification. A mechanism needs to be postulated which would continuously evaluate, across all the various parts of the developing systems, the moment at which any part of any one of them was ripe (i.e., automatized, consolidated, functioning efficiently) to be operated on metaprocedurally. A metaprocess—with the function of evaluator needs to be formulated to identify the specific quality of internal state (success, consolidation) of any procedure or set of procedures, which cues the functioning of metaprocedural processes and the rewriting of representations in Explicit_1 form. The same applies, *mutatis mutandis,* to the passage between phases 2 and 3: The organism has to recognize that its reorganizational processes generating representational changes are complete and stable and that is has predictive control over them. In several previous papers, I have submitted that it is the need for *control* rather than economy and/or success which drives the organism—control over the external stimuli, over one's internal representations and over the interactions between the two.

It is also important to stress that while the linguistic and cognitive organism uses both positive and negative feedback *within* each phase, causing behavioral change, the evaluator which operates *between* phases is cued by positive feedback mechanisms, i.e., it is *not* set into operation by failure of the system. Representational changes between phases do *not* occur because the system has detected an error or is in a conflictual, disequilibrated state—a precondition frequently invoked by stage theories. Rather, in this process-oriented account, change occurs because the system has recognized a stable success criterion which cues that it is in control and ready to pass into a new metaprocedural phase implying representational change.

I should like to speculate that metaprocedural analysis and reorganization is the specifically human way in which change takes place. It has indeed been demonstrated that chimpanzees, for instance, can be taught to use isolated lexical terms, relational terms and many problem-solving procedures. But my reading of the primate literature suggests that the animal merely generates success-oriented procedures (Karmiloff-Smith, 1983b). Thus it is not merely the lack of language which is the crucial difference. The infrahuman primate does not appear to go beyond success and work on its internal representations at a metaprocedural level in any domain. Its representations seem to remain procedural and implicit. By contrast, successful goal attainment for the human child is merely the *beginning* of a complex process of increasing control and representational change.

While I argue that metaprocedural processes generate representational change, the question of whether the underlying changed representations across language and cognition are analogous has not been addressed. It may well be that the new cross-domain representations are only possible because of previously existing, but hitherto undetected cross-domain analogies. Prior to that, a modular view of development, which necessarily incorporates domain-specific constraints, would suggest that the recurrent phases, which I aruge are common across domain, would initially operate on *different* representational structures. However, with the repeated operation of the same *processing* constraints, it is feasible to postulate their influence across domains on the reorganization of representations within domains.

While some may argue that linguistic and cognitive development are the result of the operation of totally distinct processes, my inclination is to push the process equivalence between them as far as I can, in the best of worlds to show that my initial intuition was correct, and at worst, to be able to specify explicitly why the equivalence was misfounded. In relating linguistic and nonlinguistic aspects of development within a process-oriented framework, the type of linguistic vocabulary, description, and explanatory mechanisms which are most amenable for the linguistic part of the language/cognition comparison are *functional* in nature. Structural comparisons which focus on the syntax of the language part and conceptual comparisons which focus on its semantic part do not address the process issue at all. This results in the structural and conceptual approaches positing a *unilateral* cognition/language relationship, at the expense of accounting for possible back-and-forth interactions between linguistic and cognitive development which is opened by the process-oriented approach.

I have attempted both here and in more detailed papers (Karmiloff-Smith, 1983a, 1984) to establish the plausibility of the 3-phase model. To enrichen and challenge the model further, it will be necessary to try to apply it to yet other domains, always bearing in mind the crucial difference between behavioral and representational change. It will also be essential to explore to what extent a certain number of structural and conceptual constraints must be incorporated into the model. Just as stage, structure, and concept are nonexplanatory without invoking process, so it will probably emerge that a process model requires some structural and conceptual ingredients. Strauss and Ephron-Wertheim (this volume) has argued that structuralist and information-processing approaches may well be compatible. But perhaps, ultimately, we will end up with models where the distinction between process and knowledge structures is not even the relevant one to be made.

References

Bowerman, M. (1976). Semantic factors in the acquisition of rules for word use and sentence construction. In D. M. Morehead & A. E. Morehead (Eds.), *Normal and deficient child language.* Baltimore, MD: University Park Press.

Bowerman, M. (1982). Starting to talk worse: clues to language acquisition from children's late speech errors. In S. Strauss (Ed.), *U-shaped behavioral growth.* New York: Academic Press.

Bronckart, J. P., & Sinclair, H. (1973). Tense, time and aspect. *Cognition, 2,* 107–130.

Campbell, R. N. (1979). Cognitive development and child language. In P. Fletcher & M. Garman (Eds.), *Language acquisition.* Cambridge, England: Cambridge University Press.

Cromer, R. F. (1976). Developmental strategies for learning. In V. Hamilton & M. D. Vernon (Eds.), *The development of cognitive processes.* London: Academic Press.

Dewart, M. H. (1975). *A psychological investigation of sentence comprehension by children.* Unpublished doctoral dissertation, University College, London.

Ervin, S. (1971). Imitations and structural changes in children's language. In E. H. Lenneberg (Ed.), *New directions in the study of language.* Cambridge, MA: M.I.T. Press.

Ferreiro, E. (1971). *Les relations temporelles dans le langage de l'enfant* [Temporal Relations in child language]. Paris: Droz.

Karmiloff-Smith, A. (1975, June). *Les métaphores dans l'action chez les enfants de 5 et de 12 ans: domaines spatiauz et physiques* [Metaphors in the actions of children of 5 and 12 years: Spatial and physical domains]. Paper given at the Symposium of the International Center for Genetic Epistemology, Geneva.

Karmiloff-Smith, A. (1977). More about the same: Children's understanding of child language. *Journal of Child Language, 4,* 377–394.

Karmiloff-Smith, A. (1978). Adult simultaneous interpretation: A functional analysis of linguistic categories and a comparison with child development. In D. Gerver & W. Sinaiko (Eds.), *Language, interpretation and communication.* New York: Plenum.

Karmiloff-Smith, A. (1979a). *A Functional Approach to Child Language.* Cambridge, England: Cambridge University Press.

Karmiloff-Smith, A. (1979b). Micro- and macrodevelopmental changes in language acquisition and other representational systems. *Cognitive Science, 3,* 91–118.

Karmiloff-Smith, A. (1980). Psychological processes underlying pronominalization and non-pronominalization in children's connected discourse. In J. Kreiman & E. Ojedo (Eds.), *Papers from the parasession on pronouns and anaphora.* Chicago, IL: Chicago Linguistics Society.

Karmiloff-Smith, A. (1981). The grammatical marking of thematic structure in the development of language production. In W. Deutsch (Ed.), *The child's construction of language.* London: Academic Press.

Karmiloff-Smith, A. (1983a). Language acquisition as a problem-solving process. In *Papers and reports on child language development* (Vol. 22). Stanford, CA: Stanford University Publications.

Karmiloff-Smith, A. (1983b). A new abstract code, or the new possibility of multiple codes? (Commentary on D. Premack, The codes of man and beasts). *The Behavioral and Brain Sciences, 6,* 149–150.

Karmiloff-Smith, A. (1984). Children's problem solving. In M. Lamb, A. L. Brown & B. Rogoff (Eds.), *Advances in developmental psychology (Vol. 3).* Hillsdale, NJ: Erlbaum.

Karmiloff-Smith, A. (in press). Cognitive processes and linguistic representations: Evidence from children's metalinguistic and repair data. *Cognition.*

Karmiloff-Smith, A., & Inhelder, B. (1974). If you want to get ahead get a theory. *Cognition, 3,* 195–212.

Keil, F. C. (1981). Constraints on knowledge and cognitive development. *Psychological Review, 88,* 197–227.

Lindsay, P. H., & Norman, D. A. (1982). *Human information processing.* London: Academic Press.

Newport, E. L. (1982). Task specifity in language learning? Evidence from speech perception and

American sign language. In E. Wanner & L. R. Gleitman (Eds.), *Language acquisition: The state of the art*. Cambridge, England: Cambridge University Press.

Papandropoulou, I., & Sinclair, H. (1974). What is a word? Experimental study of children's ideas on grammar. *Human Development, 17,* 214–258.

Piaget, J. (1951). *Play, dreams and imitation in childhood.* London: Heinemann.

Piaget, J., & Karmiloff-Smith, A. (in press). Un cas particulier de symétrie inférentielle, Chapter VIII, and Conflicts entre symétries, Chapter IX [A particular case of inferential symmetry, Chapter VIII, and conflicts between symmetries, Chapter IX]. In J. Piaget (Ed.), *Recherches sur les catégories* [Research on Category Theory]. Paris: P.U.F.

Siegler, R. S. (1981). Developmental sequences within and between sequences. *Monographs of the Society for Research in Child Development, 46,* (2, 189).

Sinclair, H. (1967). *Langage et opérations: sous-systèmes linguistiques et opérations concrètes* [Language and operations: linguistic subsystems and concrete operations]. Paris: Dunod.

Sinclair, H. (1971). Sensorimotor action patterns as a condition for the acquisition of syntax. In R. Huxley & E. Ingram (Eds.), *Language acquisition: models and methods.* New York: Academic Press.

Sinclair, H. (1975). The role of cognitive structures in language acquisition. In E. H. Lenneberg & E. Lenneberg (Eds.), *Foundations of language: A multidisciplinary approach.* New York: Academic Press.

Sinclair, H., & Ferreiro, E. (1970). Etude génétique de la compréhension, production et répétition des phrases au mode passif [Developmental study of comprehension, production and repetition of sentences in the passive]. *Archives de Psychologie, 40,* 1–42.

Slobin, D. I. (1973). Cognitive prerequisites for the acquisition of grammar. In C. A. Ferguson & D. I. Slobin (Eds.), *Studies of language development.* New York: Holt, Rinehart, & Winston.

Strauss, S. (Ed.), (1982). *U-shaped behavioral growth.* New York: Academic Press.

Strauss, S. & Levin, I. (1981). Commentary on Siegler's "Developmental sequences within and between sequences." *Monographs of the Society for Research in Child Development, 46,* (2, 189).

Van Kleeck, A. The Emergence of Linguistic Awareness: A Cognitive Framework. *Merrill Palmer Quarterly, 28,* 237–265.

8

*A Step-by-Step Model of Language Acquisition**

Ruth A. Berman
Department of Linguistics
Tel Aviv University

Introduction

Language acquisition is analyzed in this study as developing over time. And language development is viewed as a function of the maturation of general cognitive abilities as these combine with an expanding repertoire of linguistic data and devices. We argue that with time, children come to rely less heavily on action-based or context-anchored procedures and that these give way to more autonomous, structure-based rules of grammar. Discussion centers on children's successive construals of linguistic material as they move from a partial, item-bound knowledge of isolated elements through to end-state command of the grammar of their native tongue. The orientation adopted here can thus be identified with research in language development which highlights the reorganizations and reanalyses evinced by children in acquiring linguistic knowledge of different kinds (for instance, Bowerman, 1980, 1982a,b; E. Clark, 1980, 1981; Karmiloff-Smith, 1981, 1983, this volume; or MacWhinney, 1978, 1981), much of which evolved out of earlier ''order of acquisition'' studies pioneered by Roger Brown and his associates (Brown, 1973). This is, clearly, a perspective which departs quite radically from idealizations to an ''instantaneous'' model of acquisition, such as in principle focuses on adult grammars as the endpoint of the learning process (Chomsky, 1965, 1975). Yet like such a model, the view expounded here relies critically on the notion of linguistic *rule* as evidence for the child's having internalized a systematic organizational principle for mapping meanings onto the forms and structures which constitute the grammar of his or her language.

Our concern here centers on the acquisition of language ''as a peculiarly structured system in its own right'' (Slobin, 1981). We thus de-emphasize the nature of language as social behavior or as serving general communicative purposes in order

*This is a condensed and revised version of a paper entitled ''From Nonanalysis to Productivity: Interim Schemata in Child Language,'' presented at the First Annual Tel Aviv Workshop on Human Development, October, 1984. I am indebted to Esther Dromi and Annette Karmiloff-Smith for their enlightening comments on that paper and to Iris Levin and Sidney Strauss for their help with this revision.

to focus on the route taken by children in acquiring form–meaning relations in their native tongue. From our point of view, the task which confronts the language-acquiring child consists crucially of learning the specific linguistic forms and structures deployed by a given language to express the semantic distinctions and the discourse functions characterizing endstate proficiency in that language.

In this chapter we propose a model which specifies how children proceed from nonanalytic, nonorganized construals of isolated linguistic items at the "one-word stage" (Dromi, this volume) to knowledge of the rules and constraints which define end-state adult competence. My claims are illustrated by findings from research of my own and of others on the acquisition of Hebrew as a native tongue (reviewed in Berman, in press a.). But it will be argued that the model itself is generalizable across languages as well as across children. The constructions analyzed in this chapter all devolve upon the expression of semantic categories such as tense or number by affixing elements to a root or stem. As such, they belong in the domain of morphology rather than, say, of syntax or discourse. There is good reason why much of my own research and that of others concerned with Hebrew child language (e.g., Levy, 1983; Walden, 1982), has focused on the structure of words rather than the structure of sentences or entire texts. For it is here that Hebrew is typologically most distinct from English and other European languages for which information is available on children's language development. Much of what may be encoded syntactically, across words, in a language like English is done WITHIN words in Hebrew, as a Semitic language. This is partly because nearly all the content words of Hebrew, all verbs and most nouns and adjectives, are built up out of a consonantal root which then combines with patterns of affixation to mark notions like causative, reciprocal, or reflexive for verbs and notions like agent, instrument, or place for nouns (see Berman, 1978a; Clark & Berman, 1984). Moreover, Hebrew has a rich array of grammatical inflections which mark the categories of number, gender, and definiteness on nouns and adjectives, and of number, gender, person, mood, and tense on verbs (Dromi & Berman, 1982). We will argue, however, that not only is this model relevant to languages other than—and structurally quite different from—Hebrew, but it can be extended to various areas of the grammar (e.g., word order and subordination) and to a variety of discourse functions (e.g., expression of temporality through foreground/background relations or topicalization and object-specification).

We suggest that children pursue a series of five steps, as they proceed from initial acquaintance to full command of a given subsystem in their language. These can be briefly summed up as follows:

a. *Rote-knowledge:* initial acquisition of individual items as unanalyzed amalgams
b. *Early modifications*: a few highly familiar items are alternated constrastively
c. *Interim schemata:* a transitional, nonnormative, but at least partly productive rule is applied
d. *Rule–knowledge:* grammaticization involves strict adherence to rules coupled with some lacunae and insufficient constraints

e. *End-state usage:* rule-application is constrained by adult norms and conventions, with increasing variation of style and register

We deliberately use the neutral term *steps* to indicate segments of development which may be as short as a few days or as long as several months or more. The term further suggests that "skipping" is possible, and that children need not always pass through—or at least manifest overt behavior appropriate to—each and all of these five steps with respect to all the kinds of linguistic knowledge which they acquire. In considering the more general developmental and psycholinguistic implications of this model (see the section on "Discussion"), we will further suggest that these five steps can be reanalyzed in terms of three main *phases*—where this term is used in a sense quite close to and with much the same motivations as in recent work of Karmiloff-Smith (1983, this volume). These main phases are (a) a PREGRAMMATICAL phase, around steps a and b, where children's knowledge is largely item–bound or entails only limited formal alternations and learning is strongly anchored to real-world situations and pragmatic cues; (b) the phase of GRAMMAR ACQUISITION (steps c and d), where rules are applied productively across items in terms of linguistic structure, and items are interrelated within more general systems, categories, and paradigms; and (c) a final phase of APPROPRIATE USAGE where the repertoire of forms and rules acquired previously are deployed with increasing skill, taking account of lexical conventions, norms of usage, register distinctions, and so forth.

These phases can be viewed as age-related in only very broad, general terms. Thus, all of language learning up until the age of around 2 will be predominantly of the Phase One type, because the so-called one-word stage as well as the early-word-combinations which follow it are in essence pregrammatical. And Phase Three type learning of appropriate usage is typically a school-age phenomenon which continues on to puberty. However, none of the five steps is identifiable with any particular age or overall developmental stage. Rather, the steps may be manifested at different times for different facets of linguistic structure. Thus, for instance, grammatical inflections for verb tense and also number and gender marking on nouns in Hebrew are mastered basically by children as young as 3; but productive use of morphological devices for lexical word-formation—e.g., reciprocal or causative verbs, agent and instrument nouns—are a later development among Hebrew-acquiring children, from the fourth year and up.

A Step-by-Step Model: From Rote Via Rule

In this section, the five-step progression outlined earlier is illustrated by reference to several subsystems in Hebrew grammatical development. First, we give a brief description of each category and an account of how it patterns in acquisition along these five developmental steps (the section entitled "Development of Selected Subsystems in Hebrew"). Each step is then characterized in some detail (see the section, "Analysis of the Five Steps") as background to a more general discussion of the model as a whole (see the "Discussion" section).

Development of Selected Subsystems of Hebrew

The categories to be reviewed represent three related facets of modern Hebrew: morphophonology—the notion of a consonantal root; inflectional morphology—case-marking on pronouns, noun plurals, verb tenses; and word-formation devices. References to the studies on which conclusions are based are given for each subsection separately. These studies all focus on preschool language development, up to the age of 5 or so, and they are predominantly cross-sectional rather than longitudinal. For each subsystem, care was taken to consider findings from two different types of research data: (a) deriving from naturalistic speech output in adult–child interactions; and (b) based on specifically designed tests aimed at probing isolated features of the child's linguistic proficiency.

The Consonantal Root. Most content words in Hebrew are made up of a consonantal root which carries the basic core of the meaning of the words in which it appears (Berman, 1978a). Compare, for the root g–d–l: verbs like *gadal* 'grow up, get bigger', *gidel* 'raise, bring up', *higdil* 'enlarge, make bigger'; adjectives like *gadol* 'big, large' or *megudal* '(over)grown'; and nouns like *gidul* 'growth', *migdal* 'tower', *gdula* 'greatness', *hagdala* 'enlargement'.[1] Recognition of the consonantal root is thus an essential element in construing the MEANING of words in a Semitic language like Hebrew. It is also important for determining the GRAMMATICAL properties of a word or set of words. However, many very common words have "defective" roots. In such cases, only two or even one of the historical root elements show up in the surface form of words built up out of them. Compare, based on a single historical root, with an initial pharyngeal *ayin,* the words *ose* 'make, do', *na'asa* 'became, was made', *asiya* 'doing', *ma'ase* 'deed, act' where only the *s* is shared. Thus, current pronunciation has lost many historical distinctions; yet these still affect how words alternate grammatically, so that it is sometimes hard to identify a historical root without special formal study.

The route taken by children in learning to recognize these root elements is as described for naturalistic output in Berman (1981b) and in experimental studies with children from ages three to seven in Clark & Berman (1984) and Walden (1982). At first, (a) children have no more than a rote knowledge of words as prepackaged, unanalyzed wholes. That is, they as yet pay no language-specific attention to consonants as distinct from vowels, given that across languages, children relate to consonants as important cues to form—meaning relations from a very young age (Clark & Berman, 1984). Then, from around age two, (b) initial alternations in very

---

[1] In representing Hebrew data, a broad phonetic transcription is used. As far as possible, this is done by means of symbols familiar from English orthography, but in some cases non-English representations are necessary, thus: *c* and *ts:* for the initial sound in *tse-tse* fly or the end of the name *Katz; s* and *sh:* for the initial sound in *sugar, show,* or the end of *push;* and *x:* for the sound at the end of the words *Bach* and *loch.* Hebrew verbs are often represented in past tense (masculine singular, third person), because this is a morphologically simple form. Words are usually stressed on the last syllable. If main stress occurs on some other syllable, this will be marked in the text by an *accent aigu* on the stressed vowel. Items marked by a preceding asterisk are ungrammatical or otherwise ill-formed.

familiar items show that children recognize them as semantically connected, e.g., they know that inflectional forms such as *le-daber* 'to-talk' and *me-dabr-im* 'are-talking' are versions of the same thing. Children will then (c) apply an interim strategy of regularizing words to the full or canonic paradigm, e.g., *li-sxot* 'to swim' becomes **soxét-et* ' (she)swims' in place of required *soxa,* on the same lines as regular *li-zkor* 'to remember' alternates with *zoxér-et* '(she) remembers'. This shows that from as young as 3, children start to attend to a "consonantal skeleton" in stem–affix combinations. But they are still confined to immature reliance on the surface occurrences of consonant–vowel forms in actual words. Only from school age, at step (d), do children manifest a more abstract perception of root elements, possibly because they now know the (consonantal) orthography of Hebrew (Bentur, 1978). They also now recognize constraints on possible forms of words in root–affix combinations, providing evidence of rule-bound knowledge. In order for step (e) to be reached in this domain, formal study of "grammar" may be required to provide knowledge of the abstract historical structures and of distinctions between superficially identical forms derived from theoretically distinct roots.[2]

The path from rote learning of isolated forms to full cognizance of the grammatical structure of the radical elements underlying Hebrew vocabulary is thus quite complex. Young preschoolers are not aware of the abstract nature of this system, but as grammaticization proceeds, they do become cued to the privileged status of consonants compared with vowels in their language. And from an early age, they are able to connect words which share the same consonantal skeleton—as distinct from the more abstract, historically motivated "root"—and hence the same semantic core of meaning in Hebrew and also in Arabic (Badry, 1983).

Oblique Pronoun Inflections. In Hebrew, only nominative (subject) pronouns appear in a free, nonbound form in sentences, e.g., *ani ozev* 'I am-going', *hem azvu* 'they went'. All other pronouns occur as bound forms, fused as suffixal endings to prepositions which mark their relations to the predicate or to other nomials, e.g., *le-* + *ani* 'to + I' becomes *li* 'to-me', *bishvil+ hem* 'for + they' becomes *bishvilam* 'for-them'. The system is full of irregularities or at least of anomalies because of abstract historical distinctions not pronounced in today's Hebrew, and it thus takes children several years to master the details of this system. But because prepositions form a closed class of lexical items, because they are such high-frequency items, and because the language affords no alternative way of expressing them in the context of pronouns, learners must come to grips with these forms from very early on. These are the steps they take in doing so, based on studies of naturalistic speech output, both cross-sectional (Berman, 1981a; Dromi, 1979, Kaplan, 1983) and longitudinal (Bin-Nun, 1983), with experimental data in Rom and Dgani (in press).

At first, children use prepositions in uninflected forms, with the impersonal

---

[2] For instance, a single surface form, the past tense verb *kara* has 5 different interpretations and distinct related forms, depending on its historical root; thus: *li-kro* 'to read', *li-króa* 'to tear', *li-krot* 'to happen', *li-xrót* 'to dig (a mine)', *li-xróa* 'to kneel'.

pronoun *ze* 'it, this' as in *al ze* 'on it' or with a full noun, e.g., *al ima* 'on Mommy'. They also use a few rote-learned forms with pronoun suffixes, mainly *shel-i* 'of me = my', *lax* 'to-you', *oto* 'him'. By their third year, they will alternate a few familiar prepositions with different singular pronouns, e.g., *sheli* 'my' and also *shelxa* 'your', *li* 'to-me' and also *lax* 'to-you'. As their command of morpho-syntax expands, later on in the third year, children adopt a variety of different interim strategies for moving into the system. They may juxtapose a free, nominative pronoun to the preposition, as in English *'of they' rather than 'their'; or fuse the preposition with a pronoun form which is the same for all prepositions, analogously to English overregularization from 'mine' to 'her-n' and 'his-n,' and/or they extend the same base-prepositional form across the entire paradigm, as detailed in the studies outlined earlier. Only from around 3 do children use normative rules for selecting the appropriate suffix to suit the particular paradigm to which a given preposition belongs; and it takes until well on in school age until children learn the conventional forms of structurally or conceptually quite "difficult" prepositions when bound to pronoun endings, e.g., *bli* + *ani* 'without + I' is inflicted to *bila-day* 'without-me'.

Again, while initial entry into this system starts as early as age two, it is only by around 5 that it is mastered by children in such a way that they can extend it to prepositionals which they may never have heard being inflected, e.g., when a 5-year-old plays a game of make-believe and says things like, 'If I were spectacles, you could see *through me*' = *dark-i* from the noun *derex* 'path, way' = 'through' (Berman, 1984c). Where nonstandard forms are retained even in school-age usage (e.g., **ecle-hem* for normative *eclam* 'at-them', **ot-ex* 'you, Accusative' for normative *otax*), a more general change can be predicted in the language, as further discussed in the section on "Analysis of the Five Steps."

Plural Formation of Nouns. All Hebrew nouns and also the verbs and adjectives which agree with them occur in a singular or plural form. This changes depending on whether the noun is in masculine or feminine gender. This system is mastered early, and 3-year-olds use masculine *-im* or feminine *-ot* freely and productively in forming plurals, as shown in naturalistic cross-sectional data (Berman, 1981a; Kaplan, 1983) as well as experimental findings (Berman 1981c; Levy, 1983). Difficulties arise with two kinds of items: when the plural ending requires a change in the base form of the noun depending on its morphological class membership, and in lexical exceptions, where a masculine noun takes a feminine ending, or vice versa:

1. Masculine (=Neuter) *-im*: Feminine (final *-a* or *Vt*) *-ot*:
   kadur / kadur-im 'ball-s' *buba / bub-ot* 'doll-s'
2. Stem Changes in Plural:
   yéled / ylad-im 'child-ren' *magévet / magav-ot* 'towel-s'
3. Lexical Exceptions:
   arye / aray-ot 'lion-s' *isha / nash-im* 'woman / women'

Development proceeds as follows: At first (a), nearly all nouns are used, unanalyzed, in the singular only. A few occur, also with no analysis, in the plural form—which is then "basic" for child usage, as in words like *naaláyim* 'shoes', *bgadim* 'clothes', or the feminine nouns *kubiyot* 'blocks', *madregot* 'stairs'. By age two (b) initial alternations show up with a few highly familiar items used commonly in both singular and plural, e.g., both *yéled* 'child' and also *yladim* 'children', and *áyin* 'eye' and *enáyim* 'eyes'. As the children move into the system in their third year, (c) an interim strategy is to add the same plural ending *-im* to all words, whether masculine or feminine in form; or to use this for all but only nouns except those which end in the typically feminine final syllable *-a,* and/or to add a plural ending with no other changes in the stem noun, like those shown in (2) and (3) in the earlier listed classification. With acquisition of the normative rule (d), by late in the third year, the endings *-im* and *-ot* are added appropriately, according to the formal class membership of the noun, as feminine or masculine. Eventually, often only at late-preschool age or in elementary school, lexical exceptions are learned for nouns which have a masculine form but a feminine plural ending, and vice versa, as well as for other nonrule-bound types of alternations. Again, as in the two previous subsystems, the learning of these forms is a longish process. The initial grammar is achieved early on, between the age of 2 and 3, in what we have termed steps a through c. But full knowledge of more abstract morphophonological alternations and numerous lexicial exceptions is achieved only by late preschool age or even subsequently. And here, too, if nonnormative forms persist, we can predict that this signals a more general change in the language as a whole.

Verb-Tense Marking. Hebrew verbs have no obvious, unequivocal "basic" form analogous to English words like *sit, understand,* or *enjoy*. Instead, whenever a verb is used it must be inflected for one of five possible categories: infinitive, imperative, present, past, or future. The system is formally very complex, yet children make the initial distinctions between these five categories with great rapidity, mastering them by age 3 at latest. The path followed for these developments is traced in early case-study reports (Berman 1978b; Borochofsky 1984, Dromi 1982) as well as in detailed cross-sectional studies of naturalistic speech output (Berman & Dromi, 1984; Kaplan, 1983). At the one-word stage (a), a single nonalternating form of a verb will be used, most generally in the imperative or infinitive, with occasional rote-learned forms in past, present, or future. By age 2 and up, (b) initial alternations of two or more of the five forms occur with respect to the commonest and semantically most prototypical or general-purpose verbs such as those meaning 'go', 'do', 'make', 'give', or 'put'. Subsequently, in the third year of age, at step (c), immature construal of temporal distinctions is manifested by association of past-tense forms with activity verbs, quite generally ones with a punctual sense such as 'fell', 'broke' = 'got-broken', or 'arrived'. And nonpast durative activities or stative types of verbs may be specially marked by overextension of a present-tense type of prefix *m-* to verbs which normatively do not admit of such a morpheme (Berman, 1983). With acquisition of the productive, grammatical system at step d, from the 4th year on, children can alternate different tense/mood forms across a single verb, and the same tense/mood form is also used across verbs in different semantic classes—durative and punctual, stative and activity, etc. Again, step e is achieved only at school age, when children learn the forms of lexical exceptions which they may have misconstructed earlier—similar to the common use of English forms such as *brung* in late preschool usage.

Verb-Pattern Assignment. The final set of examples relate to the task facing the Hebrew learner in mastering the set of verb-pattern (*binyan,* 'conjugation') alternations. These are used in Hebrew to express a variety of predicate-argument relations such as: active, passive, and middle voice; reciprocal, reflexive, and inchoative change of state; and also (more marginally today) aspectual notions such as incipient or iterative. Hebrew as a Semitic language exploits verb morphology to encode semantic categories which will often be expressed by syntactic devices such as auxiliary verbs or pronouns in an analytic language such as English. Some of these alternations are illustrated below, where the notation P-n is used for the five main verb-patterns and ps = passive.

	Basic, unmarked:		
1.	*Ron yoshev* [P1]*sham*	Ingressive:	*Ron hityashev* [P4]*sham*
	Ron is sitting there		Ron sat-down there
		Causative:	*aba hoshiv* [P5]*sham et ron*
			Dad seated there OBJ Ron
		Passive:	*Ron hushav* [P5$_{ps}$]*sham*
			Ron was seated there
2.	*Ron lovesh* [P1] *me'il*	Perfective:	*Ron lavush* [P1$_{ps}$] *bi me-il*
	Ron is wearing a coat		Ron is dressed in (a) coat
		Reflexive:	*Ron hitlabesh* [P4]
			Ron got dressed-dressed himself
		Causative:	*aba hilbish* [P5]*et ron*
			Dad dressed OBJ Ron

This system has many irregularities: A single pattern can be used for more than one function (e.g., P4 is used for inchoative change of state, for reflexive, reciprocal, and for passive), and a single function may be expressed by more than one pattern. It takes children several years to gain command of these form/meaning relations as expressed through verb morphology. Moreover, acquisition of these lexical options takes place mainly after the inflectional systems of the grammar are well established as rule-governed.

At step a, children use only one, unalternating form of each verb type, i.e., a given verb-root appears in only one verb pattern (Berman, 1982). This may be a transitive form such as P1, *zarak* 'throw' but not intransitive P2, *nizrak* 'be-thrown, get-thrown', or it may be an intransitive form such as P4, *mistovev* 'turn (oneself) around' which appears before, but not together with, the transitive P3 form *sovev* 'make go-around'. Then, in the third year, at step b, children manifest initial alternations of a few familiar verbs, often using both a P1 form for self-initiated motion, say, and a causative P5 counterpart, e.g., both *yored* 'go down' and *morid* 'take-down'. Step-c interim strategies are revealed by children's "creative errors" (Berman & Sagi, 1981; Walden, 1982). From the age of 3, children innovate to fill lexical gaps—either genuine ones or ones due to inadequate vocabulary. For instance, they form causatives of verbs meaning 'swim' and 'cry' corresponding to occurrent causative forms of verbs for 'ride' or 'laugh'. A major bridge strategy from around age 4 is to mark a single major distinction along the axis of transitivity at a point where, too, children push for a one-form/one-function construal of verb-pattern alternation (Berman, 1980). For instance, they (over)use the P5 *hif'il* pattern for all and any causatives, and the P4 *hitpa'el* pattern for all and any inchoatives. With the development of end-state type, rule-governed step d, children perceive the different functions of each *binyan* pattern, and they also distinguish semantic specificities, e.g., that intransitives may be reflexive, inchoative, reciprocal, or passive, and that transitives may be agent-initiated, instrumental, and so forth. Finally, step e entails learning the normative forms of noncanonic instances such as lexical suppletions and other anomalies.

In this section, we outlined how the model we propose applies to children's acquisition of several subsystems in the grammar of Modern Hebrew. We now use these data to motivate and specify what is involved in each of the five steps which compose this model.

Analysis of the Five Steps

The overall progression from rote via rule can be characterized as follows: Initially, at step a, the child relies critically on specific linguistic input as it combines with general contextual, extralinguistic cues. By step b, attention to input is more selective, and the child starts to rely on what is most available, salient, or prototypical from language use. At step c, the learner makes initial hypotheses about structural properties and relations of the target language, and the acquisition of grammar becomes abstracted from specifics of input data—whether from situational or linguistic contexts. This type of generalization is then deepened and extended in step d to accord with the normative endstate rules. And finally, with step e, there is a *return,* as it were, to the earlier, pregrammatical attention to specific items and facts of usage. Now, the learner takes account of the constraints imposed by nonstructural factors of language usage such as lexical conventions and stylistic variation. We consider each of these steps in turn.

The Nature and Role of Rote. Our model pays account to rote learning as an important strategy in early acquisition. At this initial step, children know isolated lexical items without making any systematic connection between them, and without realizing that these initial units of acquisition may in fact have some internal structure (Bowerman, 1982b; MacWhinney, 1978, 1981). Thus, a Hebrew-acquiring child may use the plural forms *xaruz-im* 'bead-s' or feminine *kubiy-ot* 'block-s' without knowing that they have a singular counterpart, *xaruz* and *kubiya,* respectively (see "the Plural Formation of Nouns"); s/he may use both the causative verb form *le-havi* 'to bring' and the more basic motion verb *la-vo* 'to come' without knowing that they are semantically related or that they share the same abstract consonantal root (see the section "Verb-Pattern Assignment"); and the same child may use a dative prepositional marker in an unsegmented or "prepackaged" string such as the surface form *tavili* long before realizing that this is a complex structure consisting of a verb plus case-marked pronoun, *tavi?i* + *li* 'bring+Fem Imp to+me' (see the section on "Oblique Pronoun Inflections"). At this initial phase, then, straightforward imitation plays a role, and the child displays no ability to analyze surface linguistic forms (R. Clark, 1974, 1978).

The repertoire of this initial phase is made up of what are thus essentially formulaic routines or memorized chunks (Peters, 1983; Wong-Fillmore 1979). As the child does not evince any knowledge of structure, we have characterized this phase as PREGRAMMATICAL. There is as yet no alternation of a given content in more than one form. The child will not use the same verb in, say, both present and past-tense forms; and even if the verbs *come* and *go* both occur, they will not be contrasted with *come in, come out,* or with *go in, go out,* nor with each other, in any systematic way. This is not to say that the child does not have ANY linguistic knowledge or that language use is merely a parroting of what is heard. Changes in pronunciation as well as in the semantic extensions of items used at the one-word stage (Dromi, this volume) provide clear evidence that the child is being "creative" and making his or her own construals of the input data. Clearly, too, children possess a regular kind of form/meaning relationship, because the same form is used in different circumstances. A child may say *doi* once on seeing the neighbor's dog, once on hearing a dog barking somewhere else, and once when seeing a picture of a dog in a book. But this is situational or CONTEXT-BOUND alternation. As yet, the child will not alternate such an item with other linguistic forms or items which contrast with it in the grammar. That is, this initial stage of entry into a given linguistic domain displays no FORMAL reorganization of the internal structure of categories which are lexicalized and/or grammaticized in the child's native tongue.

Of special interest here is WHICH forms occur in this initial phase, as the basis on which the grammar will be built. Analysis of the initial unsegmented units documented in several studies of Hebrew-speaking children at the one-word stage (Berman, 1978b; Borochovsky, 1984; Dromi, 1982; Postol, 1982) reveals a remarkable commonality combined with a certain amount of variation in the output of these different children. Such items can thus be taken to represent the most basic, least-marked elements of the child's linguistic repertoire. Several hypotheses can be entertained as to what determines choice of such items, which are clearly only a restricted subset of everything available in the language. These include, respectively, the factors of input, of relative frequency, of formal simplicity, and of communicative function. Thus, one might argue that selection of initial, rote-acquired items derives from INPUT, the language addressed to children, mainly by their primary caretakers. In a sense this is of course correct, because children must derive exemplars from some source. But raw input data are only one factor in the process. Rather, from the very outset, children are selective regarding everything to which they are exposed, and they choose only part out of the full range of input materials for use in their own output.

A related claim would be that selection depends on FREQUENCY: The items which occur most commonly in input language will also be selected for the child's output. This also seems inadequate as an explanation. The very fact that these items have high frequency in the speech addressed to children itself suggests that they have a privileged status; they are construed not only by children but also by their caretakers as peculiarly appropriate to early verbal interchanges, hence to the initial linguistic repertoire. Besides, even if a direct correlation could be proved between input frequency and output selection (and this is unlikely in principle; see Shatz, 1982), decision making at a quite abstract level is still involved in children's choice of "basic" forms. Observational evidence is, again, provided by the fact that these are so often shared so widely across different children.

Another hypothesis relates to the notion of FORMAL SIMPLICITY (E. Clark, 1980, 1981; Slobin, 1973, 1984). The child will opt initially for those items which are linguistically unmarked and which manifest the least complexity of morphophonological form and/or syntactic structure. For instance, the masculine singular is morphologically, syntactically, AND semantically simpler, or less marked, in Hebrew than either feminine or plural (Berman, 1978a). Yet there are some words, like those meaning *pants, eyes, blocks, toys,* or *keys* which children first use in the plural form (see the section on "Plural Formations of Nouns"); and there are nouns which children use first in their feminine form, like the agent nouns meaning *baby sitter, nursery school teacher,* and, for some children, *doctor.* Thus, theoretically, the feminine nouns *para* 'cow' and *tarnególet* 'hen' are derived from their more neutral masculine counterparts—*par* 'bull' and *tarnegol* 'rooster' plus feminine form endings *-a, -et.* But in fact, Hebrew-speaking children know and use the words for cow and hen far earlier than the "formally simpler" masculine counterparts (Dromi & Berman, 1982). And there are other types of evidence which show that the very plausible factor of "formal simplicity" is not critical at this earliest phase of rote learning of isolated forms. One is a case like that of Hebrew verbs (see section on "Verb-Tense Marking"), where there is no obvious or unequivocal "basic" form available. In such instances, the moment the child selects a given surface output item, some basic structure must be incorporated for tense or mood and also root plus pattern marking. Hence, children's early overuse of infinitive forms (Kaplan, 1983) is not necessarily due to surface simplicity, because Hebrew infinitives include a prefix meaning 'to'. Rather, infinitives are a form which offer children the most mileage, so to speak. They can be used as a surrogate for requests and other directive functions so central in early language use, and they involve no claims about the temporality of an event (Berman & Dromi, 1984). Besides, morphophonological complexity in and of itself is simply ignored by children in many cases. Thus, children's early verbs in Hebrew occur in all five common *binyan* verb patterns, and children adjust the surface forms to their own output processing abilities. Similarly, many female children start by using the *feminine* form of Hebrew imperatives. This means they add an extra syllable with the ending *-i,* as in *ftexi* 'open!', *boi* 'come-here', *lexi* 'go away', and *zuzi* 'move over'. These are more "formally complex" than their masculine counterparts *ftax, bo, lex, zuz,* respectively, but that does not seem to sway the children's choice.

A final argument against formal simplicity as a primary factor in selection of initial units is that there is no necessary correspondence between linguistic simplicity, or grammatical nonmarkedness, and psychological saliency or centrality. For instance, in grammatical terms, third-person, masculine singular forms are typically unmarked and most basic in numerous unrelated languages including Hebrew. Yet this does not affect children's selection of initial pronouns, which across languages will tend to start with first-person forms (Rom & Dgani, in press).[3] On the other hand, we do not discount formal simplicity as an acquisitional principle in general. In fact, it will play an important role in language development ONCE the child has the beginnings of a grammar, especially in regulating the kind of interim schemata or bridge strategies which children adopt en route to grammatical rules (step c). But the specific grammatical structure of the child's initial units of acquisition seems irrelevant at the rote-learning stage. Here, factors other than formal simplicity will determine which units children incorporate in their early output repertoire.

My own view is that initial selection of output items depends crucially on COMMUNICATIVE factors, at a point where language is typically instrumental in function and interactionally motivated. Extralinguistic considerations of which functions or activities children seek to perform or have performed for them and which objects and events in the surroundings they attend to will largely decide what children first encode linguistically. Subsequently, however, as the child proceeds beyond rote-learned forms, such factors will become less central to the language-learning process. This claim ties in with the general conception of language development we propose: It proceeds from communicatively oriented, context-based pregrammar via increasing grammaticization on to the flexible use of grammar for communicative purposes in a rich range of discourse contexts. Next, we consider the critical role of language USE in guiding children's selection of the particular exemplars which will form the basis for their construction of a grammar.

Initial Alternations. As the child moves from initial units to early word-combinations, at the brink of morpho-syntactic acquisition, there emerge some formal

---

[3] Obviously, formal simplicity in a trivial sense is crucial right from the start: Children's initial units do *not* incorporate strings of 9 words or 19 syllables within a single prosodic contour. But this is not what Eve Clark, Slobin, and other developmental psycholinguists are referring to, nor is it what we mean here by "formal simplicity." Clearly no child produces complex strings of embedded and coordinated clauses at the stage of rote learning, just as no child strings together different words prior to the use of one unit at a time.

variations on a single theme. For instance, children may say both *man* and *men*, both *doll* and *dolls*, or in Hebrew, both *yéled* 'child' and *ylad-im* 'children', both *buba* and *bubot* 'doll-s'. The same child may also use intransitive motion verbs like P1 *yotse* 'go out', *yored* 'go down' along with their causative P5 counterparts like P5 *motsi* 'take out', *morid* 'take down'. The child now knows that there is a connection between these items, possibly some kind of "family resemblance" (Rosch & Mervis, 1975). That is, the child realizes that both *doll* and *dolls* refer to the same kind of entity, *go out* and *take out* to the same kind of event.[4] The child does not yet realize, however, at this initial entry into the system *qua* system, that there is something generalizable beyond the specific relationship between these items. That is, the child does not yet have a rule of semantic plurality so that the general notion of "more than one" is extracted as expressible by means of a specific kind of suffix. Nor will Hebrew-speaking children be aware that a certain surface shape can be attached to verbs to make them causative in content. Lacking any general principle for productively associating a given meaning with a given form beyond these specific exemplars, the child's knowledge is thus still PREGRAMMATICAL.

These initial alternations are, however, critical in that they provide children with the first bases for COMPARISON, for active working with regularly alternating forms, and hence for eventual extraction of more general form/meaning mappings across entire classes of linguistic items or structures. This is the point where learning the rules of grammar may be TRIGGERED by these initial exemplars in alternation with one another. Learning here has proceeded beyond rote, memorized chunks to what Braine (1976) terms "limited scope formulae." The same item or sequence will appear in different formal environments, but these are limited in number as well as in scope. For instance, an early type of syntactic expansion in the Hebrew child's use of language is of a finite verb followed by an infinitive—a structure used widely in Hebrew for all modal and many aspectual type verbs (e.g., ones meaning *be able to, know how to* or *begin to, continue to*). To start with, however, children will use this structure exclusively in the context of the single verb *rotse* 'want' or for girls, *rotsa*. Similarly, children learning English use forms like *don't* and *can't* in a few restricted contexts before moving on to productive, rule-bound use of auxiliary verbs in their language.

Here, too, the question arises of what determines the choice of elements which serve as "model instances." We will argue that this is not due mainly to real-world familiarity with the objects or events themselves, but rather to familiarity with these items from habitual LINGUISTIC use. And we try to show that this in turn relates to the important role of set speech routines or linguistic formulas, on the one hand, and to the nature of instances which are categorially and semantically prototypical. Clearly, to start with, children will talk about objects and events familiar to them

---

[4] This type of alternation thus goes beyond the narrower notion of *analogy* as defined for language development by MacWhinney (1978, 1981) in which a formal alternation is modeled on a single, closely corresponding, and superficially similar instance.

from their own real-world experience. Thus, Hebrew speakers use the causative verbs meaning 'make eat = feed' or 'make mad = anger' before they use the structurally identical device for words meaning 'transfer' or 'enrich', e.g., Hebrew P5 *le-ha'axil, le-hargiz* vs. *le-ha'avir, leha'ashir,* respectively. They use reflexive forms meaning 'wash' and 'dress' (P4 *le-hitraxets, lehitlabesh*) before verbs which refer to less everyday activities for the child such as *have one's hair cut, shave one's self,* again with the same surface structure. And children use intransitive, middle-voice P2 verbs like *nishpax* 'spilt,' 'got spilt', *nishbar* 'broke, got-broken' before they use this same structure to talk about less familiar events such as *nimxak* 'got erased', *nidras* 'got run-over'. Nevertheless, real-world familiarity does not suffice to explain early linguistic selectivity. Children are as accustomed to seeing things being taken, given, thrown, and caught as to seeing them being broken or spilled. Yet in Hebrew, 3-year-olds talk about the latter two events by means of P2 intransitive verbs, whereas they do NOT use these same forms to talk about taking, giving, or throwing—even though the language also has P2 forms of these verbs. Similarly, children tested for plural noun formation (Berman, 1981c) made more errors at age 3 with the Hebrew word for 'onions' than with the word for 'rabbits'. Yet both undergo the same formal changes (*batsal/betsal-im* and *shafan/shefan-im,* respectively); and there was no sociological reason to believe my subjects had had more exposure to rabbits than to onions in their real-life experience.

We suggest, instead, that familiarity in language MENTION, in LINGUISTIC use, is what determines choice of exemplars. What matters is not only familiarity with the actual entity or event but also how that entity or event is typically referred to by speakers. Thus, unlike in English, the real-world object *onions* (commonly used in salads eaten by small children in Israel) is in Hebrew usually referred to as a noncount, collective term *batsal.* Thus, the plural form *betsal-im,* although well-formed as part of the conventional lexicon, is more highly marked and less common in language usage. My study of plural nouns in Hebrew yielded results which could only be interpreted by considering how words are actually used in the linguistic conventions by members of the speech community. Otherwise, we had no way of explaining the differences within a single formal category of responses to specific items by children of different ages. Thus, 6-year-olds all gave the correct answer for the masculine nouns *davar* 'thing' and *shafan* 'rabbit' (*dvar-im, shfan-im*). Yet about half of them regularized the plurals for other nouns of exactly the same structure (e.g., *safam*—'moustache'—was pluralized as **safam-im,* not as required *sfam-im*). And, in exactly parallel fashion, for feminine nouns children gave the correct plural for *simla* 'dress, frock' as *smal-ot* and sometimes also for *giv'a* 'hill' as *gva'ot,* but they reglarized the plural of other nouns in the same formal class (e.g., *ricpa/*ricp-ot,* 'floor/s'). The various items in these two groups of nouns showed identical FORMAL alternations. Hence, the different results on specific vocabulary items is not due to structural factors. Yet again, real-world familiarity does not seem all that relevant: Children see floors as often as they see ceilings or hills (in fact, floors may be perceptually more salient than ceilings!); and they probably are just as familiar with moustaches as with rabbits in the environment.

What matters, rather, is how these items generally occur in the language, that is, as TOKENS rather than as types. There are many examples from other domains too. Thus, *shells* and *lips* are typically referred to in Hebrew in the plural, whereas *money* and the *head* are most often talked about in the singular. And that is precisely how children learn these items, even though the other option is equally grammatical (Berman, 1981a). Similarly, children know the formulaic expressions *yéled tov* 'boy good' = 'good boy' and they also hear about *ha-yeladim ba-gan* 'the-kids at-nursery-school', so *yeladim* 'boys, kids' is an early plural form. But this alternation will occur long before children learn the formally identical alternation between *mélex/melax-im* 'king/king-s', since in their storybooks, they always learn that "Once there was *a king* (or *a king* and a queen)."

This suggests that at this developmental phase, when familiarity of surface token forms in actual use is so important, a special role is accorded to FORMULAIC contexts. A particular word form may not be fully analyzed, but it will be known from a familiar ritual or set collocation. For instance, I found that children would give the plural form *dvarim* 'things' when prompted by the context of formulaic *kol miney . . .* 'all kinds of . . .'. Similarly, a small child may use the causative verb *morid* 'take down, take off' to start with in the context of *le-horid mixnasayim* 'taking off pants' as a familiar, set phrase associated with toilet rituals. That is, familiarity with the object or event in the outside world must be backed up by familiarity with their representation as LINGUISTIC items occurring as actual tokens in specific verbal contexts in order for these to constitute the basis for initial alternations.

There is still the question: What makes certain tokens better exemplars than others for these initial alternations? Why, in other words, do small children talk about things as having gotten broken, spilt, or torn, but not as having gotten taken, given, or thrown? This is not merely a matter of formal simplicity, since these items all manifest similar grammatical and semantic permutations. Rather, the initial exemplars which will trigger the rules of grammar represent PROTOTYPICAL instances of a particularly set of form/meaning relations. The notion of prototype is taken here from work in cognitive psychology (Rosch, 1977, Rosch & Mervis, 1975) and in linguistic semantics (Fillmore, 1975), as well as in research into child language (Bowerman, 1980; Slobin, 1984). It represents an interplay between mental categorization of objects and events from the real-world, extralinguistic context; token occurrences in language use of how objects and events are typically referred to in speech; and the concept of linguistic markedness, where priority is with those members of a grammatical paradigm or a lexical subsystem which are least marked, most basic, or universal.

This set of claims can be illustrated by early verb-pattern alternations of Hebrew-speaking children (see the section on "Verb-Pattern Assignments"). It seems to me that the forms which are the first to be varied by children are representative of the more general kinds of predicate-argument relations encoded by different languages and in specific linguistic categories. Thus, the first verbs used in past tense will typically represent punctual, endstate events ("Verb-Tense Markings") and they will often occur in the passive-intransitive P2 pattern in Hebrew, as in *nishbar* 'broke = got broken', *nishpax* 'got-spilt', or *nikra* 'tore'. These are prototypical of the "event" type of scene or situation, where speakers first comment on the fact that something has happened to a given object (Savasir & Gee, 1982). Moreover, children will soon also encode these typical situations in "activity" type scenes (Slobin, 1984), and will use the P1 transitive verb to talk about the fact that someone has broken, spilled, or torn something. By contrast, predicates which have the same form in Hebrew but refer to transfer of possession or location—e.g., those meaning *give, take,* or *throw, put*—are first used as transitive verbs in P1, as actions that someone performs. And these will NOT be formulated in the middle or passive-voice P2 forms until much later in children's usage, because these activities are typically not viewed as "events," as things which happen, but rather as activities which people perform.

The important role of prototypical scenes in how children first encode situations is discussed by Slobin (in press) in terms of a "basic child grammar" which is shared across children and languages, prior to the emergence of language-specific, typologically motivated knowledge (corresponding to what I have characterized as pregrammatical (Berman, 1984b). It is also implicit in much of the literature on early child language. For instance, Maratsos (1983) notes that children's initial word-combinations widely favor Animate–Actor plus Action (e.g., *baby cry*) or else Action plus Inanimate Object (e.g., *bite finger*). That is, children are selective about which structures are used in referring to situations which in theory could be viewed from a quite different perspective. In sum, the initial alternations of forms characterizing step b are not haphazard at all. Rather, they represent instances which are the most salient, basic, unmarked, frequent, or "normal"—that is, the most *prototypical*—for a given set of form-meaning relations. This prototypicality is anchored not only in cognition, in perception, and in real-world experience, it is also bound to linguistic usage and the universals of natural language grammars. These prototypical exemplars are restricted in scope, but they perform an important function. They go beyond mere rote learning and show that the child realizes that two or more surface forms may be instances of "the same thing." And they provide learners with an opportunity to practice and compare instances as well as to extract generalizations and extend them to new or unfamiliar exemplars in the next phase, as discussed below.

Interim Schemata. This is perhaps the most interesting phase from a psycholinguistic point of view, because it marks the beginning of clearly nonroutinized learning and the child's creation of grammar. Children's "creative errors" provide what is probably the single most important source of evidence of the onset of rule-bound, formal processing of linguistic material. Typical errors include: regularization of anomalous forms or lexical exceptions to the paradigm cases—as in the classic examples of English *mans* or *swimmed* as well as *brung;* innovative filling-in of real or assumed lexical gaps—as when an Israeli child coins nonexistent causative verbs like **mavke* 'make-cry' or **masxe* 'make-swim', and overextension of a given grammatical device to inappropriate contexts—as when French-learners overuse *que* and Hebrew-learners overuse *še-*, both meaning roughly 'that', in all types of embedded clauses, not only where they are required by the grammar.

The factor of PRODUCTIVITY first begins to play a role at this interim phase, when "the learner evidently prefers a bad generalization to none at all" (Gleitman & Wanner, 1982, p. 7). It is now that children use the kind of intermediate strategies characteristic of the "waiting-room" period (Slobin, 1982) while they are busy figuring out both what aspects of a particular notion are encoded in the language *and* what means are used to do so. We noted several examples of such interim generalizations in the section on "Development of Selected Subsystems of Hebrew": Use of a single pronominal ending without regard for the forms required by a given preposition, and/or use of the same form of a preposition without regard for a particular suffix; adding plural endings with no change in the stem; overmarking of nonpast-tense verbs by the prefix *me-*, even where it does not occur in the grammar, much as English-speaking children overmark a verb like *cut* in the past-tense form **cutted;* and overextension of a given verb-pattern *binyan* form to express a given semantic content—such as nonconventional use of P5 *hif'il* to mark causatives or P4 *hitpa'el* to mark inchoatives—along with neutralization of distinctions between the entire set of transitive versus intransitive types of predicates, respectively.

The interim schemata adopted by children en route to the normative rules of end-state grammar depend heavily on quite general psycholinguistic principles of acquisition of the kind formulated, for instance, by E. Clark (1980, 1981), MacWhinney (1978, 1984), and Slobin (1973, 1984). These include such factors as semantic TRANSPARENCY, expressed as the need for distinct surface marking of each semantic category, and formal SIMPLICITY, in the sense of relative lack of structural permutations and rearrangements in alternating between one form and another of the same element or set of elements. In this phase, children will tend to seek one-to-one mappings of form/meaning relations, so that a single form expresses only one function, and a given function is expressed by only one form. Reliance on such principles gives rise to quite specific strategies in children's construals of form–meaning relations. One is the device of OVEREXTENSION—noted earlier in the broad use of a single subordinating conjunction across different kinds of clauses by means of Hebrew *še-* 'that' or French *que* 'that' (E. Clark, in press); in overuse of prefixal *me-* to mark the present-tense of Hebrew verbs; and in overextension of the P5 *hifil* verb pattern (which is often causative in function) to P1 pattern verbs which are already transitive, to yield the equivalent of "stop making-push me" for "stop pushing me," or "I'm making it turn inside out" (e.g., a hat) for "I'm turning it inside out" (Berman & Sagi, 1981). A related interim procedure is REGULARIZATION: Here, the principles of formal simplicity and of semantic transparency may combine, as children attach a formal device to a single, invariant word stem or base structure. This strategy then enables them to apply a rule without regard for structural or lexical factors which constrain its application. For instance, children may simply add the ending *-er* in English or *-an* in Hebrew to form new names for agents—a *caker* for someone who makes cakes—or for instruments—a *liner* for something that we draw lines with (Clark & Berman, 1984).

These transitional strategies manifest considerably more VARIABILITY than the pregrammatical phases which precede and the acquisition of endstate grammar to follow. The specific devices employed, and the domain to which they apply, tend to be confined to certain items or classes of items and also to certain children. We have noted, for instance, that in acquiring oblique pronouns, Hebrew learners will often use a single prepositional stem across the entire paradigm. For example the preposition *mi-* 'from' yields well-formed *mimeni* 'from-me' and *mimenu* 'from-him'; and children may regularize these to other contexts, such as **mimenax* 'from-you' and **mimenam* 'from them' (cf. normative *mimex, mehem*). But not all children will use this particular strategem for this particular paradigm. They may either regularize OTHER prepositional stems, but not this one. Or they may favor another kind of interim strategy in moving into this complicated set of formal alternations. For instance, they may regularize the pronoun endings rather than the prepositional stem. And some but by no means all children will use free, nominative pronouns with a preposition (e.g., **al hem*—'on they'—for *aleyhem*—'on them').[5]

To some extent, of course, the particular strategies adopted will depend on the nature of the subsystem being learned. In some cases, a superordinate semantic category may form the basis for (spurient) generalizations. An example is use of "one/more than one" for Hebrew plurals, disregarding the distinction between formal classes of nouns as masculine or feminine; or use of the same morphological form for agent and instrument nouns across the board (Clark & Berman, 1984). Form-based cues may underlie other kinds of generalizations, as were noted earlier for oblique pronouns in Hebrew. In other words, the types of strategies available to children, e.g., regularization or overextension, like the acquisitional principles which underlie them—such as maximal distinctiveness and formal simplicity—are shared ACROSS children and also across linguistic systems. But the specific paths selected in relation to given lexical items or structures will differ at this stage. If this were not the case, these interim devices would ultimately become canonized as the grammar of that language.

A central fact about such strategies is, then, precisely that they are "interim" procedures, and hence relatively shortlived. They are by nature TRANSIENT phenomena. In fact, where such juvenile errors or deviations from the end-state grammar do persist, or where they apply across the board for an entire paradigm or subsystem, then general change can be predicted for the language in coming generations.[6] Thus, this interim step c raises a further question of general psycholinguistic

[5] An interesting question which lies outside the present discussion is what *determines* these individual differences. One pertinent idea is that of de Villiers (1980) regarding the scope of the generalizations that children are able to make at the time of their early word combinations. She suggests that "children who go through a prolonged holophrastic period and who are comparatively old when they begin combining words, might introduce rules that act over broader categories than children who leap early into syntax and proceed by the accretion of a great many more idiomatic rules" (de Villiers, 1980; pp. 15–16).

[6] Examples where child language can be seen to herald changes in the adult grammar are discussed in some detail in other work of mine (Berman, 1981b, 1984c). These include: (a) reanalysis of imperative forms as the future stem without a person-marking prefix; (b) leveling of the distinction between first-

significance which, like the one noted in footnote 5, lies beyond the present study: What determines why some interim rules are abandoned, whereas others eventually lose their status as partial and transitory—hence as immature—and over time also become absorbed into the normative, endstate grammar of older speakers?

These interim schemata form an important, perhaps necessary, bridge between pregrammar and endstate rules. They differ from what has preceded, since they go beyond the rote-learned and imitative material of step a, and they are no longer confined to familiar, often formulaic contexts as in step b. Rather, these strategies are GRAMMATICAL in at least two senses: They are creative, since they extend beyond familiar items to new instances, and they are structure-dependent rather than item-bound or context-based as formerly. In other words, in applying a step c procedure, children reveal that they have extracted critical features of form/meaning relations in their language. But these procedures are not the same as step d rules of grammar. Learners still need to interrelate specific generalizations into an overall system, and they need to express them in ways which conform to the constraints that govern the adult grammar.

Rule Acquisition. Perhaps the most important contribution of psycholinguistics to the study of language development is the recognition, based largely on Chomsky's early claims about the nature of linguistic competence, that the child's task entails internalizing the rules of grammar. These rules are in essence productive; they are extendable to novel instances and thus account for the ability to produce and interpret utterances which one may not previously have encountered. At the point of rule acquisition, learners no longer operate with specific collocations or restricted sets of items or even structures. Rather, in applying the rules of grammar, they have recourse to quite abstract linguistic categories—such as of noun or verb, of grammatical subject or direct object, of completive versus ongoing aspect.

The nature of the rule will depend on the specific subsystem on which it operates. It may be a purely formal rule, with no semantic correlate—e.g., that stressed final *-á* is a marker of feminine gender in Hebrew (cf. *korá/korót* 'rafter-s', but *kóva/kova'ím* 'hat-s'; it may be a categorial rule—e.g., that in English particles must follow pronoun objects (cf. *wake up the kids* but not **wake up them*); it may be semantic—as in languages where action-versus-state verbs differ with respect to aspect marking (as in English), or passivization (as in Russian). The kind of rules may also be typologically governed by the nature of the target language; e.g., the distinction between languages whose word order is grammatically determined compared with those where word order may be alternated for pragmatic purposes; between verb-initial languages with prepositions and postposed relative clauses compared with verb-final languages which have postpositions and preposed relative

---

and third-person marking in future-tense verbs; and (c) extension of the subordinator *še-* 'that' not only syntactically, as already noted, but also lexically, to prepositions and conjunctions where it is normatively disallowed.

clauses; or between languages which have a largely analytic, isolating kind of grammar compared with those which have a rich bound morphology and numerous inflections (Berman, 1984b; Slobin, 1982, 1984).

I suggest that there are two properties critical to rule acquisition as the crux of grammatical development. First, rule application entails a release from various "supportive" strategies. Children start out with context-bound, action-based procedures for producing and interpreting utterances, such as relying on the real-world plausibility of an event when verbally encoded. And they proceed to consideration of surface factors of linear succession of elements, ordering of items, and transparency of form-meaning relations (Berman, in press b; Cromer, 1976). Then, as learning is freed of such props, knowledge becomes increasingly STRUCTURE-DEPENDENT, applying not to words, but to classes of words, not to strings of specific elements but to construction types (Chomsky, 1975). An example is the fact that English interrogatives are formed by inverting the first auxiliary element and the subject noun phrase, irrespective of the form or substance of the auxiliary or of the subject nominal in question, and that this applies to matrix but not to embedded questions (cf. *Why is the baby crying* vs. **I wonder why is the baby crying*). Or, also in English, indirect-object (dative) shifting to postverb position is mandatory when both objects are pronouns (cf. *I gave the book to him/I gave him the book* vs. *I gave it to him,* but not **I gave him it*). There is evidence that such formal constraints are acquired along with the general rules which govern these constructions at step d, and that children do not make errors in these across-the-board, categorially based operations (Baker, 1979; Maratsos & Chalkley, 1980).

A second property of more advanced grammar-learning of the step d kind is what I have elsewhere termed the "typological" thrust (Berman, 1984c). As children accumulate knowledge and experience in their language, they come to recognize the options available and appropriate to that language. For instance, in a relatively free word-order language like Hebrew, Hungarian, or Turkish, children realize that constituents can and should diverge from the canonic, or unmarked, order of simple sentences—a kind of knowledge which is not accessible to the child learning English or French. Similarly, children learning a language like Hebrew will surprisingly early master the intricacies of its agreement system, which marks verbs for number, gender, and person to agree with the subject noun, while adjectives and demonstratives agree with head nouns in number, gender, and definiteness. The reason why they can do this so rapidly is because the system is so central and pervasive to the grammar of their language that it sets up certain expectations for learners, once they have a large enough repertoire of exemplars and rules, of both input and output materials in their language. True, "basic child grammar" may be largely shared and universal in the early phases (Slobin, 1984), but once rule acquisition becomes established as a predominant mechanism of language acquisition, the particular grammar of the mother tongue will assert its sway over the learner. Take, for instance, Hebrew-learning children's gradual acquisition of the abstract notion of a consonantal root. It is feasible that to start with, all children attend to the consonantal elements occurring in the stream of speech in much the same way. As they move into the grammar of their native tongue, however, children learning a Semitic language will start to pay special attention to consonants, as root elements which carry the semantic core of words and which have important structural consequences for the forms of words. By this point, learners manifest a structure-dependent, rule-governed knowledge which is delicately attuned to the kind of language they have learned. Such knowledge is shared by all learners of a given language and marks the onset of important differences in the knowledge of speakers of language X compared with language Y.[7]

Acquisition of the rules of grammar is thus a very general, highly abstract kind of learning which is not dependent on particular items or elements, nor is it bound to usage in particular circumstances. Step d in this model thus relates to grammar at its most autonomous, as freed from the communicative factors which motivate pregrammar. But this very autonomy also means that it is insufficiently constrained by the demands of convention and by considerations of usage factors that will impinge on the grammar at step e.

Usage and Convention. This final step highlights the integrative nature of our model because it interlocks in rather special ways with the earlier, pregrammatical phase. At this point, learning is no longer primarily a matter of abstraction and generalization, as in steps c and d. Rather, attention again focuses on specific items, on the one hand, and on the context of use, on the other—and in this we see its link to steps a and b. Here, learning is again largely ''item-based,'' (MacWhinney, 1984), and it is once more dependent on behavioral as well as structural considerations. But now, learning is anchored in a large body of prior knowledge, of items AND of rules. For example, in the acquisition of lexical classes like agent and instrument nouns, knowledge of the semantics of such nouns and of the formal devices for coining new ones is rule-bound. But the learner must also discover that in English *cook* refers to a person, *cooker* to a thing, while *drill* is the name of a thing, *driller,* the name of a person. And a Hebrew-speaker learns to refer to male and female agent nouns through the grammar, e.g., *pasal–pasélet* 'sculptor/sculptress', *shofet* 'judge' *shofétet,* 'woman judge'; yet the femine noun *ganénet* is 'nursery school teacher', and is not used as the feminine form of the related *ganan* 'gardener'. That is, what the learner must know is what the CONVENTIONAL associations are for using a given device in expressing a given meaning in the well-established, current lexicon (Clark & Berman, 1984). Similarly, children must acquire a generalized repertoire of word-formation devices in the form of verb patterns used to encode such notions as causativity or inchoative change-of-state in Hebrew. But they also need an extensive stock of actual exemplars specifying for them which verbs form the inchoative in which *binyan* pattern (compare P2 *nexlash* 'weaken = grow weak,' P4 *hitayef* 'tire = grow tired,' P5

---

[7] This is not to deny the existence of a UNIVERSAL GRAMMAR. On the contrary, crosslinguistic constraints of form and substance are taken to underlie and hence be presupposed by children's ability to learn the grammar of a PARTICULAR language.

hivshil 'ripen = turn ripe'), and which verbs or adjectives form their inchoative by syntactic means of auxiliaries, e.g., *nihya tipshi* 'become stupid'. In like fashion, English-learning children must learn that the causative can be expressed by a special, suppletive verb—e.g., *feed the doll, teach the doll to swim;* by the same verb used in a transitive construction—e.g., *move the doll, dress the doll;* or by a syntactic use of auxiliaries—e.g., *make the doll sing, put the doll to sleep.*

Another type of item-sensitive, post-rule learning is the case of lexical exceptions. Thus, the child will first acquire an entire set of rules for number and gender agreement in Hebrew, before learning which specific items in the language happen to violate these generalizations, as with irregular past-tense verbs or noun plurals like *children* and *teeth* in English. Many of these irregular items are common in early child language, and so may be rote-learned prior to rule acquisition. Others may be inappropriately regularized to give deviant forms at the stage of rule learning to yield a U-shaped learning curve (Bowerman, 1982a). Still others will be learned, once again BY ROTE, after the system qua system has been internalized. Thus, an interim strategy adopted by many learners in acquiring oblique pronouns in Hebrew is to regularize the same ending *-ex* 'you + Feminine' to all prepositions. Only later will the choice of ending be adjusted to the specific item, to yield: *mim-ex* 'from you' but also *it-ax* 'with-you' and *al-áyix* 'on-you', respectively.

We noted earlier that if the interim strategies of step c persist across time and across children, language CHANGE may be indicated. In fact, it is at step e that more general change can most reliably be diagnosed. For if children continue to regularize across exceptions, if they fail to take account of conventions of usage and norms of the adult speech community at this point, when the rule-bound grammatical facts are already known, then change would seem particularly likely. And there is evidence that it is the usage of older children, who already "have" the grammar, so to speak, that will determine the course of modifications in the language over time (as shown for English past-tense forms in Bybee & Slobin, 1982).

Step e learning is not only item-based but also context-sensitive, relating to language USE rather than to language structure. One manifestation is recognition of which devices are favored for a particular function in the context of discourse. For instance, in word formation, colloquial, everyday usage is as important as are structural considerations in determining the options which are preferred for coining new words (Clark & Berman, 1984). More generally, in talking about events, the speaker can select different grammatical structures to depict the selfsame situation from a particular perspective or point of view. For instance, one can use either active or passive voice and numerous other devices for focusing on one part of an utterance or in order to topicalize one argument rather than another (Bates & MacWhinney, 1982). Thus, Hebrew-learning children will acquire numerous structural devices which function for downgrading an agent so as to focus on the patient or some other participant in the event. But they need to learn more than simply how to form the relevant constructions. Once they know a given formal structure, it is at step e—when attention can be given to DISCOURSE requirements—that children will learn how to make use of and how to vary these devices effectively for communicative purposes (Karmiloff-Smith 1981, 1983). Use is then made of the rules of grammar in ways which accord with what is preferred in actual language use and with different discourse styles or registers of usage. It is this kind of learning which enables speakers to take account of the situation in which language is being used—including the status and relationship of participants in the speech event, shared knowledge, stylistic appropriateness, and so forth.

Step e knowledge also interacts in special ways with formal language STUDIES. Thus, children's learning of orthography has an impact on how root consonants are construed in Hebrew (Bentur, 1978). Formal study of school grammar and general level of education affect deployment of rules for Hebrew morphophonemics as well as in other parts of the grammar (Schwarzwald, 1981). And literacy gives rise to the diglossia associated with the differences between casual, colloquial spoken usage and more formal, expository styles of expression. The knowledge added at step e is thus very different from grammar acquisition. The rules of grammar characteristic of step d are shared across speakers of a given language, by adults and children, by educated and less-educated speakers, by highly facile, verbal types and those who lack powers of self-expression. They represent the general, abstract, structure-based *faculté de language* for which a Chomskian "language acquisition device" aims to account. Yet learners also need to discover how these rules are deployed in language use, in the conventional, well-established repertoire of forms and words, and in the alternation of different structures according to speaker perspective and to discourse needs. It is such knowledge that turns learners from competent language knowers into proficient language users.

This points to another feature of this endpoint step: Here, too, as was formerly noted for step c, INDIVIDUAL DIFFERENCES play an important role. It is at this point that, rather than the structure-based factors which determine the course of acquisition at steps c and d, LEARNER-BOUND variables emerge as critical. I refer here to such factors as general world experience; exposure to different registers and styles of usage; amount of formal school-study; size and range of vocabulary; and social and affective elements relevant to the individual personality, powers of expression, and personal verbal skills. It is thus step e that accounts for the fact that while all learners of a language acquire basically the same grammar, no two speakers use it quite the same.

Discussion

The model earlier presented is very different indeed from a view of developmental stages as "discrete time slices" (Marshall, 1980). Nonetheless, its perspective is that of developmental psycholinguistics rather than of the field of inquiry that has come to be termed "language acquisition" (Keil, 1981; Wexler, 1982). The latter focuses on what is termed "the logical problem of language acquisition," that is, on formal characterizations of the END-STATE grammar which must meet the requirement of learnability. Such an orientation often relies on an idealization of acquisition as instantaneous. Here, instead, we have tried to characterize the steps taken by children en route from initial to final knowledge of their language. Underlying this model is the claim that the child does not proceed from nonanalysis to full-blown knowledge, but that there may be several interim levels of analysis on the way. The knowledge of the language-acquiring child is thus not construed as a matter of "all or nothing." Rather, different kinds and levels of PARTIAL knowledge play a role en route to the adult language.

An important facet of this partial knowledge is that some types of knowledge may be fully organized and certain rules of the grammar fully established, well before others. Thus, for Hebrew acquisition, the inflectional systems of the grammar (e.g., plurals and verb tense) are well established before children acquire command of derivational word-formation in their language. Learners acquire subject–verb agreement in Hebrew before noun–adjective agreement, for instance (Kaplan, 1983). And across languages, they gain command of the syntax of simple clauses before mastering coordination or embedding. This staggering of language-development may derive from different sources in each case. For example, productive rules of grammar such as inflectional morphology which applies across the lexicon may in general be established earlier than less rule-based, more convention-bound facts of lexical structure. Children use subject–verb strings extensively before age 3, at a time when they have only a small repertoire of adjectives to which agreement rules might apply. And there is a logically cumulative relationship between simple clauses on the one hand, and their co-occurrence in conjoined or subordinated structures, respectively. Yet these distinct examples all demonstrate that the steps outlined in our model may be taken by children across different developmental stages, at different ages, and at various points in the acquisition of their mother tongue. The only exception would seem to be the "one-word stage" which in the terms outlined here is essentially pregrammatical—perhaps, in fact, the only true "stage" in language development (see Dromi, this volume). Apart from this, the five-step model presented here will apply and reapply to different kinds of material and domains of language learning, from the time of the child's early word-combinations and on through to late childhood (see Karmiloff-Smith, this volume).

A second general point is that the behaviors noted for each of these steps are not in principle exclusively juvenile or immature. They remain part of the adult repertoire, but as knowledge becomes restructured and less partial, there is increasingly more reliance on procedures associated with steps d and e, less on steps a through c. Yet rote memorization also continues to play a role in our linguistic knowledge as adults. Thus, idioms and other formulaic kinds of expressions are known and used by adult speakers as prepackaged chunks, often with no possible analysis into parts (consider a familiar expressions such as *hoist by one's own petard* in English). Similarly, along the lines of our step b, familiarity with set forms and expressions seems to play an important role in the application of certain grammatical rules which may no longer apply productively across the current end-state grammar at all. For instance, Schwarzwald (1981) found that Hebrew-speaking adults as well as school children performed better—in the sense of observing the prescriptive, school-book grammar norms—on familiar verbs than with rare or highly literary words. And my research on plural formation in Hebrew nouns showed that adults as well as younger speakers made the normative stem changes for just those nouns which are known to them from set, formulaic contexts by contrast with other nouns belonging to the same form–class (Berman, 1981c). Even the kinds of interim strategies we noted as step c do occur in adult usage. This is shown by the misconstruals of individual speakers in producing or interpreting utterances; by nonnormative usages characteristic of a speaker's idiolect or a few isolated items in a person's repertoire; and by immature syntactic strategies (e.g., violations of agreement constraints) to which speakers may resort in the course of talking, under the exigencies of rapid ongoing processing. Besides, there are indications that certain interim schemata and nonnormative generalizations occurring in the speech of middle-class children are also found among speakers of less standard dialects or less-educated background (Ravid, in progress; Schwarzwald, 1981).

The GENERALITY of the models I propose, as of any such model, can be queried along several dimensions—for instance, whether it applies across children, across languages, and across domains. With respect to the first issue, a more carefully controlled and comparable data-base than was available to me might be needed to meet rigorous demands of generality. I have, on the one hand, indicated how this model can accommodate individual differences in language acquisition—most particularly through the application of different interim procedures to different areas at step c and again at step e in the way speakers deploy their endstate grammatical knowledge. Yet I have, on the other hand, suggested that there is a shared core of purely linguistic, grammar-based knowledge which is common to all (and only) children learning a natural language. Elsewhere I have argued in some detail that this builds upon a prior foundation of what Slobin (in press) terms "basic child grammar" and that it emerges out of a shared, crosslinguistic phase of pregrammatical entry into language use (Berman, 1984b). It is along these lines, then, that this model incorporates linguistic universals as the core or base-frame within which children come to focus on the grammar of their own specific language, increasingly from step c onward.

As for the generality of this model across linguistic subdomains, as was noted in the introduction, my focus has been largely on morpholexical facets of language structure. Yet the subsystems described in the section on "Development of Selected Subsystems of Hebrew" in themselves involve the manipulation of a wide range of different types of linguistic materials. Moreover, current research indicates that these steps can be shown to account for the acquisition of syntactic categories such as word order and the distinction between form–classes such as nouns, verbs, and adjectives (as reviewed in Berman, 1984a, 1984b). And the work of Karmiloff-Smith (1979, 1981, 1983) shows highly analogous developmental phases to be relevant to the acquisition of anaphoric reference, on the one hand, and of narrative discourse structure, on the other. I cannot here address the question of whether this model is further generalizable beyond the domain of LANGUAGE learning. Yet I see no a priori reason why similar steps and phases should not apply to the development of other systems of knowledge as well.

The overall scheme I have argued for thus views learning as a continuous process of integration and reintegration. Initially, at steps a and b, communicatively anchored, extralinguistic or pragmatic factors are critical for pregrammatical entry into the system, and language learning is embedded in general patterns of behavior. As grammar becomes more autonomous, and language development more structure-bound, release from reliance on context yields a pure kind of "grammaticization" of knowledge, which may be too rigidly rule-bound with insufficient attention to nonstructural constraints. Finally, the rules of grammar are reintegrated with communicative motivations, in such a way as to take account of discourse conventions, conversational appropriateness, stylistic variation, and the knowledge and needs of interlocutors are considered along with those of the speaker. Thus, the commonalities of the first and final phases as both "communication-oriented" are clearly of only a superficial nature. Initially, verbal communication is closely bound up with the nonverbal behavior of the child in interaction with people and objects in specific situational contexts. End-state communication can be linguistically quite autonomous and minimally interactional—through the audio medium of the radio or in the visual channel of books and newspapers.

To conclude, consider the status of each of the five steps in interrelation with the others. It is not necessarily the case that all five steps have equal status across all of language development or with respect to all types of linguistic subject matter. True, each step presupposes the preceding one and leads into the next. Yet steps b and c seem less autonomous, and they may be less preconditional than either the initial entry of step a or the acquisition of grammar proper in step d. Thus, acquaintance with and manipulation of familiar exemplars in (b) may be viewed as merely an extension of step a, whereas initial attempts at rule formation in step c are clearly only bridges to the grammar. The view here is, rather, as noted at the outset, that these steps each play a role in the three major phases we envisage en route from the one-word stage to mature language knowledge and language use.[8] The phases can be schematically represented as follows:

Pregrammar Grammaticization End-state Language

Steps a........> b.......> c........> d..............e

This scheme presents the steps taken en route to language knowledge as nondichotomous and nondiscrete. One leads into another in the form of a CONTINUUM, with much overlap within and across language domains. As noted earlier, a child may be at step c with regard, say, to use of word-order devices but at step d or beyond in the use of inflectional morphology. And children may continue to rely partially on rote-learned items and familiar sequences even when they can already apply step c- or d-type rules to other instances of the same type. Nor does the learner follow a strictly linear path in progressing along this continuum from nonanalysis to productive rule-application and then on to further constraints or elaborations.

[8] It might even make sense to continue the dots in the chart BEYOND step e, which we depict there as an "end-state," because in time both language knowledge and language use may deteriorate with age.

Rather, there is a constant reintegration and reembedding of prior procedures and prior information into more mature and more general knowledge as well as use of the learner's linguistic repertoire.

References

Badry, F. (1983). *Acquisition of lexical derivation rules in Moroccan Arabic*. Unpublished doctoral dissertation, University of California, Berkeley.

Baker, C. L. (1979). Syntactic theory and the projection problem. *Linguistic Inquiry, 10,* 533–581.

Bates, E., & MacWhinney, B. (1982). Functionalist approaches to grammar. In E. Wanner and L. Gleitman (Eds.), *Language acquisition: The state of the art*. Cambridge, England: Cambridge University Press.

Bentur, E. (1978). *Some effects of orthography on the linguistic knowledge of Hebrew speakers*. Unpublished doctoral dissertation, University of Illinois, Urbana.

Berman, R. A. (1978a). *Modern Hebrew structure*. Tel Aviv: Universities Publishing.

Berman, R. A. (1978b). Early verbs: How and why a child uses her first words. *International Journal of Psycholinguistics 5,* 21–39.

Berman, R. A. (1980). Child language as evidence for grammatical description: Preschoolers' construal of transitivity in Hebrew verbs. *Linguistics, 18,* 677–701.

Berman, R. A. (1981a). Regularity versus anomaly: The acquisition of Hebrew inflectional morphology. *Journal of Child Language, 8,* 265–282.

Berman, R. A. (1981b). Language development and language change: Evidence from the acquisition of Hebrew morphophonology. *Journal of Child Language, 8,* 609–626.

Berman, R. A. (1981c). Children's regularizations of noun patterns. *Papers and Reports on Child Language Development, 20,* 34–44.

Berman, R. A. (1982). Verb-pattern alternation: The interace of morphology, syntax, and semantics in Hebrew child language. *Journal of Child Language, 9,* 169–191.

Berman, R. A. (1983). Establishing a schema: Children's construal of verb-tense marking. *Language Sciences, 5,* 61–78.

Berman, R. A. (1984a). On the study of first language acquisition. In A. Guiora (Ed.), *Language learning*. Detroit, MI: Wayne State University Press.

Berman, R. A. (1984b). A crosslinguistic perspective: morphology/syntax. In P. Fletcher & M . Garman (Eds.), *Language acquisition* (2nd ed.). Cambridge, England: Cambridge University Press.

Berman, R. A. (in press a). Acquisition of Hebrew. In D. I. Slobin (Ed.), *Crosslinguistic study of language acquisition*. Hillsdale, N.J.: Erlbaum.

Berman, R. A. (in press b). Cognitive components of language development. In C. Pfaff (Ed.). *Crosslinguistic studies of language acquisition processes*. Rowley, MA: Newbury House.

Berman, R. A., & Dromi, E. (1984). On marking time without aspect in child language. *Papers and Reports on Child Language Development, 23,* 23–32.

Berman, R. A., & Y. Sagi. (1981). Word-formation processes and lexical innovations in childhood. *Hebrew Computational Linguistics, 18,* 31–62 (in Hebrew).

Bin-Nun, Y. (1983). *On an interim phase in the acquisition of inflected pronouns: Data from three- to four-year-old Hebrew usage*. Unpublished manuscript, Tel Aviv University.

Borochovsky, E. (1984). On the phenomenon of imitation in early speech. *Hebrew Computational Linguistics, 22* (in Hebrew).

Bowerman, M. (1980). The structure and origin of semantic categories in the language-learning child. In M. Foster & S. Brandes (Eds.), *Symbol and sense*. New York: Wiley.

Bowerman, M. (1982a). Starting to talk worse: Clues to language acquisition from children's late speech errors. In S. Strauss (Ed.), *U-shaped behavioral growth*. New York: Academic Press.

Bowerman, M. (1982b). Reorganizational processes in lexical and syntactic development. In E. Wanner & L. Gleitman (Eds.), *Language acquisition: The state of the art*. Cambridge, England: Cambridge University Press.

Braine, M. D. S. (1976). Children's first word combinations. *Monographs of the Society for Research in Child Development,* 41 (Serial No. 164).

Brown, R. W. (1973). *A first language: The early stages.* Cambridge, MA: Harvard University Press.

Bybee, J., & Slobin, D. I. (1982). Rules and schemas in the development of the English past tense. *Language, 58,* 265–289.

Chomsky, N. (1965). *Aspects of the theory of syntax.* Cambridge, MA: MIT Press.

Chomsky, N. (1975). *Reflections on language.* New York: Panther.

Clark, E. V. (1980). Convention and innovation in acquisition of the lexicon. *Papers and Reports on Child Language Development, 19,* 1–20.

Clark, E. V. (1981). Lexical innovations: How children learn to create new words. In W. Deutsch (Ed.), *The child's construction of language.* New York: Academic Press.

Clark, E. V. (in press). Acquisition of Romance, with special reference to French. In D. I. Slobin (Ed.), *Crosslinguistic study of child language.* Hillsdale, N.J.: Erlbaum.

Clark, E. V., & Berman, R. A. (1984). Structure and use in the acquisition of word formation. *Language, 60,* 542–590.

Clark, R. (1974). Performing without competence. *Journal of Child Language, 1,* 1–10.

Clark, R. (1978). Some even simpler ways to learn to talk. In N. Waterson & C. Snow (Eds.), *The development of communication.* New York: Wiley.

Cromer, R. (1976). Developmental strategies for languages. In V. Hamilton & M. Vernon (Eds.), *The development of cognitive processing.* London: Academic Press.

de Villiers, J. (1980). The process of rule-learning in child speech: A new look. In K. Nelson (Ed.), *Children's language (Volume 2).* New York: Gardner Press.

Dromi, E. (1979). More on the acquisition of locative preposition: An analysis of Hebrew data. *Journal of Child Language, 6,* 547–562.

Dromi, E. (1982). *In pursuit of meaningful words: A case-study analysis of early lexical development.* Unpublished doctoral dissertation, University of Kansas, Lawrence.

Dromi, E., & Berman, R. A. (1982). A morphemic measure of early language development: Data from Modern Hebrew. *Journal of Child Language, 9,* 403–424.

Fillmore, C. J. (1975). An alternative to checklist theories of meaning. *Berkeley Linguistic Society, 1,* 123–131.

Gleitman, L. R., & Wanner, E. (1982). Language acquisition: The state of the state of the art. In Wanner & Gleitman (Eds.), *Language acquisition: The state of the art.* Cambridge, England: Cambridge University Press.

Kaplan, D. (1983). *Order of acquisition of morpho-syntactic categories by Hebrew-speaking children aged 1;9 to 3;6 years.* Master's thesis, Tel Aviv University (in Hebrew).

Karmiloff-Smith, A. (1979). *A functional approach to child language.* Cambridge, England: Cambridge University Press.

Karmiloff-Smith, A. (1981). The grammatical marking of thematic structure in development of language production. In W. Deutsch (Ed.), *The child's construction of language.* New York: Academic Press.

Karmiloff-Smith, A. (1983). Language development as a problem-solving process. *Papers and Reports on Child Language Development, 22,* 1–22.

Keil, F. C. (1981). Constraints on knowledge and cognitive development. *Psychological Review, 88,* 197–227.

Levy, Y. (1983). The acquisition of Hebrew plurals: The case of the missing gender category. *Journal of Child Language, 10,* 107–121.

MacWhinney, B. (1978). The acquisition of morphology. *Monographs of the Society for Research in Child Development, 41,* 1–2.

MacWhinney, B. (1981). Basic syntactic processes. In S. A. Kuczaj (Ed.), *Language, thought, and development, Vol. 1: Syntax and semantics.* Hillsdale, N.J.: Erlbaum.

MacWhinney, F. (1984). Hungarian language acquisition as explempifying a general model of grammatical development. In D. I. Slobin (Ed.), *Crosslinguistic study of child language.* Hillsdale, N.J.: Erlbaum.

Maratsos, M. (1983). Some current issues in the study of the acquisition of grammar. In P. M. Mussen (Ed.), *Carmichael's manual of child psychology* (4th ed.). New York: Wiley.

Maratsos, M., & Chalkley, M. (1980). The internal language of children's syntax. In K. Nelson (Ed.), *Children's Language,* (Vol. 2.) New York: Gardner Press.

Marshall, J. (1980). On the biology of language acquisition. In D. Caplan (Ed.), *Biological studies of mental processes.* Cambridge, MA: MIT Press.

Peters, A. (1983). *The units of language acquisition.* Cambridge, England: Cambridge University Press.

Postol, S. (1982). *Language at the single-word stage: A longitudinal study of a child acquiring Hebrew.* Unpublished manuscript, Tel Aviv University.

Ravid, D. (in progress). *Transient and fossilized phenomena in the usage of different groups of Hebrew speakers.* Doctoral dissertation, Tel Aviv University.

Rom, A., & Dgani, R. (in press). Acquiring case-marked pronouns in Hebrew: Semantic and morphophonological factors. *Journal of Child Language.*

Rosch, E. (1977). Human categorization. In N. Warren (Ed.), *Advances in cross-cultural psychology.* London: Academic Press.

Rosch, E., & Mervis, C. (1975). Family resemblances: Studies in the internal structure of categories. *Cognitive Psychology, 7,* 573–605.

Savasir, I., & Gee, J. (1982). The functional equivalents of middle-voice in child language. *Berkeley Linguistic Society, 8,* 607–616.

Schwarzwald, O. (1981). *Grammar and reality in the Hebrew verb.* Ramat-Gan, Israel: Bar-Ilan University. (in Hebrew)

Shatz, M. (1982). On mechanisms of language acquisition. In E. Wanner & L. Gleitman (Eds.), *Language acquisition: The state of the art.* Cambridge, England: Cambridge University Press.

Slobin, D. I. (1973). Cognitive prequisites for the acquisition of grammar. In C. Ferguson & D. Slobin (Eds.), *Studies of child language development.* New York: Holt, Rinehart, Winston.

Slobin, D. I. (1981). Psychology without linguistics = linguistics without grammar. *Cognition, 10,* 27–35.

Slobin, D. I. (1982). Universal and particular in the acquisition of language. In E. Wanner & L. R. Gleitman (Eds.), *Language acquisition: The state of the art.* Cambridge, England: Cambridge University Press.

Slobin, D. I. (in press). Crosslinguistic evidence for the language-making capacity. In D. I. Slobin (Ed.), *Crosslinguistic study of child language.* Hillsdale, N.J.: Erlbaum.

Walden, Z. (1982). *The root of roots: Children's construction of word-formation processes in Hebrew.* Unpublished doctoral dissertation, Harvard University.

Wexler, K. (1982). A principle theory for language acquisition. In E. Wanner & L. R. Gleitman (Eds.), *Language acquisition: The state of the art.* Cambridge, England: Cambridge University Press.

Wong-Fillmore, L. (1979). Individual differences in second-language acquisition. In C. J. Fillmore, D. Kempler, & W. S.-Y. Wang (Eds.), *Individual differences in language ability and language behavior.* New York: Academic Press.

9

The One-Word Period as a Stage in Language Development: Quantitative and Qualitative Accounts*

Esther Dromi
School of Education
Tel Aviv University

Introduction

When Keren, my daughter and the subject of the present investigation, was age 24(14)[1] she spent an afternoon visiting her grandmother. During the visit, Keren was introduced to Sharon, who was about her age and whom she had not met before. The two girls played together. After returning home I recorded the following conversation between Keren and myself:

M: *kereni, sixakt im sharon?*
kereni play with Sharon
[present, 2nd person / feminine, singular]
'kereni, did you play with Sharon?'

K: *at yoda'at, sharon tinoket*
you know Sharon baby
[present, 2nd person / feminine, singular]
'(Do) you know, Sharon (is a) baby!'

*This chapter is based on a reanalysis of the data collected for my doctoral dissertation which was submitted to the University of Kansas, 1982. I would like to thank Melissa Bowerman, my adviser, for her help throughout my dissertation project. Thanks are also extended to the participants of the 1983 Tel Aviv University Annual Workshop on Human Development who encouraged me to reconsider some earlier claims. Especially I would like to acknowledge the invaluable reviews and discussions of Annette Karmiloff-Smith, Liliana Tolchinski-Landsman, and I. M. Schlesinger. I express my deepest gratitude to Ruth Berman, Eve Clark, Iris Levin, and Sidney Strauss for their very careful and helpful comments on earlier drafts of this chapter. This is Working Paper no. 30 of the Tel Aviv University Study Group on Human Development and Education.

[1] In this chapter, children's ages are given in months and days; hence 24(14) is to be read as 24 months and 14 days. All Hebrew examples appear in italics and are followed by English gloss. Numbers in parentheses following an example word indicate the number of the word in the subject's cumulative lexicon. Thus (023) means that this word was the 23rd word acquired by the subject.

hi *lo* *medaberet.*
she no talk
[present, feminine
singular]
'She (can) not talk,'

hi *omeret* *rak* *milim.*
she say only words
[present, feminine
singular]
'She says only words.'

This chapter reports on my detailed investigation of the extended period of time during which Keren produced only "one word at a time" (Bloom, 1973).

The term *the one-word stage* is commonly used to label the quite lengthy period when the child's linguistic productions are strongly constrained to only single words (e.g., Bloom, 1973; Greenfield & Smith, 1976; McNeill, 1970; McShane, 1980; Rodgon, 1976; Scollon, 1976, 1979). Given the purposes of this book, I consider here a set of criteria which seems to me essential for determining whether a given slice in a sequence of behaviors within a specific domain does, in fact, constitute a developmental stage. Following Wohlwill (1973) and other theorists (e.g., Flavell, 1971, 1982; Karmiloff-Smith, this volume), I criticize the readiness of writers to loosely apply the term *stage* to any observable change in child's behavior. In turn, I suggest that the term *stage* be reserved to only those constructs within a structurally defined systems (e.g., the linguistic system), which show a certain amount of underlying uniformity and that have identifyable boundaries.

The mere emergence of a new behavior, it seems to me, cannot in itself be taken as a single proof that the child has moved from one developmental stage to the next. Thus, the observation that at some point the child begins to utter multiword linguistic constructions does not conclusively imply that a radical qualitative change in the underlying organization of the child's linguistic knowledge had occurred.

A much stronger evidence than the identification of overt addition of components is required before one can conclude that the one-word period is a stage in language development. Such a claim should be based on findings indicating that during the one-word period the child acquires a unique characteristic of the language that was missing before, and that it will be hierarchically integrated into the developing system in later stages. A qualitative change in the underlying organization of the linguistic system must be detected before we can conclude that the move from one-word productions to early word combinations reflects the emergence of a new developmental stage. To state it more generally, it is claimed here that even in a specific domain (just as across different domains) only clear changes in the underlying characterization of the behavior studied justify the application of the stage concept.

In the present chapter I describe a longitudinal detailed study of a developing linguistic system of a single child. In this study I carefully recorded changes over time in the child's lexical behavior during the one-word period. This period spanned for my subject 7 months and 11 days. Throughout this time the child has acquired 337 different Hebrew words which she used repeatedly in various contexts. Changes in several aspects of using these words are examined here and compared, in an attempt to reveal some underlying characteristics of the system.

For the purpose of data collection I employed a diary-study methodology which incorporated three independent sources of evidence on each lexical item that was acquired by the child, and subsequently used by her (see Dromi, 1982, for a detailed description of methods, and also Braunwald and Brislin, 1979). The complete data base consisted of 337 different lexical items (i.e., Hebrew words) and a number of entries on each word depicting all the changes that were noted for using it over time. The analysis of such a rich data base helped me to reach the conclusion that the one-word period is indeed a stage in language development, a stage which can be divided into two distinct phases: a preparatory phase and a second phase of consolidation.

In the first section, I discuss the utility of the concept of stages for the specific domain studied—language acquisition. In this section I review earlier claims about the status of the one-word period in relation to subsequent stages of production. Several estimates on quantitative measures—such as the total number of words learned, the length of period, and the rate of early lexical growth, which are based on a number of different studies—are given in the second section. In the third section, I briefly describe the procedures of data collection and data analyses applied in the present investigation. The fourth section contains the findings from my own study and are organized according to the following topics: (a) the measurement of vocabulary growth; (b) developments in phonology and morphology; (c) the development of word reference; and (d) the development of word meaning. In the fifth and final section the main findings are integrated and are invoked for supporting the claim that the one-word period is a stage in language development because it meets such criteria as novelty, qualitative change, and distinct boundaries. It is shown that during the one-word stage the child is preoccupied with the single task of learning how to correctly map words onto their conventional range of meaning.

Is the One-Word Period a Stage in Linguistic Development?

The utilization of the concept of stage raises an immediate association with Piaget's theory. In this theory, stages are suggested as a heuristic for seeking links across widely differing domains: spatial, social, physical, linguistic, conceptual, etc. In the Piagetian theory the question of whether a sequence of behaviors within a single domain manifests a stage-like development is not of direct interest. In a number of chapters in this volume (e.g., those by Keil and by Case, Marini, McKeough, Dennis, & Goldberg), important issues and as yet unresolved problems about the meaning of a stage model and its usefulness for a general cognitive developmental theory are raised and thoroughly discussed (see also Brainerd, 1976, 1978; Flavell,

1971; Wohlwill, 1973). The question of whether the notion of a stage is functional for the study of a sequence of attainments in a particular domain is central to the present chapter. A proposal is made that the stage concept is useful, especially when one studies the development of a highly complex system (e.g., language) which is hierarchically organized. The term, however, should be reserved for only specific units which clearly meet a set of criteria for being considered a stage in development.

In order for a slice of development to be considered a stage in a particular domain, it is not enough to define its starting and ending points. It is necessary to make explicit which are the qualitative features that differentiate it from the preceding and subsequent slices in that sequence. A stage must usually span for a recognizable stretch of time (see Karmiloff-Smith, this volume, for the distinction between micro and macro developments), and the emergent behaviors during a given stage should constitute a novelty. Some underlying uniformity must be found in a number of events that superficially may appear heterogeneous, before one can conclude that these behaviors evolve from the same underlying process, and hence belong to the same stage. Finally, a developmental stage in a specific domain usually constitute a necessary condition for the transition towards the subsequent stages.

A survey of contemporary studies on the one-word period reveals that very little attention has been paid to the question of whether the one-word period is a developmental construct that meets some of the suggested criteria for being termed a stage in language development. Some relevant, though indirect claims regarding the status of this period in relation to the subsequent period of word combining have, however, been made.

McNeill (1970) who is a proponent of an extreme nativist view, applies the term *holophrastic stage* for the early period of production during which children use single-word utterances. The term *holophrase* refers to single-word utterances which correspond to complete adult sentences (Ingram, 1971; Rodgon, 1976). According to McNeill (1970) (and other nativists; e.g., Fodor, 1975, 1980), children are born endowed with the underlying representations of the universal grammatical rules of languages. Therefore, an underlying representation of a fully grammatical sentence is present in the child's mind even when, in fact, the child produces only a single-word utterance. The kind of limitations hypothesized by McNeill to account for holophrastic speech are reduced short-term memory, low attention span, and severe programming deficiencies. Such constraints are gradually removed through maturational processes, as reflected in the child's growing ability to construct longer utterances, and finally, grammatical sentences. McNeill makes a strong claim that only quantitative differences exist between holophrastic speech and later speech, because he considers it impossible for the child to develop knowledge of syntax on the basis of previous experiences with single words. Since McNeill claimed that there exists no qualitative differences in the child's underlying characterization of his/her linguistic knowledge during the different phases of production (i.e., the one-word period, the period of early syntax, the period of late syntax), it seems to me that he had not attributed to the holophrastic speech the status of a real developmental stage.

Bloom (1973) was the first to dispute McNeill's arguments. She claimed that the reason one-word–stage children do not produce sentences is that they do not know the syntactic rules of their language. According to Bloom, single-word utterances are not *holophrases;* rather they are simple linguistic entities that the child produces while focusing on single aspects of situations. During the early phases of the one-word period, children are unable to deal with more than one aspect of a situation at a time, and they therefore produce single words one at a time. Toward the end of the one-word period, children start to differentiate among different aspects of the same situation, and they also refer to them verbally. However, they still do not know syntax, and as a consequence, they now utter sequences of single words. Bloom has argued that there is a qualitative difference between the one-word stage and later stages of linguistic development. According to her, the difference lies in the underlying knowledge of syntactic rules which is present only when the child produces multiword utterances.

The two contradictory opinions regarding the relationship between the one-word period and the early period of word combining generated a controversy among researchers as to whether *discontinuity or continuity* exists between the one-word period and subsequent periods in language acquisition. Bloom's (1973) proposal that there is discontinuity between the two has not gained much support and was objected by Greenfield and Smith (1976), and Scollon (1976) on different grounds.

Greenfield and Smith's investigation of the different ways children use the same single-word utterances throughout the one-word period led them to the conclusion that, although this period seems to be static in syntactic form, it is remarkably dynamic in terms of an underlying growth in semantic content. They do not assume that the concept of a sentence is present in the mind of children who produce only one-word utterances. Rather, they suggest that in the course of the one-word period, children develop their ability to convey complex messages by means of combining words with nonlinguistic elements. Since they found that the semantic relations expressed by single words (e.g., agent, action, location, possession, etc.) are also conveyed by later speech, they argued that: "Between early language and matured language. . . the basic organization may be the same" (Greenfield & Smith, 1976, p. 223).

Scollon (1976, 1979) highlights still other aspects of continuity between the one-word period and the early phases of multiword speech. He identified in the speech of his subject *vertical constructions* and argued that these sequences of utterances were constructed according to a topic comment organization to convey a single message. Since such constructions were present in speech samples recorded during both the one-word period *and* the multiword period, he concluded that the discourse structure during the single-word stage and the multiword stage is similar. According to Scollon, during the prolonged single-word period, the child practices the main prerequisites for syntax which are directly transferred to multiword structures.

These arguments indicate that there is no one predominant view about the status of the single-word period in the sequence of linguistic growth. Most researchers argue on different grounds, however, for continuity and gradual transition into the more advanced level of speech. Bloom vigorously objects to this view. One goal of the present investigation is to provide input to the very interesting question of whether the one-word period is indeed a distinct stage in linguistic development, which manifests unique characteristics with respect not only to subsequent stages but also to preceding ones.

Some Developmental Measures of the One-Word Period

Most naive observers who watch the linguistic behaviors of infants will testify that the single-word period is the first step in language development. Around the world, parents eagerly await the day when their infant starts to speak (cf. Dale, 1976; de Villiers & de Villiers, 1978). Most charts that describe the important milestones of early child development note that during the second year of life infants begin to produce their first words (e.g., Nelson, 1973).

The scientific investigation of the one-word period, however, is quite recent. It is only in the last decade that intensive efforts have been focused on the investigation of the linguistic behaviors that precede the emergence of syntax (e.g., Bloom 1973; Bowerman, 1978, 1980; Braunwald, 1978; Carter, 1978, 1979; Ferrier, 1978; Greenfield & Smith, 1976; McShane, 1980; Scollon, 1976, 1979). As recently as only 12 years ago, Brown (1973) called into question the possibility of carrying out a valid investigation on single-word corpora. He argued that single-word utterances are closely tied to their nonlinguistic contexts and that it is therefore extremely difficult to interpret their meanings without overcrediting the child with too much knowledge.

The difficulties involved in interpreting early words and in classifying them into distinct categories are intrinsic to the study of early speech (see Nelson, 1973; Greenfield & Smith, 1976; and Dromi, 1982). At the outset of production, words may be highly idiosyncratic. Sometimes it is even difficult to differentiate a meaningful word from the countless sounds which the child produces in the course of the day (Braunwald, 1978; Veneziano, 1981). A close examination of both the word and its context and a careful and continuous effort at collecting data are required in order to identify significant developmental attainments of the child.

Most modern psycholinguistic studies of the one-word period are observational studies in which parent-investigators anecdotally describe specific aspects of the linguistic growth of their subjects. Therefore, very little is known about the quantitative measures of the one-word period. It is not at all clear *when* children start to produce meaningful words, for *how long* they produce only one word at a time, nor *how many* lexical items children's lexicons contain prior to the first evidence for syntax. From a methodological point of view, it does not seem desirable to collapse the findings of diverse studies in order to arrive at a picture of the quantitative characteristics of the one-word period. Data deriving from a variety of investigations are not comparable because researchers have employed different definitions of what constituted a meaningful word, different sampling procedures, different methods of recording data, and various techniques of data analysis. Despite this limitation, some very general outlines do emerge from existing studies. These outlines are presented and will be used later to test the generality of my exact measures on the quantitative characteristics of the one-word period in one Hebrew-speaking child.

According to Nelson (1973), Bloom (1973), Halliday (1975), and Bowerman (1978), children start to utter meaningful words around their first birthday. Children continue to produce only single-words for a period that ranges between 10 and 12 months (Greenfield & Smith, 1976). Initially, new words enter the child's lexicon slowly (Braunwald, 1978; Fischer & Corrigan, 1981; Nelson, 1973), and at some point the rate of acquiring new words change radically (Bloom, 1973; Corrigan, 1976; Halliday, 1975; McShane, 1980). When exactly the spurt in lexical growth is noted does not appear to be precisely documented. Investigators also differ in their views about the relationship of this spurt to the attainment of specific cognitive or social skills (Bloom, 1973; Corrigan, 1976; Fischer, 1980; Fischer & Corrigan, in press; Menn & Haselkorn, 1976). McShane (1980), for example, has recently argued that this change in rate of acquisition is a direct reflection of the main achievement of the one-word stage, defined as the sudden recognition of the *naming* function of words. According to McShane, very shortly after recognizing this function, the child moves into the multiword stage of constructing sentences.

A goal of the present study is to directly examine the quantitative aspects of the one-word period in one child. An attempt is made here to find out whether clear boundaries of the period can be identified. Other aspects that are examined include the rate of lexical growth during this period as well as the total number of words that are learned by the child before she moves into word combining. The study also traces changes in the form of words (i.e., phonological and morphological modifications) and their extensional properties.

Methods of Data Collection and Data Analysis

My own daughter Keren was the subject of this investigation, and I followed her linguistic behaviors between the ages of 10(12) and 17(28). A rich body of data was collected over the period of 32 weeks that spanned the entire one-word period of the child's production.

Three independent means were employed for collecting natural speech data:

1. A continuous handwritten diary which included extensive descriptions of the phonological forms together with the linguistic and nonlinguistic contexts of production. This record contained detailed information about new words and repeated uses of old words.
2. Nine audio tapes ranging in length from 55 to 180 minutes that were recorded no more than three weeks apart. These samples provided a representative picture of Keren's ordinary speech at specific points in time.
3. Four 30-minute videotaped play sessions of the subject interacting with one of her parents.

The combination of data collection procedures was adopted to compensate for the drawbacks inherent in any one procedure alone. It also provided a means to conduct five different measures of reliability and validity of the data, which were found to be reasonable high.

Interjudge agreement on what constituted a meaningful word for the child was .95. A point-to-point comparison between mother's description of the contexts of each utterance and a trained independent judge's description of the same contexts revealed a high level of agreement, ranging from .80 to 1.00, with a mean of .90. The proportion agreement between two judges on the phonetic transcription of the audio tapes was 98% and 99% for two different recordings, the first and last, respectively. Comparisons conducted between mother's handwritten notes and transcribed audio recordings revealed that 66% of the words that were noted in the diary were recorded at least once in one of the 9 audio tapes. The percentage agreement on the date of recoding a new word in the written diary was .92 (see Dromi, 1982 for details on data collection techniques, transcription, as well as for the statistical procedures of validity and reliability).

In the course of data analysis, each of the subject's productive words was repeatedly scored on a weekly basis on three independent dimensions: (a) morphophonemic changes in the form of the word; (b) the development of the word's reference; and (c) the development of the word's extension. The analysis of word reference was carried out in order to find out whether the child's word was an object word, an action word, a modifier, a social word, or an indeterminant word which named both action and object or modifier and object. The analysis of a word's extension was aimed at describing the range of application of a word compared to its conventional meaning in adult Hebrew. Here a distinction was made among four different categories: underextension, overextension, regular extension, and unclassified.

The procedure of repeated scoring of each word was obligatory in order to allow the measure of developmental processes over time. A one-week interval of time was chosen as the basic unit of measurement, because it appeared to provide an optimal level of description for the dynamic developmental processes noted. The fact that calendar weeks were used is significant, because it provides a clear indication of real-time chronology. Since every change in the child's lexical development was identified in the framework of real time, the processes observed could be evaluated from two perspectives: (a) by looking at all the behaviors recorded during any given week, it was possible to evaluate developments across the entire lexicon; and (b) by looking at one word across time, it was possible to evaluate the developmental history of a given word with reference to the time that had passed since it emerged in the child's speech (i.e., the "age" of the word).

In most cases, judgments on the category membership of a word for any given week were based on several entries. Whenever conflicting evidence was encountered, the word was assigned to the indeterminant or unclassified categories. If during a single week only one entry was recorded for a word, this entry was taken as sufficient evidence upon which to make judgments about reference or extension.

Results and Discussion

Before turning to the presentation of my findings, I wish to reconsider the difficulties in determining what functions as a meaningful word in the growing linguistic system of the child. The need for a workable definition of a meaningful word is obvious because early words very often do not resemble adult words in form or in meaning. In the present investigation a meaningful word was defined as *any sequence of sounds which were consistently produced by the child in repeated similar contexts*. In all cases these sequences were comprehended by an adult listener who interacted with the child and/or by an observer who watched the child or listened to her recordings. There were several criteria which helped in making a decision about the emergence or use of a meaningful word. Some words resembled adult forms because they were phonological approximations of conventional Hebrew words. Very often the fact that the child consistently used the same form in similar contexts (referring to closely related referents) helped to identify a new word. Frequently the child used gestures in order to indicate to the listener what the referent of her word was. Finally, intonation patterns and repetitions indicated that the child's construction was meaningful and that she expected either a verbal acknowledgment or an active response to her word. (On this issue see also Keenan, 1977; Keenan & Klein, 1974; Scollon, 1976; and Veneziano, 1981.)

Keren's first comprehensible word was recorded when she was 10(12). On this particular day I identified her first word *haw* 'bow wow' (001), which she uttered on a walk when we passed a little white dog who was barking at us. Keren was pointing to the dog when she said *haw*. The appearance of Keren's first word *haw* (001) was a novelty. It signified the first event in which the child uttered a specific sound which she, the speaker, and I, the hearer, related to the same real-world referent. Prior to this time, I was never able to match Keren's infinite number of speech sounds with specific extralinguistic referents. Therefore, I argue, that the difference between a child who produces a single meaningful word and a child who produces a number of different speech sounds (i.e., sensorimotor morphemes in Carter's (1978) sense, or babbling in the more traditional interpretation), is not a simple matter of degree.

On the last day of recording, Keren was 17(23). This cut-off point was selected 6 days after I realized that the child was producing two and multiword utterances quite frequently and with great ease.[2] Keren moved from the one-word period to the multiword period very sharply. When she was 17(16) I noted in her diary:

> Today it has become clear to me that Keren no longer faces 'programming-span' limitations. Most of her utterances are two and multiword expressions. It happened quite suddenly.

---

[2] It is important to note that prior to this occasion Keren sometimes produced 2- or 3-word constructions. These constructions, however, were unanalyzed amalgams (MacWhinney, 1978) such as *maze* 'what this' or *abalada* 'daddy went to work'. These expressions were, of course, modeled as such to the child and therefore cannot be considered as productive multiword expressions.

Out of the 39 speech events that were recorded in the diary on that day 17 were productive multiword expressions. On the last day of study, the child produced during 30 minutes of audio-recording 279 utterances, 36% of which were multiword expressions. Some characteristics of Keren's linguistic behaviors throughout the one-word period will be subsequently described. In particular, I present data about: (a) the rate of vocabulary growth; (b) changes in the form of words; (c) account on the development of class–word membership; and (d) description of the development of word meaning.

Measurement of Vocabulary Growth

The subject's complete productive lexicon of single-word utterances included 337 meaningful words (see Dromi, 1982, for a list of these words). In order to determine the rate of the child's lexical development, I calculated the total number of words in

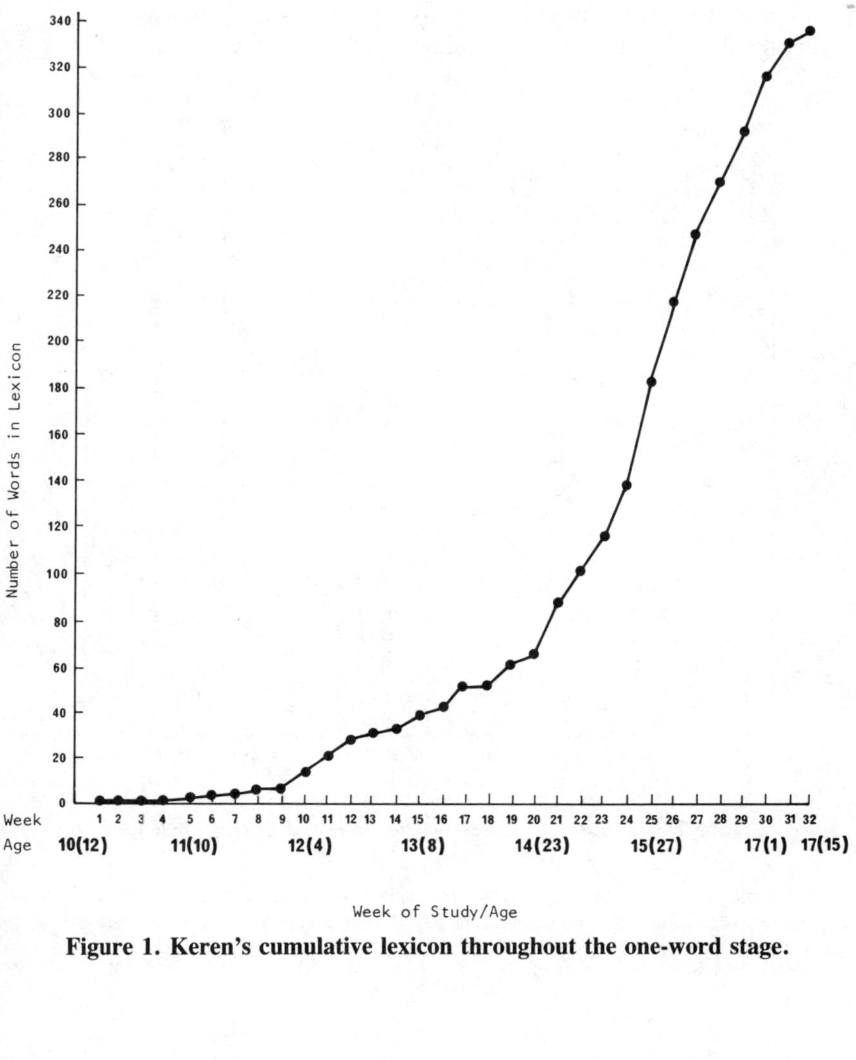

Figure 1. Keren's cumulative lexicon throughout the one-word stage.

the cumulative lexicon (see Fig. 1) and the number of new words acquired during each week (see Fig. 2). The latter is plotted to show the slope of the acquisition curve for new words.

Figures 1 and 2 indicate that in the early phases of production new words entered the child's vocabulary very slowly. The rate increased gradually until it reached an average of about 18 new words per week between the 21st and 24th weeks of study.[3] During the 25th week of study, when Keren's ages were 15(27)–16(3), I observed a striking increase in the number of words that she acquired. A peak figure of 44 new words was recorded for that single week. This was at least twice as many words as were recorded in any previous week and also more than in any subsequent week. The high rate of acquiring new words continued for the following two weeks. During this block of three successive weeks the child acquired a total of 111 new words, which constituted one third of her productive lexicon of one-word utterances. The sudden change in the rate of acquisition was observed about eight weeks before the subject started to produce two- and multiword combinations.

After three weeks of intensive learning the rate of acquisition started to decline.

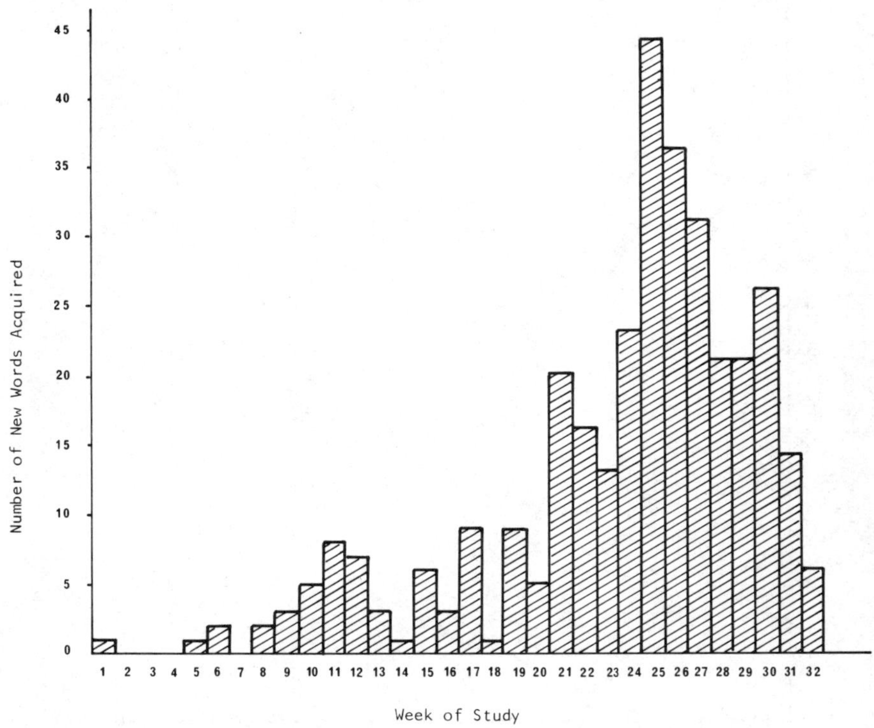

Figure 2. The number of new words acquired during each week.

[3] This average is lower than was estimated by Carey (1978) who suggested that the average rate of learning new words during the early years and up to age 6:0 is about 9 words a day (see also Clark, 1983, p. 811).

During weeks 28, 29, and 30 the rate was similar to that observed for the weeks that preceded the sudden increase. During weeks 31 and 32 there was a considerable further decline. In the last two weeks of the study the subject acquired only 14 and 7 new words, respectively: this corresponds to the point at which she started to combine words. To summarize, the rate at which Keren acquired new words was neither linear nor exponential, as could be expected, but rather it was curvilinear.

Changes in the Form of Words

Throughout the period of investigation, all the overt changes in a word's form were noted. These changes are presented according to two types of underlying processes: (a) changes in the phonological manifestations of words, and (b) morphological markings on existing lexical items. Although I present phonological and morphological modifications in separate categories, I wish to note here that the two classes that are usually treated as distinct in adult grammars may not be such for very young speakers. It is possible that for the child who acquires Hebrew as a first language, the addition of either prefixes or suffixes on some stems is not a sufficient evidence for morphological productivity (see Berman, this volume; Berman & Dromi, 1984; and Dromi & Berman, 1982, for a more explicit account of Hebrew grammar and the implications it might have on the child who acquires a language that employs a highly rich bound morphology). As is indicated, my results show that during the one-word period phonological and morphological processes are highly correlated (see Fig. 3). This finding indicates that both processes may function for the child as similar means for shaping her own productions. An improvement which may be motivated by growing communicative needs as well as by the increasing awareness that words are conventional tools in any given language (Clark, 1983).

Table 1 shows the distribution of 276 child's words in three mutually exclusive classes of phonological forms by the months of investigation.[4]

Table 1. The Distribution of Phonological Classes by the Months of Investigation.[a]

	Months								
Word type	**1**	**2**	**3**	**4**	**5**	**6**	**7**	**8**	**Total *N* of Words**
Conventional words		1 (20)	2 (9)	2 (17)	6 (26)	8 (12)	20 (19)	17 (45)	56 (20)
Phonological approximation			16 (70)	10 (83)	16 (70)	54 (83)	86 (79)	21 (55)	203 (74)
Nursery words	1 (100)	4 (80)	5 (21)		1 (4)	3 (6)	3 (3)		17 (6)

[a]Numbers in parentheses indicate the relative frequency of the class in percentages.

[4]In the present analysis and in the two following analyses a total of only 276 words were scored. Of Keren's words 61 were not recorded frequently enough to allow analyses of changes over time; 16 out of the 61 words were personal names.

In each column the distribution of the new words acquired in this particular month is presented. There are three classes of form: (a) conventional words, (b) phonological approximations, and (c) nursery words. While the first two classes refer to the child's attempts to produce conventional adult Hebrew words, the third category consists of nonconventional words some of which are found in Hebrew "baby-talk," others of which were idiosyncratic to my subject. Of all the new words acquired by the child throughout the one-word period, 74% were initially produced as phonological approximations of adult input forms. Only 20% of the words were articulated correctly from their outset. The group of nursery words was relatively small, and it consisted of only 17 words (6% of all the words acquired).

During the first two months of study nursery words constituted the major category of words in the chid's lexicon. The relative proportion of nursery words declined sharply toward the fourth month of study, and during the last month of study no nursery words were learned. The relative proportion of phonological approximations and conventional forms is constant throughout time. It is only during the last month of study that an increase is noticed for conventional forms, while a decrease is observed in the relative number of phonological approximations of new words. In general it seems then, that the phonological form-class membership of a word (a, b, or c) is significantly related to age of acquisition ($\chi^2 = 148$, *df* 54 $<.001$).

Since most of the child's words changed their phonological and morphological manifestations across time, it seemed desirable to measure the dynamic properties of the system. For this purpose the overall number of phonological changes that included: additions of previously omitted phonemes, modifications of consonants and/or vowels, substitution of a distorted phoneme by a more approximated one, and a complete replacement of baby words by conventional words were calculated. Morphological markings that were mainly inflections for number, gender, and diminutive on nouns, and number, gender, and tense on verbs, were also plotted for this analysis. Figure 3 presents the distribution of scores by week of study. Each bar

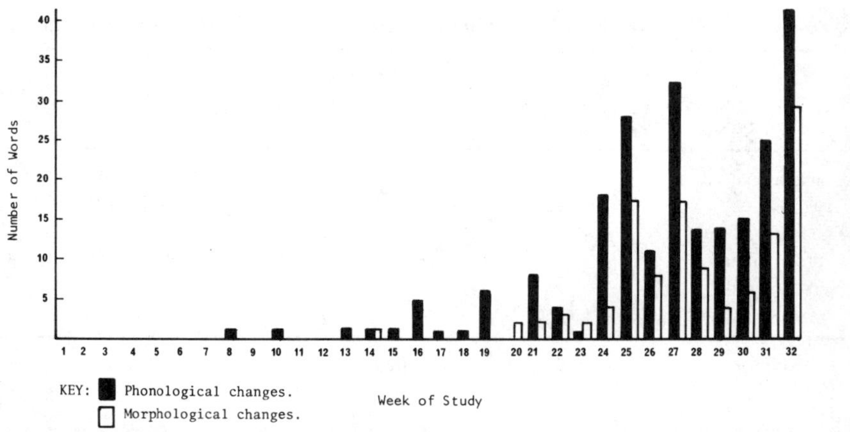

Figure 3. The distribution of words which underwent phonological or morphological changes according to the week of study.

in this figure represents the absolute number of words that *underwent change* during a particular week.

As mentioned earlier, there is a striking similarity between the rate of phonological and morphological acquisitions. The two parallel and fluctuating changes that are observed indicate that most phonological and morphological processes were recorded toward the end of the period, starting at the 21st week of study. The fluctuating rates of changes during the second half of the period indicate that short periods of intensive modifications were followed by short periods of rest (compare, for example, weeks 24–27 to weeks 28–30). The last two weeks of study show an increase in the improvement of the system. This finding is significant, mainly because during this time the total number of new words acquired by the child decreased greatly (see Fig. 2). I therefore conclude here that the child's efforts to improve the form of her words was inversely related to the rate of her lexical acquisition. Towards the end of the one-word period, the intelligibility of words improved considerably while the size of the lexicon remained more or less unchanged.

The Development of Reference

In the existing literature on the one-word period, the difficulty of establishing whether a word refers to an object or an action has rarely been considered. In most cases researchers either did not specify how they made such decisions, or they simply treated words classified as nouns and verbs in adult language as object words and actions words, respectively (see Bowerman, 1976; Dromi, 1982).

In the present analysis, each of Keren's words was assigned to one category of reference on the basis of a close examination of the contexts in which the child supplied it. Every effort was made to minimize the influence of the observer's knowledge of the referential properties of the same word in adult speech.[5] Scores were given to each word on a weekly basis, and it was possible to assign a word to different categories in two adjacent weeks. Figure 4 shows how the words acquired by the child during each month of study were distributed, according to their behavior during the first week of use.

Figure 4 indicates that words belonging to one of five classes have been acquired and used by the child during the one-word period. Some words designated clearly either objects or actions (e.g., *oto* 'car' (011) was never used to point to nonobject references, *alax* 'walk' (226) on the other hand, refered to actions alone). Other words were used as social words (e.g., *kuku* 'pikaboo' (034) produced while playing, or *ken* 'yes' (154) used as an acknowledgment). Still other words were modifiers (e.g., *ol ol gadol* 'large' (173) for describing properties of objects or events). Finally, I identified a class of words which could not be assigned to any of the above mutually exclusive classes, because they had multiple referential functions (e.g., *xam* 'hot' (071) was used as the child's name for heaters and ovens, and at the same

[5] A set of operational definitions as well as a number of examples to illustrate how the words were assigned to the categories of reference are given in Dromi (1982, chap. 4).

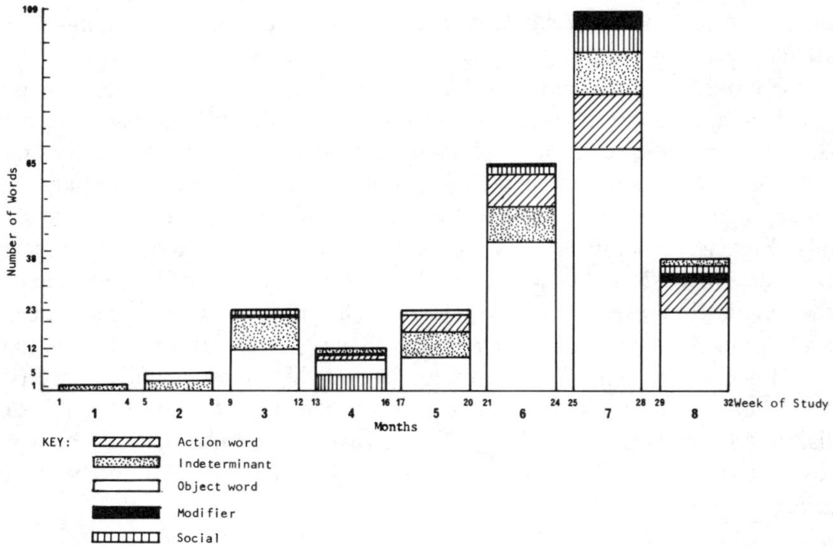

Figure 4. The distribution of the child's categories of reference by months.

time was uttered in some contexts as a modifier denoting the property of being either hot or cold; *dio* 'giddiup' (015) was used while pointing to pictures of animals as well as when pointing to the action of riding). In the present analysis the indeterminant category was generated for words which manifested double or multi-referential behaviors.

In the sample as a whole, the majority of words (59%) were object words. The four other categories were far less well represented: indeterminant, 16%; action words, 14%; social words, 7%; and modifiers, only 4%. Words for objects were first recorded in Keren's lexicon during the second month of study. During all subsequent months (except for the fourth) they constituted the major class of new words learned. The first action word entered the child's lexicon during the fourth month of study. This word was *toda* 'thank you' (038), and it was learned during the fifteenth week of study and was subsequently used to request desired objects. Starting with the fourth month, action words were learned continuously and during the seventh month (weeks 25–28) they constituted the second to largest group of new words acquired. Modifiers and social words first entered the child's lexicon during the third month of study and showed a peak during the following month (weeks 13–16).

Of special interest to us is the category of indeterminant words used for both actions and objects or for objects and modifiers. During the first two months of study, most of the words acquired by the child were used for more than one function. Between the third and seventh months, 5 to 10 words showed indeterminant behaviors. The relative number of words in this category declined significantly toward the beginning of the seventh month. During weeks 26, 28, 29, and 30 only one indeterminant word per week was recorded, and during the last two weeks of study Keren did not acquire any indeterminant word. The finding that with time the category of indeterminants becomes smaller until it completely disappears is important. It shows that as the overall linguistic experience of the child increases, his/her referential abilities improve. Toward the end of the one-word period most words in the child lexicon have a single, clearly defined reference. It seems that as Keren got closer to syntax, her linguistic system became much more differentiated. That is, the mapping of new words onto either object concepts or relational concepts is much sharper than it was at the beginning of the one-word period.

The Development of Word Meaning

Learning the conventional meaning of words is a dynamic, long-term, and covert process that involves complex interactions between two developing systems in the child, the cognitive system and the linguistic system. Despite the extensive theoretical efforts that have been expended over the last few years on the question of how children learn the meanings of words (e.g., Bowerman, 1978, 1980; Clark, 1973, 1975, 1983; Nelson, 1974; Schlesinger, 1982), the state of the art is quite confusing (Carey, 1982). This confusion means that strong predictions cannot be made about initial extensional behaviors of newly acquired words and changes in the behaviors over time.

I argued elsewhere (Dromi, 1982, 1983) that different words in the child's lexicon take different routes as their meanings develop. An examination of the extensional behaviors of each of the first words acquired by Keren revealed that some words were initially used by her in very restrictive settings only (underextended), while others were applied too broadly from the outset (overextended). A surprisingly large number of words (126, or 45% of the 276 words analyzed) showed regular uses from the very first week of their production. These words were applied to a set of referents all included in the adult's category of referents for the same word. A fourth category was that of unclassified words. A word was assigned to this category if it showed an unclear pattern of extension that could not be identified as either underextension, overextension, or regular extension. In my sample, some unclassified words were "empty" words uttered by the child without a clear indication that she attached some constant meaning to them (e.g., *avoda* 'work'), other unclassified words were idiosyncratic, for which an adult meaning could not be attributed. A number of unclassified words were undifferentiated words that had been used as cover terms for *whole situations,* or *scenes.* These words were thus concurrently extended to objects, actions, and relations (e.g., *niyar* 'paper' used for pencils, papers, pictures, paintings, written pages, or the action of writing and drawing). Many of the unclassified words had been acquired early by Keren and were used undifferentiatedly for a period of several months (see Dromi, 1984, for a more detailed examination of this phenomenon).

Not only that different words initially showed different extensional behaviors, but when all the changes over time in a category membership of a word were considered (i.e., the word's developmental history), the patterns of meaning acquisition became even more diverse. The first 276 words acquired by the subject exhibited as many as 58 different combinations of the different possibilities of extension. Only few words, and largely those acquired as regular extensions, showed a stable behavior over time. Because different words exhibited different extensional properties, it seemed crucial to test the effect of time on the child's general strategies of meaning acquisition. Is it the case that throughout the one-word period the child applies the same underlying rules for learning the meanings of words? Does the child alter his/her strategies at a specific point in time? If, in fact, the child changes his/her way of treating the relationship between a word and its meaning, when does this change occur?

Data addressing these questions on the relationship between real time and the development of word meaning are found in Fig. 5. In this figure, the relative proportion of underextended, overextended, regular extended, and unclassified words acquired by my subject in each of the 32 weeks of the study is presented. The distribution of only newly acquired words for any given week is given in Fig. 5.

A close inspection of Fig. 5 reveals that during the one-word period the child does not employ a single strategy of meaning acquisition. It seems that the "stretch" of time in question can be divided into two phases, each showing unique characteristic of mapping words onto their meanings. During the first phase (weeks 1–18), the patterns of extension noted for new words are quite irregular: In some weeks all the words acquired were unclassified, in others most or all of them showed regular extension or underextension. The unpredictable picture, however,

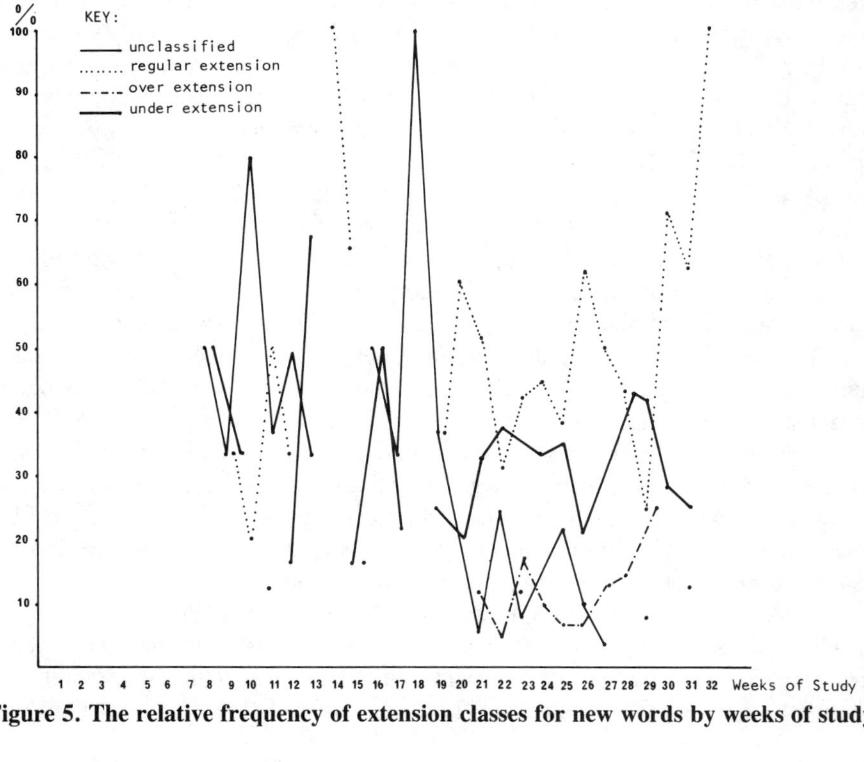

Figure 5. The relative frequency of extension classes for new words by weeks of study.

changes towards the 19th or 20th week of study. It becomes obvious that during the second phase of the single-word period, (weeks 19–32), most new words learned (70% and more) corresponded somewhat to their conventional meanings. In the majority of the cases, when the child started to produce a new word, it was either attached to a class of equivalent referents all belonging to the corresponding adult category for the same word (regular extension), or it referred to a single referent which belonged to such a category (underextension).

The application of a new word to one or to a subset of correct referents clearly indicates that the child has succeeded in establishing a one-to-one relationship between a word and its conventional meaning. The move from underextension to regular use of a word depends on the child's generalization abilities, and, in fact, takes place in a relatively short time. In Dromi (1982), I reported that most of Keren's underextended words stayed in this category for a period of about 2 weeks only (*M* = 2 weeks, *SD* = 2 weeks, *MODE* = 1, *RANGE,* 1–10 weeks), this finding clearly suggests that underextension and regular extension are closely related categories.

The tendency of new words to exhibit unclassified behaviors is significantly lower in the second phase of the stage than it was earlier. During the first phase there were several weeks in which only unclassified words were learned (e.g., 1st, 6th, and 18th weeks). This was not the case for the second phase. Between the 20th and 29th week of study, unclassified words constituted only 20% of the new words learned. The relative proportion is even further reduced during the last five weeks of study.

The category of overextension emerges only in the second phase of study. This suggests that evidently the application of a new word for referents that lie outside the adult category requires relatively mature processing abilities. It is plausible that only after having some experience with language can the child start to carry out detailed componential analyses of meaning. And it is then that the child will extend the range of application of a word beyond its original scope to referents that share components with the standard referent (Bowerman, 1978). Another explanation can be that, after gaining some knowledge about the nature and function of words, children start to use existing words to express meanings for which they have not yet acquired more appropriate terms (Clark, 1983). Given the amount of attention devoted to overextensions in the literature (e.g., Bowerman, 1978, 1980; Clark, 1973, 1975; Nelson, 1974; Rescorla, 1980, 1981), it is surprising that the relative proportion of initial overextension is so low, and that this category emerges so late.

The fact that words tend to change their extensional properties over time makes it necessary to examine not only new words but also how all the words were used by the child in any given week. Figure 6 depicts the relative distribution of words in the four classes of extension according to week of study.

The interrelationships among the four curves—representing the four classes of extension—are quite interesting. Throughout the 32 weeks of study (the one-word period), the two curves for underextension and overextension remain parallel and relatively low. In contrast, the curves for unclassified and regular extension cross

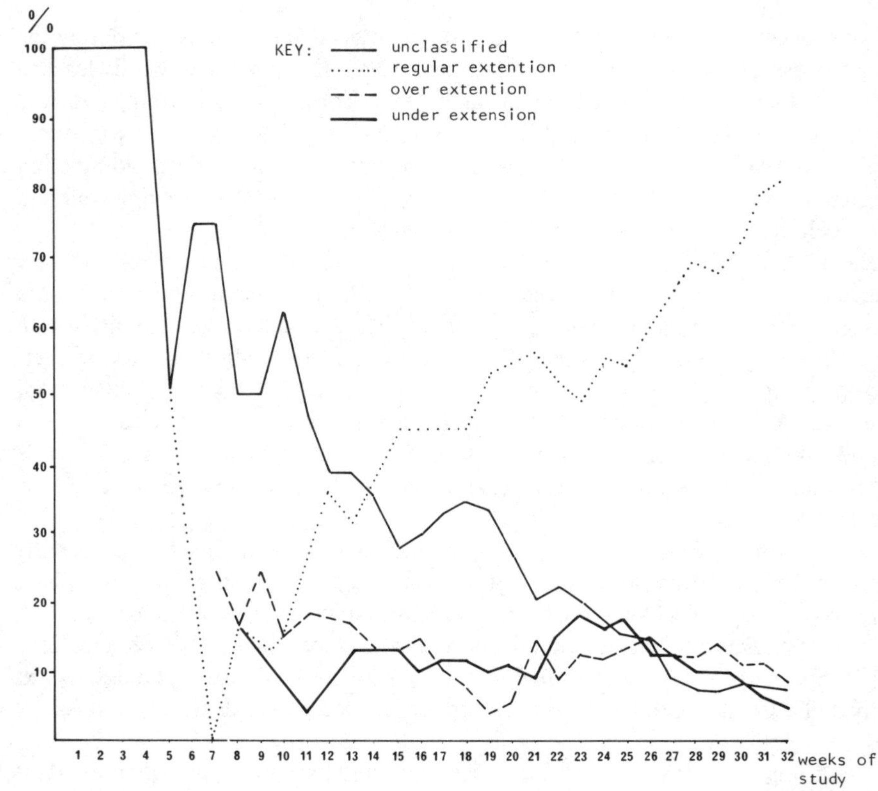

Figure 6. The relative frequency of extension classes according to the week of study.

each other and are negatively correlated. It is obvious that with time fewer and fewer words showed unclassified behaviors just as more and more words were used regularly by the child. The gradual and continuing increase in the number of regularly extended words is important. It clearly indicates that the child's linguistic system has undergone major change during the one-word period. It seems that the more words that Keren learned, the more she became aware of their symbolic properties, their relationship to a class of referents (or an underlying category), and their main feature of being conventional.

The move from the more initial strategy of using words more loosely to encode general intentions, inconsistent meanings, and personal experiences to the more conventional usage does not happen all at once. The child does not suddenly stop uttering unclassified words. As is shown in Fig. 6, the relative number of unclassified words decreases constantly with time, whereas the number of regular extended words shows linear growth. The number of words applied regularly by the child increases considerably during the second half of the one-word period. In weeks 19–21 the number of regular extended words equals the sum of all the other three categories, and toward the end of study it doubles or even triples it.

It seems to me that great difference exists between the beginning and the end of the one-word period. Whereas at the beginning of the period each word is learned as a special case (and thus shows an idiosyncratic pattern of extension), at the end of the period most words that are learned are immediately attached to correct underlying categories (and thus are used flexibly in a variety of accepted contexts). During the latest weeks of study most words are used by the child in a manner quite similarly to adult usage. On the basis of this finding, I conclude that the acquisition of lexical meaning is the most obvious, identified underlying developmental process in the one-word period. During this period, Keren did not simply add more words to her vocabulary. Rather, she seemed to be actively involved in the exploration of the main property of words: that of being associated with clear conventional meanings.

During the one-word period the child learns how meanings are being conveyed by words and which are the main principles to be followed in order to correctly apply them in the right contexts. I hypothesize that among these principles one would identify the following: (a) words convey consistent meanings; (b) words for objects are usually different than words for actions or relations; (c) words are conventional means in that they must correspond with adults' productions in both form and meaning; (d) words in most cases are associated with more than one specific referent; and (e) the meaning of two words contrast with each other, etc. My data show that the acquisition of these principles and possibly others is not instantanious. Lexical meanings are being learned in an ongoing developmental process which begins with the emergence of the first word and extends throughout the several months of producing one-word at a time or even longer than that.

Integration of Findings and Conclusions

In order for a slice of development to be considered a stage in a particular domain we first have to define its starting and ending points. This, however, is not enough; it is further necessary to make explicit which are the qualitative features that differentiate this stage from preceding and subsequent ones.

The emergence of the first meaningful word is an emergence in development. Keren's *haw haw* (001) was a novelty because it signified her first successful attempt to refer to some underlying meaning by means of producing a consistent, recognizable sequence of phonemes. Even those psychologists who favor an additive explanation of development would agree that the difference between a child who playfully utters a variety of sounds and a child who purposefully produces a single word is not simply a matter of degree. The growing ability to intentionally produce similar sounds in repeated contexts make it possible for the child to verbally communicate with primary caretakers.

The emergence of the first meaningful words, however, is only a first step in learning how to use words meaningfully. The present study indicates that it takes up to eight months and may be even more until the child gradually grasps and actively applies the main underlying principles of using words appropriately. It is argued here that by the end of the one-word period the child becomes much more successful in the direct mapping of words onto their conventional meanings. At this point a clear change in the child's linguistic behavior is noticed. The number of new words learned during a single week decreases, and multiword productions start to emerge. Keren's transition into the multiword period was sharp, and it was accompanied by an intensive effort of improving the intelligibility of existing words in her productive lexicon (see Fig. 3).

In what ways do multiword constructions differ from single words? Is this difference a qualitative one? I hypothesize that the emergence of two and multiword constructions is the first linguistic evidence for the child's ability to convey basic semantic relations such as agent, location, possession, patient, etc. (Brown, 1973; Schlesinger, 1971, 1982, and personal communication, October 10, 1983). According to this proposal, combinational or relational meanings are not conveyed through single words. Rather, it is argued that single words function as linguistic labels, and as such they convey referential meanings only. Note here that the last proposal is in contrast to Greenfield and Smith's (1976) thesis that during the one-word period, children express relational meanings by combining words with nonlinguistic elements (e.g., intonation and gesture). I question the last proposal, mainly because I find it hard to accept Greenfield and Smith's (1976) premise that "an element that phonetically resembles an adult vocabulary item also semantically resembles it" (p. 45). Without assuming that children's meanings are closely related to adult's meanings, these authors could not have performed their analysis. In their system, a distinction was made between the referential use of a word (e.g., *Daddy* said, while pointing to the entity FATHER) and the relational use of the same word (e.g., *Daddy* said, while pointing to the entity BRIEFCASE). It was mandatory to hypothesize that the child uttering *Daddy* in the latter context knows exactly what this word means in adult speech in order to conclude that the word was used to express a relation of possession between DADDY and A BRIEFCASE. I argue that it is very often too complicated to make a decision about whether the child overextends an existing word to refer to an irrelevant referent or whether the child refers to some possible relation that s/he is able to notice. I suspect that the adoption of a single alternative is too biased.

So far I have argued that the one-word period is distinct because it has identifiable boundaries and because its underlying characteristic is that of acquiring lexical–referential meanings. To these two considerations I wish to add a third one—which provides support to the claim that the one-word period is indeed a stage. Here I wish to refer to the relatively long stretch of time throughout which children produce only "one word at a time." My subject produced only single words for 32 weeks or 8 months. This figure is consistent with estimates and reports on other children (e.g., Braunwald, 1978; Greenfield & Smith, 1976; and Halliday, 1975). A period of 8 to 12 months is quite impressive especially if one considers the relatively short span of time during which normal children achieve a full command over the basic forms, structures, and rules of their mother tongue (i.e., at about the age of 3:0 to 3:6).

To the best of my knowledge, the actual length of the two- and multiword period have not yet been studied. My observations of Keren's linguistic development in the periods that followed the emergence of word combinations were less systematic than the ones I described in this chapter. In Keren's case, however, the length of the two-word period was surprisingly short, lasting for six weeks only. From several attempts to collect speech samples from other children at the two-word level, I have also learned that usually young speakers stay at this level for a few weeks and then quickly move to the phase of producing simple sentences. This partial evidence on the considerable difference that exists between the one- and multiword periods may indicate that there is a real qualitative difference between the two. I hypothesize that whereas the one-word period is a real stage in language acquisition, the two- or multiword period is a transition into early syntax (see Feldman, 1980; Berman, this volume). Partial and indirect support to my claim comes from a study on the language-learning process of developmentally language delayed (DLD) populations. Allen and Rapin (1980) and Rapin (1982) have reported that DLD children who remain in the one-word stage for long periods, despite intensive training, have a much worse prognosis than those DLD children who move with training into the two-word stage. With additional intervention only the latter group eventually reaches the more advanced level of learning syntax.

Since the one-word period seems to be a stage in language development, a much broader examination of its underlying characterization is needed. The information reviewed and discussed so far is based on my inspection of the *within-stage developments* of only a single child. Therefore, I will make an attempt to test the generality of my findings.

In several previous studies on the one-word stage, researchers have anecdotally noted an increase in the rate of lexical acquisition shortly before the onset of word combination (e.g., Bloom, 1973; Halliday, 1975; Nelson, 1973). In two more recent studies this increase was measured directly during recorded experimental sessions (Corrigan, 1976, 1978, 1979; McShane, 1980). Keren's continuous record provides strong evidence that a spurt in lexical behavior is indeed observed. In Keren's case this spurt was recorded six months after the emergence of the first meaningful word, and two months before the end of the one-word period. Bloom (1973) and Halliday (1975) reported a sudden increase in the child's vocabulary between the ages of 16 and 18 months. I am not convinced that chronological age should be used as a measure for comparison among children, but I find it interesting that Keren's age was 15(27) in the 25th week of study when the radical change in the rate of her lexical growth was noted.

McShane (1980) has argued that the sharp increase in the number of words is immediately followed by the emergence of word combination. He attributed this finding to a sudden insight into the "naming" function of words. My data only partially support McShane's claim. It was not the case that during the 6th or 7th month of study (weeks 21–28) the new words that were acquired by the child suddenly showed distinct pattern. Rather, what I observed was a gradual change in the relative frequency of all the categories of reference throughout time (see Fig. 4). It was only in the last month of study that a clear change was noted in the distribution of words in terms of the five categories of reference. During this month, the rate at which Keren acquired new words decreased and very few indeterminant words were recorded.

One might suspect that the low number of new words recorded in Keren's lexicon during the last few weeks of study is an artifact of measurement. I doubt that this was the case, because during this time many more conversations than ever before between Keren and me were held and recorded. In spite of the slowdown in Keren's acquisition of new vocabulary in the last month of study, I was still very busy collecting new data for the diary. At that time new entries described new uses of old words, their phonological improvements, and a few acquisitions of morphological markings.

To the best of my knowledge, there is only one research in which the relationship between the growing lexicon of single words and the emergence of word combinations was considered.[6] Anisfeld (1984, pp. 107–109) reported that in five different English-speaking subjects who were studied longitudinally by McCune-Nicolich (1981), as MLU values exceeded 1.0 morpheme per utterance, the rate of learning new words slowed down sharply. I think that the slowdown in the rate of learning new words toward the end of the one-word stage indicates that the stage is coming to an end. I suspect that other researchers have failed to document this slowdown for mainly methodological reasons (see Fischer & Corrigan, 1981). When the child's lexicon exceeds 50 or 100 words, it is extremely hard to study it carefully (Nelson, 1973). Only in a study involving systematic daily recordings does one obtain the amount of information needed in order to develop an exact and complete picture of interactive processes that are changing throughout time.

Of course, my systematic attempt to look at fine changes over time must be replicated with other children in order to completely rule out the possibility that my findings are idiosyncratic to my daughter Keren. However, on the assumption that my findings are representative, I conclude that the one-word period is a distinct stage in language development. This stage is characterized by the acquisition of lexical referential meanings, and it has identified boundaries.

The one-word stage seems to be divided into two phases. The first phase is a transitional phase which spans a period of several months (5 months in Keren's case), and may be considered a preparatory phase (e.g., Feldman, 1980; Menn & Haselkorn, 1976). During this first phase new words are learned by the child very slowly and inefficiently, as if the child laboriously learns every new word as a special case. During the second phase of the one-word stage, which is possibly shorter than the earlier one (and spanned 3 months in Keren's case), the child becomes much more efficient. The number of new words learned rises sharply and the child's ability to master the relationship between a word and its conventional meaning considerably improves. During the last few weeks of the one-word period new words are learned less often, and the child is very busy improving her repertoire of existing words in terms of both their forms and meanings. I hypothesize that the last few weeks of the one-word stage are, in fact, the period of consolidation and of equilibration, and as such they constitute the end of a distinct stage and the prelude to the transition into syntax.

---

[6] I wish to thank I. M. Schlesinger for bringing to my attention this discussion in Anisfeld's publication.

References

Allen, D. A., Rapin. I. (1980). Language disorders in preschool children: Predictors of outcome a preliminary report. *Brain and Development, 2,* 72–80.

Anisfeld, M. (1984). *Language development from birth to three*. Hillsdale; NJ: Erlbaum.

Berman, R., & Dromi, E. (1984). On marking time without aspect. *Papers and reports on child language development.* Stanford, CA: Stanford University Dept. of Linguistics.

Bloom, L. (1973). *One word at a time*. The Hague: Mouton.

Bowerman, M. (1976). Semantic factors in the acquisition of rules for word use and sentence construction. In D. Morehead & A. Morehead (Eds.), *Directions in normal and deficient child language*. Baltimore, MD: University Park Press.

Bowerman, M. (1978). The acquisition of word meaning: An investigation into some current conflicts. In N. Waterson & C. Snow (Eds.), *Development of communication.* New York: Wiley.

Bowerman, M. (1980). The structure and origin of semantic categories in the language learning child. In M. Foster & S. Brandes (Eds.), *Symbol as sense*. New York: Academic Press.

Brainerd, C. J. (1976). "Stage," "structure," and developmental theory. In G. Steiner (Ed.), *The psychology of the twentieth century*. Munich, Germany: Kindler.

Brainerd, C. J. (1978). The stage question in cognitive developmental theory. *Behavioral and Brain Science, 1,* 173–213.

Braunwald, S. R. (1978). Context, word and meaning: Toward a communicational analysis of lexical acquisition. In A. Lock (Ed.), *Action, gesture and symbol: The emergence of language*. London: Academic Press.

Braunwald, S. R., & Brislin, R. W. (1979). The diary method updated. In E. Ochs & B. B. Schieffelin (Eds.), *Developmental pragmatics*. New York: Academic Press.

Brown, R. (1973). *A first language: The early stages*. Cambridge, MA: Harvard University Press.

Carey, S. (1978). The child as word learner. In M. Halle, J. Bresnan, & G. Miller (Eds.), *Linguistic theory and psychological reality*. Cambridge, MA: MIT Press.

Carey, S. (1982). Semantics and development: State of the art. In L. Gleitman & G. Wanner (Eds.), *Language acquisition: The state of the art*. Cambridge, England: Cambridge University Press.

Carter, A. L. (1978). From sensorimotor vocalization to words: A case study of the evolution of attention-directing communication in the second year. In A. Lock (Ed.), *Action, gesture and symbol: The emergence of language*. New York: Academic Press.

Carter, A. L. (1979). Prespeech meaning relations: An outline of one infant's sensorimotor morpheme development. In P. Fletcher & M. Garman (Eds.), *Language acquisition: Study in first language development*. Cambridge, England: Cambridge University Press.

Clark, E. V. (1973). What's in a word? On the child's acquisition of semantics in his first language. In T. E. Moore (Ed.), *Cognitive development and the acquisition of language*. New York: Academic Press.

Clark, E. V. (1975). Knowledge, context and strategy in the acquisition of meaning. In D. Dato (Ed.), *Georgetown University round table on languages and linguistics*. Washington, DC: Georgetown University Press.

Clark, E. V. (1983). Meanings and concepts. In P. H. Mussen (Ed.), *Handbook of child psychology (Vol. 3)*. New York: Wiley.

Corrigan, R. L. (1976). *Patterns of individual communication and cognitive development*. Doctoral dissertation, University of Denver, CO.

Corrigan, R. L. (1978). Language development as related to stage 6 object permanence development. *Journal of Child Language, 5,* 173–189.

Corrigan, R. L. (1979). Cognitive correlates of language: Differential criteria yield differential results. *Child Development, 50,* 617–631.

Dale, P. S. (1976). *Language development: Structure and function*. New York: Holt, Rinehart, & Winston.

deVilliers, J., & de Villiers, P. (1978). *Language acquisition*. Cambridge, MA: Harvard University Press.

Dromi, E. (1982). *In pursuit of meaningful words: A case study analysis of early lexical development.* Doctoral dissertation, University of Kansas, Lawrence.

Dromi, E. (1983). On the identification of the extensional behaviors of new words. *Rassegna Haliana di Linguistica Appliata, 15,* 235–246.

Dromi, E. (1984). The word-context production strategy in the early acquisition of meaning. In C. E. Johnston & C. L. Thew (Eds.), *Proceedings of the Second International Congress for the Study of Child Language* (Vol. 2, pp. 127–141). Lanham, MD: University Press of America.

Dromi, E., & Berman, R. (1982). A morphemic measure of early language development: Data from modern Hebrew. *Journal of Child Language, 9,* 403–424.

Feldman, D. H. (1980). Stage and transition in cognitive–developmental research: Getting to the next level. *The Genetic Epistemologist, 9,* 1–6.

Ferrier, L. J. (1978). Some observations of error in context. In N. Waterson & C. Snow (Eds.), *The development of communication.* New York: Wiley.

Fischer, K. W. (1980). A theory of cognitive development: The control and construction of hierarchies of skills. *Psychological Review, 87,* (6), 477–531.

Fischer, K. W., & Corrigan, R. (1981). A skill approach to language development. In R. Stark (Ed.), *Language behavior in infancy and early childhood.* Amsterdam: Elsevier.

Flavell, J. H. (1971). Stage related properties of cognitive development. *Cognitive Psychology, 2,* 421–453.

Flavell, J. H. (1982). On cognitive development. *Child Development, 53,* 1–10.

Fodor, J. A. (1975). *The language of thought.* Cambridge, MA: Harvard University Press.

Fodor, J. A. (1980). On the impossibility of acquiring ''more powerful'' structures. In M. Piatelli-Palmarini (Ed.), *Language and learning.* Cambridge, MA: Harvard University Press.

Greenfield, P. M., & Smith, J. H. (1976). *The structure of communication in early language development.* New York: Academic Press.

Halliday, M. A. K. (1975). *Learning how to mean: Explorations in the development of language.* London: Edward Arnold.

Ingram, D. (1971). Transitivity in child language. *Language, 47,* 888–909.

Keenan, E. O. (1977). Making it last: Repetition in children's discourse. In S. Ervin-Tripp & C. Mitchell-Kernan (Eds.), *Child discourse.* New York: Academic Press.

Keenan, E. O., & Klein, E. (1974). *Coherency in children's discourse.* Paper presented at the Linguistics Society of America meeting.

MacWhinney, B. (1978). The aquisition of morphology. *Child Development, 43,* 1–103.

McNeill, D. (1970). *The acquisition of language: The study of developmental psycholinguistics.* New York: Harper.

Menn, L., & Haselkorn, S. (1976). Now you see it, now you don't: Tracing the development of communicative competence. Proceedings of the 7th annual meeting. North-East Linguistic Society.

McShane. J. (1980). *Learning to talk.* Cambridge, England: Cambridge University Press.

Nelson, K. (1973). Structure and strategy in learning to talk. *Monographs of the Society for Research in Child Development, 38* (1–2, Serial No. 149).

Nelson, K. (1974). Concept, word, and sentence: Interrelations in acquisition and development. *Psychological Review, 81,* 267–85.

Rapin, I. (1982). *Children with brain dysfunction.* New York: Raven Press.

Rescorla, L. A. (1980). Overextension in early language development. *Journal of Child Language, 7,* 321–335.

Rescorla, L. A. (1981). Category development in early language. *Journal of Child Language, 8,* 225–238.

Rodgon, M. (1976). *Single word usage, cognitive development, and the beginnings of combinatorial speech: A study of ten English-speaking children.* Cambridge, England: Cambridge University Press.

Scollon, R. (1976). *Conversations with a one year old.* Honolulu: University of Hawaii Press.

Scollon, R. (1979). A real early stage: An unzippered condensation of a dissertation on child language. In E. Ochs & B. B. Schieffelin (Eds.), *Developmental pragmatics.* New York: Academic Press.

Schlesinger, I. M. (1971). Learning grammar: from pivot to realization rule. In R. Huxley & E. Ingram (Eds.), *Language acquisition: Models and methods.* New York: Academic Press.

Schlesinger, I. M. (1982). *Steps to language.* Hillsdale, NJ: Erlbaum.

Veneziano, E. (1981). Early language and nonverbal representation: A reassessment. *Journal of Child Language, 8,* 541–563.

Wohlwill, J. F. (1973). *The study of behavioral development.* New York: Academic Press.

10

The Ambiguity of Stage and Structure in Behavior: Person and Environment in the Development of Psychological Structures*

Kurt W. Fischer
Richard L. Canfield
Department of Psychology
University of Denver

Stage and structure in behavior are fundamentally ambiguous concepts. An inescapable predicament for developmental scientists is that any single behavior is ambiguous, equivocal, or uncertain with respect to its underlying structural properties. That is, a single behavior observed in a single child in a single context can be analyzed in terms of a number of alternative structural formulations. So long as the range of observations of behavior is not widened, it is difficult to determine which of the alternative structures provides a better account of the organization of the behavior.

At times, some scholars have reacted to this fact with dismay, suggesting that the equivocal nature of stage and structure in behavior spells doom for theories of the development of psychological structures (e.g., Flavell, 1983; Strauss & Ephron-Wertheim, this volume). A closer look reveals, however, that theories of stage and structure are not unique with regard to the ambiguity issue. The same issue pervades all knowledge. Science itself is a process of devising methods to reduce the ambiguity of knowledge (Anderson, 1978; Bullock, 1981; Dewey, 1948; Hanson, 1961; Peirce, 1931–1958; Wittgenstein, 1953).

The issue has been magnified in cognitive development, however, by the way scholars have approached the study and analysis of stage and structure. Cognitive-developmental theories and research methods have tended to increase the problem rather than reducing it, because they have encouraged a narrow focus of explanation and investigation. A broader focus is necessary: First, developmentalists need to heed the general scientific dictum that a wide range of converging observations should be made to test any structural explanation. Second, a structure should be

---

* The authors would like to thank the following individuals for their contributions to various parts of this chapter: Daniel Bullock, Helen Hand, Charles Reichardt, Louise Silvern, and Michael Westerman. Preparation of this article was supported by grants from the Carnegie Corporation of New York and the Spencer Foundation. The statements made and views expressed are solely the responsibility of the authors.

seen not as a fixed, unchanging characteristic of a person but as a characteristic that varies. These variations can occur not only over long time periods but from moment to moment, and they arise not only from factors in the person but also from factors in the environment.

With this broader focus, the uncertainty or ambiguity of stage and structure becomes manageable, and it is possible to move toward building a nonarbitrary scientific basis for a structural approach. In this chapter we will attempt to point the way toward such an approach. Generally, we will outline how scholars can reduce the ambiguity or uncertainty of stage and structure by treating them as characteristics of the person and the environment conjoined. More specifically, we will describe a methodological orientation that can be used to deal with some of the most problematic sources of ambiguity.

The Fundamentally Equivocal Nature of Knowledge

Ambiguity arises whenever experience is interpreted (Broughton, 1981; Westerman, 1980; Wittgenstein, 1953). Whenever someone builds a model, proposes a theory, or gives an explanation of a particular phenomenon, the model, theory, or explanation will necessarily capture only some aspects of the phenomenon and ignore others. Because so much is always left out, we can never be sure that our model or explanation is correct. Indeed, we can be certain that it is incorrect, at least in the sense that it is incomplete. There is no way finally and totally to prove a theory in developmental psychology or in any other field of science (Campbell, 1974; Cook & Campbell, 1979; Duhem, 1962; Hanson, 1961; Reichardt, 1983). Someone can always point out a significant aspect of the phenomenon that has been neglected, or someone can propose another explanation that is based on a set of observations which overlaps only in part with those on which the first model is based.

Of course, scientists often act as if a theory has been confirmed. Suppose that a theory predicts that children's thinking changes systemically from behavior A to behavior B to behavior C. When observations support that prediction, the conclusion is usually drawn that the theory is true (Hanson, 1961; Campbell, 1969). In the extreme, this conclusion is not justified: Because so many other relevant observations are possible, the theory can be only *supported* by the confirmatory observations. It cannot be *proven.*

Diagnosis of the Difficulty for Approaches to Stage and Structure

Because of this limitation, scientific theory progresses in two ways—by the pitting of rival theories against one another and by the expansion of the range of observations encompassed by a successful theory. When two theories are pitted against one another, one can be judged to be better than the other, even though neither one can be proven. When a theory is expanded to include a wider range of observations, the expanded version (if successful) can be judged to be superior to the narrower version.

Pitting Rival Theories Against One Another

For rival theories to be pitted against one another, they must encompass some of the same aspects of a phenomenon. If there is no real overlap between them, then they cannot be used to make differential predictions, and so they cannot be pitted against one another. Many rival theories in cognitive development do not clearly encompass the same aspects of developing behaviors. Often, they address fundamentally different questions.

One of the most basic reasons for this absence of overlap between cognitive-developmental theories is that investigators typically take one of two fundamentally different, dichotomized approaches to development. One approach is based on a functionalist, mechanistic orientation such as learning theory. This approach focuses on how the environment molds behavior, and it neglects how characteristics of the child contribute to behavior. Another approach is based on a structural, organismic orientation such as Piaget's (1970) theory. It focuses on how characteristics of the child (especially developmental status) produce consistency across tasks, and it says little or nothing about how the environment molds behavior (Fischer & Bullock, 1984; Fischer & Silvern, 1985; Reese & Overton, 1970).

Because of the different focuses of the two approaches, it has proved to be difficult for researchers to derive specific contrasting predictions so as to pit the approaches against each other. There is so little overlap in the aspects of development addressed by the two approaches that it has been difficult to contrast them fruitfully. In most studies purporting to contrast the approaches, the proponents of one approach have failed to take the other approach seriously: They have demonstrated that the variables emphasized by their approach (environmental or organismic variables, respectively) do indeed have an effect on the behavioral characteristics measured by proponents of the other approach. But too often, they have concluded from these findings that the other approach was invalid. They have tended to dismiss the other approach rather than taking it seriously.

A more balanced orientation is required, acknowledging the importance of both environmental and organismic factors. From this orientation, a different conclusion can be drawn from the research based on the dichotomized structural and functional approaches: Environmental factors sometimes affect stage of performance, and the developmental status of the child sometimes predicts performance across environmental variations. Both person and environment affect development (Fischer & Silvern, 1985). This conclusion is an important starting point for understanding stage and structure, although it is hardly surprising.

The problem of lack of overlap between theories that superficially seem to deal with the same phenomenon extends beyond research growing out of the dichotomized structural and functional approaches. Even when two theories are both dealing with changes in the same general class of factors (person or environment), they cannot be pitted against one another unless they address some of the same aspects of behavior. For example, Piaget and his colleagues have made general predictions about the broad course of developmental change during childhood, including large-scale changes in reasoning about proportionality (Inhelder & Piaget, 1955/1958; Piaget & Szeminska, 1941/1952). Siegler (1981) has made much narrower, more specific predictions about the patterns of success and failure on a limited class of proportionality tasks. Unless specific overlapping areas can be found between the aspects of behavior on the proportionality tasks predicted by these two approaches, it is not possible to pit them against each other. It appears that such areas of overlap are scanty at best (Strauss & Ephron-Wertheim, this volume).

Expanding the Range of Observations

A second way of testing the adequacy of theories, models, or hypotheses is by expanding them to include a wider range of data. A small sample of data is open to more rival explanations and thus more ambiguity than a larger sample. Campbell (1974, p. 11) compares the task of the scientist to that of a person trying to identify the same dot in two prints of a single newspaper photograph (which are composed of a dot matrix). If each picture is covered with a large sheet of paper that allows the person to examine only one dot at a time, the task is extremely difficult. The sample of data is so small that the range of possible interpretations for each dot is enormous. However, as the size of the hole is enlarged, the task becomes more manageable because patterns become identifiable. Up to a point, the larger the portion of the picture shown, the easier it is to identify the same dot in each print. (Of course, beyond a certain point, enlarging the hole further only produces a mass of irrelevant data—portions of the newspaper page that are not relevant to the identification of the dot.)

A common weakness of much research on stage and structure in cognitive development is reliance on only one or a few types of data to build a theory that claims broad scope. To promote the construction of less ambiguous structural models, investigators need to increase the range of their data base. They need to envisage a number of different consequences of any model, hypothesis, or theory, analyzing what happens across several types of data. Then they will be likely to observe patterns rather than merely isolated points (Fischer & Bullock, 1981), and it will be more difficult for other investigators to offer alternative explanations to their theories.

The extension of a model to a broader range of data is a natural part of the scientific process, and it promotes more differentiated and powerful models. Such an extension process has occurred in research on conservation of number, for example. The original model of conservation of number proposed by Piaget and Szeminska (1941/1952) was challenged with several new types of data, and gradually the formulation has been extended and refined to fit the old data as well as the new.

Piaget and Szeminska presented children with two lines composed of equal numbers of items, and then they lengthened or shortened one of the lines without changing the number. Before 6 or 7 years of age, children believed that lengthening an array increased its numerosity and shortening it decreased its numerosity; that is, they confounded number with length or density. Preschool children thus did not understand that number remained constant in the face of these irrelevant transformations. After 6 or 7 years, on the other hand, children stated that lengthening or shortening the lines did not change the number. Piaget and Szeminska concluded that beginning at 6 to 7 years, children achieved the construction of a logic for dealing with number concepts. This conclusion was based on tasks using several different materials for assessing conservation of number (buttons, coins, and so forth), but in other ways the tasks and behaviors considered were restricted: First, small numbers such as 2 and 3 were excluded, and tasks were generally limited to numbers of at least 8 to 10 in each line. Second, only two basic aspects of the children's behavior were assessed—whether they gave the correct response (that the number of items did not change) and whether they could verbally explain why the number did not change.

The challenge to Piaget and Szeminska's formulation has featured a lively debate over when children can first understand that the number of items is not changed by irrelevant transformations such as alterations in the spatial layout of items. Gelman (1972) challenged the conclusion that children did not understand conservation until 6 or 7 years. She assessed conservation with a task and a response measure that were different from Piaget and Szeminska's: The task was limited to small numbers (2 or 3) in each of the two arrays, and the measure of whether children noticed a change was their surprise reaction to surreptitious transformations in the number of items. In these tasks, children as young as 3 years of age were more surprised by changes in number than by changes in length and density. Gelman (1978) concluded that the basic structure of conservation of number was present by 3 years of age.

A major rival hypothesis for explaining Gelman's results provided a less radical departure from Piaget and Szeminska's original interpretation; it also explained the results of not only Gelman but also Piaget and Szeminska (Cooper, 1984; Starkey & Cooper, 1977). Preschoolers can produce a conservation response via a different strategy from older children. They use number-estimation skills such as subitizing or counting to quantify small numbers such as 2 and 3. In this way they can make conservationlike judgments without actually considering the nature of the transformations of the arrays (addition or subtraction of an item vs. changes in length and density). When children see the initial two arrays, they count or subitize them and arrive at a number for each array. Then after one of the arrays is transformed, the children count or subitize each array again. With this strategy, the children do not need to consider the nature of the transformations in the arrays. They need only to estimate the number of items in each array before and after the transformations. Piaget and Szeminska's tasks, on the other hand, are much harder to pass by simple counting because of the larger numbers. Not until 6 or 7 years can children pass these tasks by understanding conservation—that transformations in the spatial layout do not change the number of items.

This hypothesis was tested in several experiments that added new tasks and types of data to the original research. Starkey and Cooper (1977) used the tasks of both

Gelman and Piaget and Szeminska but added assessment of another aspect of behavior—reaction time. They reasoned that with a conservation strategy, there should be little or no relation between the time to give a response and the size of the array, because children would know that there had been no change in the number. With a counting strategy, on the other hand, arrays with larger numbers of items would take longer to count. The results supported their hypothesis. Preschool children showed a correlation between reaction time and the size of the number. But at a later age, when children began to give correct conservation explanations, the correlation went away. At the same time, on number-estimation tasks as opposed to conservation tasks, these same children took longer at all ages to count large arrays.

Several other investigators changed the conservation tasks so as to prevent children from simply counting the number of items. Silverman and Briga (1981) added a second transformation to the conservation task: In addition to lengthening one line, they also covered two items in the line so that children could not simply recount them. Under these circumstances, 3-year-olds did not show conservation. Halford and Boyle (in press) used arrays in which children could not easily count the number of items in each line. Conservation responses were again absent in 3-year-olds, but 6- and 7-year-olds showed conservation of number despite the difficulty of counting the number of items.

As long as the argument was restricted to the limited range of tasks and behaviors assessed by Piaget and Szeminska and by Gelman, it was difficult to resolve the disagreement between their alternative hypotheses about conservation. Also, their respective studies did not even use the same tasks nor assess the same behaviors. However, when a wider sample of tasks and behaviors was used, a sensible overall pattern emerged: By 3 or 4 years of age, children can show something approximating conservation of number so long as they are allowed to count the number of items. But they cannot show conservation when they are not allowed to count and so are forced to base their judgment on the understanding of certain kinds of transformations. That type of conservation of number first emerges at 6 or 7 years.

In the same way, many issues concerning stage and structure can be illuminated by widening the range of tasks and behaviors considered. Indeed, we believe that many controversies about stage and structure will be resolved only when researchers broaden their range of observations.

Including Both the Person and the Environment in Research: The Child-in-a-Context

Unfortunately, cognitive-developmental researchers often use a limited set of observations—one or a few tasks and one or two response measures. This problem has been accentuated by the dominant influence of the structural and functional approaches. One group of researchers, the functionalists, limit themselves primarily to tests of environmental effects on behavior and so neglect the broad class of organismic variables. The other group, the structuralists, limit themselves primarily to tests of organismic effects and so neglect the broad class of environmental variables. If developmental science is to resolve the controversies between these two approaches about the nature of stage and structure, then researchers must study the effects of both environmental and organismic factors in the same studies.

With this broader range of observations, the theories put forth by both groups will necessarily change, so that concepts of stage and structure will become collaborational (Fischer & Bullock, 1984) or transactional (Sameroff & Chandler, 1975). It is not the child alone that moves through stages, nor is it the environment alone that induces all changes in behavioral organization. It is the collaboration of child and environment that leads to structural change. What develops is the child-in-a-context. When developmental researchers systematically include factors from both child and environment in each study, then it becomes possible to begin to construct sensible, workable theories of stage and structure. We feel that some progress has already been made toward this collaborational approach.

Sources of Developmental Variation Other Than Age and Learning History

Some research already exists that suggests important environmental and organismic variables to be included in research. The vast literatures based on the structural and functional approaches suggest, of course, that age and learning history have substantial effects on the maturity and complexity of behavior. As children grow older, they generally become smarter, and children typically mold their behavior to fit contingencies of reinforcement and punishment. It makes sense, therefore, to investigate how both age and learning history relate to developing skills in the same children. As interesting and important as age and learning history are, however, investigators need to cast a wider net. If scholars are to build better theories of stage and structure, research paradigms cannot be limited to the standard questions engendered by structural and functional approaches.

Factors that are of special importance for inclusion in the research net are those that raise questions about traditional concepts of stage and structure. One such set of factors calls into question the unity and stability of stage and structure: For virtually any measure of stage, level, or step, children are not fixed at a given point in a developmental scale. Instead, their behavior varies routinely over a zone in the scale (Fischer & Elmendorf, in press; Vygotsky, 1978). Two factors that have been shown to have significant effects on that variation are degree of environmental support and task complexity. A second challenge to traditional concepts of stage and structure comes from evidence of substantial individual differences in developmental sequences. People do not appear to follow monolithic, universal sequences such as those proposed by Piaget (1970) and Kohlberg (1969).

Environmental Support: Functional Levels

The existence of a developmental zone provides an important challenge to traditional stage theory. Even within a given context, people seem to evidence a range of developmental steps. When children are asked, for example, to make up a number of stories about some material that they have been tested on, such as doctor roles, they spontaneously produce stories at several different steps on a developmental scale for understanding the role of doctor (Hand, 1981/1982; Watson & Fischer, 1980).

Table 1 shows a developmental sequence that children evidence for the understanding of the doctor role in pretend play. The left-hand column lists the cognitive-developmental levels for the steps in the sequence, based on skill theory (Fischer, 1980). The middle column describes the general type of role skill at each step. The right-hand column presents the behaviors that were actually modeled by the experimenter in one of the experimental conditions, which will be described (Watson & Fischer, 1980).

The first three steps represent precursors to the doctor role. At Step 1, *self as agent,* children pretend to carry out at least one action, such as drinking or eating, but they do so only with their own body, not with a doll or other separate agent. At Step 2, *active other agent,* they make a doll carry out such an action. Step 3, *active substitute agent,* brings the extension of the action to a substitute object, as when children make a block act as if it is walking or eating.

Table 1. A Developmental Sequence of Social Role Playing.

Cognitive Level	Step	Role-Playing Skill	Modeled Behaviors
Sensorimotor systems	1	*Self as Agent:* A person pretends to carry out one or more behaviors, not necessarily fitting a social role.	Experimenter pretends to drink from an empty cup.
Single representations	2	*Active Other Agent:* An agent performs one or more behaviors, not necessarily fitting a social role.	Experimenter pretends that a doll is talking, walking, eating, and washing, as if it were actually carrying out the actions itself.
	3	*Active Substitute Agent:* An object substituting for an agent performs one or more behaviors, not necessarily fitting a social role.	Experimenter pretends that a block is walking, eating, talking, and going to sleep, as if it were a person or a doll.
	4	*Behavioral Role:* An agent performs several behaviors fitting a social role (in this case, doctor).	Experimenter pretends that a doctor-doll uses a thermometer and an otolargyngoscope on another doll.
Representational mappings	5	*Social Role:* One agent behaving according to one social role (doctor) relates to a second agent behaving according to a complementary social role (patient).	Experimenter pretends that a patient-doll is sick and a doctor-doll examines her and gives her some medicine. The patient-doll makes appropriate responses during the examination.

(*continued*)

Table 1. (*Continued*)

Cognitive Level	Step	Role-Playing Skill	Modeled Behaviors
	6	*Social Role with Three Agents:* Two complementary roles (patient and nurse), instead of just one, are simultaneously related to the first role (doctor).	Experimenter pretends that a doctor-doll examines a sick patient-doll and is aided by a nurse-doll. Both patient and nurse respond appropriately.
Representational systems	7	*Intersection of Social Roles for Two Agents:* Two separate, agent-complement role relations are coordinated so that one agent is in two roles simultaneously and relates to another agent in two complementary roles (doctor-father with patient-daughter).	Experimenter pretends that a doctor-doll examines a sick patient-doll and also acts as father to the patient, who is his daughter. The patient-doll responds appropriately as both patient and daughter.
	8	*Intersection of Social Roles for Three Agents:* Three separate agent-complement role relations are coordinated so that one agent in two or three roles simultaneously relates to two other agents each in the relevant complementary roles (doctor-father-husband with patient-daughter and mother of patient-wife).	Experimenter pretends that one doll is a doctor, father, and husband relating to two other dolls. The second doll is a sick patient and the first doll's daughter. The third doll is the patient's mother and the first doll's wife.

Note: The stories and exact scoring criteria can be obtained from the authors upon request. From "Development of Social Roles in Elicited and Spontaneous Behavior During the Preschool Years" by M. W. Watson and K. W. Fischer, 1980, *Developmental Psychology, 16,* p. 485. Copyright 1980 by the American Psychological Association. Reprinted by permission.

The first behavior that demonstrates a social category for doctor is Step 4, *behavioral role,* where children make a doll act out several actions fitting the category of doctor. The child may have a doll say, "I'm a doctor," give another doll medicine, and use a thermometer. The patient doll does nothing active but merely serves as a passive recipient of the doctor doll's actions. With Step 5, *social role,* children make the patient doll an active participant in the interaction, as both doctor and patient carry out several actions fitting their roles. Step 6, *social role with three agents,* adds a nurse to the interaction.

The last two steps involve understanding that individual agents can occupy two or more roles at once. At Step 7, *intersection of social roles for two agents,* children make a man doll act simultaneously as doctor and father to a girl doll, who acts as his patient and his daughter. A third agent is added to the intersection at Step 8, *intersection of social roles for three agents,* as a woman doll acts as the mother of the patient and the wife of the man.

When children were asked to make up stories about doctor roles, they spontaneously produced several different steps from the sequence in Table 1. For example, one pattern for a 7-year-old girl was to act out five stories in a six-minute period—two at Step 4, two at Step 2, and one at Step 1. Research to date suggests that this sort of spontaneous behavior shows a natural upper limit for any particular child in any specific context, which is called the functional level. The functional level for the 7-year-old girl was Step 4; she produced several stories at Step 4 and several at lower steps. In fact, her behavior represents a common pattern for 7½-year-olds, who in a study by Watson and Fischer (1980) typically showed a functional level of Step 4.

One important method for investigating functional levels is to assess performance across testing conditions that vary the degree of environmental support for high-level performance. In high support, the context provides clear information about the task demands for high-level performance, and it keeps memory demands to a minimum. In low support, people are left to their own devices to produce a behavior. A reasonable hypothesis is that functional levels will vary directly with degree of environmental support.

In several studies in the Cognitive Development Laboratory at the University of Denver, we have tested this hypothesis. Children were tested under at least two conditions. In the high-support condition, called elicited imitation, an experimenter demonstrated a story for each step being tested, as illustrated in Table 1 in the column "Modeled Behaviors." After each story was modeled, the child was asked to act out a similar story. The low-support condition occurred after the high-support condition and was called free play. The child was left alone in the room with the toys used in the modeled stories and asked to make up several stories of his or her own dealing with the same topic as the modeled stories. (The 7-year-old girl described earlier was performing under this condition.) Most of the studies included a second low-support condition, called best story, in which the experimenter returned to the room and asked the child to show her the best story he or she could.

For every study, degree of environmental support had a potent effect on functional level (Fischer, Hand, Watson, Van Parys, & Tucker, 1984). At most ages, high support produced a much higher functional level than low support. With the sequence in Table 1, for example, 7½-year-olds typically showed a functional level of Step 4 in both low-support conditions but Step 6 in the high-support condition. In other studies, the size of the gap between high- and low-support conditions has been even larger (Hand, 1982; Van Parys, 1983/in press). Also, in current studies we have added another condition, providing an intermediate degree of environmental support. As predicted, this condition produced an intermediate functional level, in between the levels of the low-support and high-support conditions (Purcell, 1983).

Within the same general context, then, the stage or structure that children demonstrated varied greatly as a function of degree of environmental support. Even though they were tested with the same materials, in the same room, and on the same developmental scale, their functional level was much greater under high-support conditions than under low-support conditions. In effect, their competence changed with alterations in the degree of support (Fischer & Elmendorf, in press; Rubin, Fein, & Vandenberg, 1983).

Task Complexity: The Need for Multiple Tasks

Another environmental factor that seems to have potent effects on the variation of behavior along a developmental scale is task complexity. An implicit assumption in the Piagetian literature is that a single task can be used to assess not only the stage that it was designed for but also earlier stages (Kohlberg, 1969). Consider a classification task designed to assess the concrete-operational ability of multiplication of classes, such as multiplying three shapes and three colors to classify objects on both dimensions simultaneously in a 3 × 3 matrix. It is assumed that this task can also be used to assess the preoperational development of classification (Inhelder & Piaget, 1959/1964).

A reasonable alternative hypothesis is that task complexity affects the developmental level that a child can demonstrate (Roberts, 1981). If the complexity of a task places it well beyond a child's highest functional level for a given context, then the child's performance on that task may seriously underestimate his or her functional level. There have been a few empirical tests of whether task simplification improves level of performance. For example, in several studies, simplified versions of Kohlberg's (1969) moral dilemmas produced more advanced moral stages than the original, highly complex dilemmas (e.g., Kuhn, 1976; Lee, 1971). However, hardly any studies have systematically investigated the general relation between task complexity and level of performance—how different degrees of task complexity affect level of performance.

To test how task complexity relates to level, researchers need to assess behavior with multiple tasks that vary in complexity. The high-support condition from our research employed one especially useful type of multiple-task method, the scalogram method: For every step in a predicted developmental sequence, a separate task was administered, so that there were independent assessments of all steps. In most cases, the task for a later step subsumed many of the skills assessed by tasks for earlier steps. Consequently, it was possible to determine how success on earlier tasks related to the type of error on later tasks. If the Piagetian assumption is correct that the nature of the errors on a complex task reflects children's developmental level, then children's errors on later tasks should generally show the same step as their functional level (the highest step that they passed) on the earlier tasks. If the alternative hypothesis is correct that performance on complex tasks underestimates most children's level, then children's errors should generally show a lower step than the highest step passed.

We have analyzed the errors in two sets of studies—one set investigating the development of the doctor role (Roberts, 1981; Watson & Fischer, 1980) and another set investigating the development of multiplication of classes (Fischer & Roberts, 1984). In both sets, the Piagetian hypothesis was not supported. Children's errors on tasks beyond their functional level did not typically reflect their functional level.

The alternative hypothesis was supported, but with an interesting twist. When children failed a task that was much more complex than their functional level, their errors typically were at a lower step than their functional level. However, when they failed a task that was only slightly more complex than their functional level, they frequently showed their functional level in their errors.

For example, in the research on the development of the doctor role, 11 children passed Step 5 (social role with two agents) and failed Step 6 (social role with three agents). Of those 11, 6 demonstrated Step 5 (their functional level) in their failing performance for Step 6, and 5 performed at a lower step. That is, with this task that was only slightly more complex than their functional level, approximately half the children showed their functional level in their errors.

When the task was much more complex than the children's functional level, the errors on the task were typically far below functional level. The children used a fall-back strategy, a simple behavior that allowed them to do something with the materials but did not demonstrate their functional level. In the doctor role research, the most common fall-back strategy was simply using the doctor's instruments on oneself or on the dolls, without making the dolls do anything; on the developmental scale, this behavior was either at Step 1, self as agent, or intermediate in complexity between Step 1 and Step 2. Children of all ages commonly used this strategy when they were faced with a task much more complex than their functional level.

Thus task complexity, like environmental support, seems to cause substantial variation in level of performance. In the face of environmental effects of these types, investigators have sometimes concluded that the concept of stage should be abandoned. We do not find such pessimism to be justified. Certainly, variations in stage induced by environmental factors are powerful and systematic. But variations in stage related to changes in age or developmental maturity are equally powerful and systematic (Fischer & Silvern, 1985).

State and structure should not be treated as fixed characteristics of a child independent of context, but should be seen as characteristics of a child-in-a-context. Then the order underlying variations in stage can be analyzed as a function of characteristics of both the environment and the person.

Individual Differences in Sequences

The variation of level over short time periods is one of the fundamental phenomena that needs to be treated in a broadened concept of stage. Another equally important phenomenon that needs to be treated is individual differences in developmental sequences. It appears that people do not all develop through identical stages or levels but show substantial individual differences. (The present argument relies heavily on the review by Fischer and Silvern, 1985, of the literature on stages and individual differences in cognitive development.)

There have been few direct empirical tests of whether there are individual differences in cognitive development. Most studies testing for a developmental sequence have been designed in such a way that it has not been possible to detect individual differences in the sequences. In general, these studies have used methods that tested a single sequence, allowing only confirmation or disconfirmation of that sequence and precluding detection of any other sequence. At best, they have been designed to allow the detection of individual differences in the speed of movement through a sequence but not in the nature of sequences (Feldman, 1980; Wohlwill, 1973).

On the other hand, a large amount of indirect evidence indicates strongly that individual differences in sequences are common. A host of environmental factors have been shown repeatedly to produce developmental variations (e.g., Biggs & Collis, 1982; Flavell, 1983; Levin & Simons, this volume; Turiel & Davidson, this volume). Because individuals routinely encounter diverse environments, the potent effects of the environment will necessarily lead them to differ developmentally. Similarly, a host of organismic factors have been shown repeatedly to produce developmental variations (e.g., Feldman, 1980; Fischer & Elmendorf, in press; Silvern, 1984; Vaillant, 1977). Because individuals manifest organismic diversity, the powerful effects of organismic factors will necessarily lead them to differ developmentally.

Fortunately, several researchers have tested directly for individual differences in developmental sequences. The evidence clearly shows individual differences in a diverse array of skills, including standard cognitive skills as well as language and social-emotional skills (Block, 1982; Bloom, Lightbown, & Hood, 1975; Corrigan, 1980; McCall, Eichorn, & Hogarty, 1977; Peters, 1977; Sroufe, 1979).

A powerful method for detecting individual differences is a version of scalogram analysis in which children are tested on multiple tasks that can potentially form different sequences. Scalogram statistics can then be used to determine empirically which of the possible sequences actually occur (Krus, 1977). One of the most interesting tests of individual differences used this method. Knight (1982) used seven separate tasks to assess the development of early reading skills in first through third grades. Children showed three different developmental sequences for these tasks, with each sequence evidencing a distinct relation to reading problems. The children who showed the most common sequence were normal readers with no evident reading problems, whereas those who showed the other two sequences had problems with spelling or rhyming, respectively. The three sequences seemed to reflect different strategies for reading. For example, the children with rhyming problems appeared to have general difficulties with the sounds of words, and consequently they apparently learned to read many words by learning the word as a whole rather than analyzing it into its sounds.

Evidence suggests that emotional factors may be especially potent at inducing individual differences in sequences (Block, 1982; Silvern, 1984; Sroufe, 1979; Vaillant, 1977). One extreme emotional experience that seems to induce major variations in development is child abuse, which appears commonly to induce severe dissociative disorders such as hysteria and multiple personality (Fischer & Pipp, 1984a). In a recent study of 100 patients with multiple personality, 97 of them had suffered severe abuse as children (Putnam, Post, Guroff, & Silberman, 1983). Also, the more types of abuse they had suffered, the more personalities they had as adults. In multiple personality, the individual constructs distinct, mutually incompatible personalities that control consciousness at different times, whereas in the normal personality, the individual constructs a relatively integrated personality. The different sequence followed in multiple personality seems to be evident by 4 to 6 years of age, by which time the child has clearly begun to use the strategy of constructing different personalities to deal with emotional experiences (Bliss, 1980).

In summary, stage and structure do not appear to be monolithic, unitary, fixed characteristics of a child. Instead, they seem to be dynamic, variable properties of the child-in-a-context. Different children show different stages in a given context, and different contexts produce different stages in the same child. To deal with the ambiguity of stage and structure, investigators need to use methods that allow them to detect variations in stage and to expand their range of observations to include likely sources of such variation. These sources include not only the age and learning history of the child, but also factors that specifically relate to short-term variations in stage such as degree of environmental support and task complexity, and factors that are likely to produce long-term individual differences in developmental sequences such as emotional experiences and strategies of approaching tasks.

A Case Study of Research on Stage and Structure: Optimal Levels

In the face of so much developmental variation, is it possible to articulate workable concepts of stage and structure? If a child's stage will not stay still for researchers to examine it, how can it be studied empirically? Must researchers simply operate on the faith that eventually they will find some kind of structural order?

We find substantial bases for optimism in the first-order consistencies that previous studies have already found. Age has systematic effects on children's general developmental level (Case, this volume; Fischer & Pipp, 1984b). Variations in environmental support produce orderly variations in functional level. When children fail a complex task, the developmental level they show in their errors seems to relate to how far the complexity of the task surpasses their functional level in that domain. Individual differences in developmental sequences appear to arise in sensible ways from organismic characteristics and experiences.

The nature of stages and structures, we believe, will be evident when researchers study how these first-order consistencies combine in a collaborative analysis of psychological development (Bullock, 1981). A case study from our laboratory demonstrates how stage and structure can be fruitfully investigated in this way.

This study investigated the hypothesis that development proceeds through a series of optimal levels. That is, at any point in development, children evidence an optimal level, an upper limit on the structural complexity of skills they can control. The limit increases sharply at certain points in development, when children enter a new developmental level. That is, a criterion for the emergence of a new stage or level is a sharp increase or spurt in the upper limit of children's skills. A number of studies document the occurrence of such spurts at various points in development (e.g., Corrigan, 1983; Dromi, this volume; Kenny, 1983; Tabor & Kendler, 1981).

According to skill theory, people develop through 10 such levels between 4 months and 30 years of age. The levels are defined structurally in terms of sets of variations in behavior and relations between those sets. For the purposes of this chapter, it is not necessary to review the structural definitions for all the levels; they are described in detail in Fischer (1980). The concept of optimal level is elaborated by Fischer and Bullock (1981) and Fischer and Pipp (1984b).

Optimal level is not a characteristic of the child alone but of the child and the environment collaborating. For children to function at their upper limit, a whole host of organismic and environmental conditions must be met. Here are a few of them. The children must be in an emotional and motivational state that facilitates optimal performance. They must be familiar with the content of the skills to be assessed. They must have had the opportunity to practice the skills. The context for performing the tasks must provide support for high-level performance. The tasks presented to the children must be at a level of complexity that facilitates use of their optimal level rather than interfering with it. The children must adopt the strategy for approaching the tasks that is assumed by the investigator's structural analysis of the tasks. And so forth.

Depending on whether or not children are functioning at optimal level, the developmental patterns they manifest will differ greatly. At optimal level, children will show a spurt or discontinuity in performance as they enter a new cognitive-developmental level. When not at optimal level, they will not show a spurt but will demonstrate slow, gradual, continuous improvement in performance. (For a discussion of factors other than optimal level that can produce spurts, see Fischer and Bullock, 1981.) In the developmental literature, a common controversy between structuralist and functionalist approaches is whether developmental change is discontinuous or continuous. According to the optimal-level hypothesis, both approaches are correct: Development is both discontinuous and continuous. At optimal level, children show discontinuous change upon the emergence of a new level. Under ordinary circumstances, however, children seldom function at optimal level, and so they show continuous change.

In skill theory, most of the controversies between structural and functional approaches are recast so that the controversies disappear. For example, are individual differences the norm in cognitive development, as functionalists often argue? Or are there universal developmental sequences shown by all people? According to skill theory, both arguments are true. When behavior is examined at a high degree of abstraction, where each behavior can be fit into a highly general type of structure, all people develop through the same ten levels. On the other hand, when behavior is examined at an intermediate level of abstraction that captures the specific contents and strategies of the behavior, individual differences in sequences are the norm.

The study testing the optimal-level hypothesis focused on 2 of the 10 hypothesized levels (Fischer, Hand, & Russell, 1984; Kenny & Fischer, 1984). Six tasks dealing with arithmetic concepts were constructed to assess each level. At Level 7, called single abstractions, children can coordinate two or more concrete examples to form a general concept such as intention, responsibility, addition, or subtraction. Each of the six tasks required children to demonstrate an understanding of one of three general arithmetic operations—addition, subtraction, or multiplication. Correct performance required provision of a general definition of the operation and application of that definition to concrete arithmetic problems.

At Level 8, abstract mappings, children can construct simple relations between two or more abstractions, such as intention and responsibility or addition and subtraction. Each of the eight tasks required children to demonstrate an understanding of one of four general relations between arithmetic operations—addition with subtraction, addition with multiplication, multiplication with division, and subtraction with division. Correct performance required provision of a general explanation of the relation and application of that explanation to concrete arithmetic problems.

The design for the study varied characteristics of both the people and the environment. Subjects were eight students (four of each sex) at each age/grade from third grade through the beginning of college. They were tested individually under four different environmental conditions, involving low and high environmental support and low and high practice. Two conditions were administered in each of two sessions. In the initial session, subjects were first given each of the tasks and asked to give an answer (low-support, low-practice condition). Second, they were shown a good answer for each task and given some time to examine the answer; then the answer was taken away, and they gave an answer of their own again (high-support, low-practice condition). At the end of the first session, subjects were encouraged to work on the problems at home. Two weeks later, they were tested again with the same procedures as in the first session. Because the first session had provided practice on the tasks and the intervening two weeks provided more opportunity for practice, the two conditions in the second session both involved high practice: a low-support, high-practice condition, and then a high-support, high-practice condition. Predictions were that with low support and low practice, performance for both levels would show gradual, continuous development; whereas with high support and high practice, performance would show abrupt, discontinuous development. The other two conditions would produce patterns of data intermediate between these two extremes. The age/grades predicted for the appearance of spurts were fourth to sixth grade for Level 7 and eighth to tenth grade for Level 8.

The predictions were strongly supported. In the low-support, low-practice condition, performance increased slowly with age/grade, and even in the college students it did not reach more than 50% correct. In the high-support, high-practice condition, performance increased abruptly at a certain age/grade for each level. For the six Level 7 tasks, performance spurted from near zero in third grade to over 60% correct in fifth grade. For the eight Level 8 tasks, it spurted from near zero in ninth grade to over 80% correct in tenth grade. The predictions applied to individual subjects as well, in that in the high-support, high-practice condition, virtually all subjects showed performance consistent with the upper-limit criterion: They either passed most or all the tasks at a given level, or they failed all or most of them (Fischer, Pipp, & Bullock, 1984).

This study illustrates, we believe, that research based on a collaborational approach can illuminate concepts of stage and structure. The study included a sample of tasks, a range of ages, and a range of environmental conditions. The optimal-level hypothesis was used to predict the conditions under which certain of the tasks would demonstrate an abrupt developmental change in a specific age region. The results of the study supported two hypothesized levels (stages) of cognitive development based on a structural-developmental theory, but the stages were not unitary or monolithic. They appeared and disappeared depending on the particular environmental condition examined.

The arithmetic study is only one of several investigations that indicate the promise of methods that cast a wide net including both organismic and environmental variables. The other studies have generally used different specific methods to determine the conditions under which stagelike development occurs, but they all share an approach that emphasizes inclusion of a range of variables and behaviors. McCall (1983) and Fischer, Pipp, and Bullock (1984) provide reviews of some of these methods.

Conclusion: Guidelines for Doing Research on Stage and Structure

In part, this article is a response to a certain malaise—even dismay—that has infused some developmental circles in recent years. Because Piaget's stage theory does not seem to have worked and because stages and structures are so difficult to pin down, some developmentalists seem to have concluded that the problems of stage and structure are insoluble. Even at the conference on which this book is based, a few of the discussions showed this malaise. Of course, it is not possible to offer a simple recipe for doing significant research on stage and structure. If there were a simple recipe, the problems of stage and structure would have been solved long ago. But we do believe it is possible to do research in a way that is likely to produce illuminating findings.

In this chapter, we have attempted to outline a framework for doing research to illuminate stage and structure. First, we placed the difficulties with stage and structure in the context of epistemology: The difficulties are in no way unique but are merely one instance of a general problem in human knowledge. Knowledge is equivocal, ambiguous, or uncertain, and although its uncertainty can be substantially reduced, it cannot be eliminated.

Second, procedures that work for reducing the uncertainty of scientific knowledge in general can be applied productively to investigating stage and structure. Indeed, they lead to a set of guidelines for doing research on stage and structure. The guidelines are necessarily global, general, and incomplete, because they are rules of thumb, not step-by-step recipes.

1. Sample a range of tasks, environmental conditions, organismic characteristics, and aspects of behavior. Some particular types of sampling have proved especially useful in previous research.

a. It is useful if the tasks can be fit into an a priori structural analysis. For example, they can be placed into a developmental ordering, or they can be categorized into a few specific stages or structures. Such orderings and categorizations provide a tool for analyzing the variable effects of tasks, and therefore they can provide great power in analysis of developmental patterns such as continuities and discontinuities (Fischer, Pipp, & Bullock, 1984). It is also useful if the study is designed so that the tasks allow the detection of individual differences in developmental patterns.

b. It is useful if the environmental conditions provide variable degrees of support for high-level performance so that the researcher can examine the zone of developmental stages or levels that naturally appear in a given context.

c. It is useful if subjects vary along at least one or two dimensions that are likely to produce structural or stage differences such as age, cognitive style or strategy for the domain(s) under investigation, and social-emotional state or background. When these characteristics can be ordered or categorized systematically a priori, they too can provide great power in analysis of developmental patterns.

d. It is useful if several different aspects of behavior can be observed. Consideration of a broad range of behaviors tends to greatly reduce the number of possible alternative hypotheses or theories for interpreting a set of results.

2. When possible, pit rival theories against one another, or expand the range of observations to include those from more than one theory or approach.

a. If rival theories are to be pitted against one another, they need in fact to have substantial overlap. In particular, they should be based on similar tasks, testing conditions, and aspects of behavior. The study of cognitive development is full of controversies between rival theories that evaporate as soon as researchers recognize the different types of observations on which the theories are based. The controversy over conservation of number illustrates a common scenario: Researchers argue about when a skill first really develops, and further research indicates that their different findings arose from the fact that they were not assessing the same skill.

b. One of the surest ways to make theoretical progress is to expand the range of observations used for testing a theory. A typical result of such an effort is that either the theory is extended and improved or it is abandoned. Two theories that appear to be rivals but that do not deal with the same types of observations can often be integrated by including in a single study the types of observations from both theories. The controversy over conservation of number seems to have been resolved by studies of this type.

3. Work with well-articulated hypotheses that clearly specify particular patterns of data, and use methods that directly assess those patterns of data. Early in a research effort, hypotheses may not be well articulated, and then movement toward clear predictions of patterns of data should be a major goal of the research effort. Although we have not directly addressed this issue in the chapter, it has informed all parts of our argument. For example, in the discussion of individual differences, we pointed out that most cognitive-developmental research testing for universal se-

uences (such as Piaget's or Kohlberg's) has been designed in such a way that individual differences in sequences can not be detected. Two of the most fundamental problems in cognitive-developmental research have been that (a) the patterns of data specified by a hypothesized stage or structure have never been stated and (b) when patterns have been specifiable, the methods used have not allowed the investigator to detect those patterns. Fischer and Bullock (1981, 1984) provide guidelines for moving toward specifying patterns of data for stage and structure and then using appropriate methods to detect those patterns.

References

Anderson, J. R. (1978). Arguments concerning representations for mental imagery. *Psychological Review, 85,* 249–277.

Biggs, J., & Collis, K. (1982). *A system for evaluating learning outcomes: The SOLO taxonomy.* New York: Academic Press.

Bliss, E. L. (1980). Multiple personalities. *Archives of General Psychiatry, 37,* 1388–1397.

Block, J. (1982). Assimilation, accommodation, and the dynamics of personality development. *Child Development, 53,* 281–295.

Bloom, L., Lightbown, P., & Hood, L. (1975). Structure and variation in child language. *Monographs of the Society for Research in Child Development, 40* (2, Serial No. 160).

Broughton, J. M. (1981). Piaget's structural developmental psychology V. Ideology-critique and the possibility of a critical developmental theory. *Human Development, 24,* 382–411.

Bullock, D. (1981). On the current and potential scope of generative theories of cognitive development. In K. W. Fischer (Ed.), *Cognitive development. New Directions for Child Development, No. 12.* San Francisco, CA: Jossey-Bass.

Campbell, D. T. (1969). Prospective: Artifact and control. In R. Rosenthal & R. L. Rosnow (Eds.), *Artifact in behavioral research.* New York: Academic Press.

Campbell, D. T. (1974, September). *Qualitative knowing in action research.* Kurt Lewin Award Address, Society for the Psychological Study of Social Issues, Meeting with the American Psychological Association, New Orleans.

Cook, T. D., & Campbell, D. T. (1979). *Quasi-experimentation: Design and analysis issues for field settings.* Chicago: Rand McNally.

Cooper, R. G., Jr. (1984). Early number development: Discovering number space with addition and subtraction. In C. Sophian (Ed.), *Origins of cognitive skills.* Hillsdale, NJ: Erlbaum.

Corrigan, R. (1980). The use of repetition to facilitate language acquisition. *Journal of Psycholinguistic Research, 3,* 231–241.

Corrigan, R. (1983). The development of representational skills. In K. W. Fischer (Ed.), *Levels and transitions in children's development. New Directions for Child Development, No. 21.* San Francisco, CA: Jossey-Bass.

Dewey, J. (1948). *Reconstruction in philosophy* (enlarged ed.). Boston: Beacon.

Duhem, P. (1962). *The aim and structure of physical theory.* New York: Atheneum.

Feldman, D. H. (1980). *Beyond universals in cognitive development.* Norwood, NJ: Ablex.

Feshbach, S. (1970). Aggression. In P. H. Mussen (Ed.), *Carmichael's manual of child psychology (Vol. 2).* New York: Wiley.

Fischer, K. W. (1980). A theory of cognitive development: The control and construction of hierarchies of skills. *Psychological Review, 87,* 477–531.

Fischer, K. W., & Bullock, D. (1981). Patterns of data: Sequence, synchrony, and constraint in cognitive development. In K. W. Fischer (Ed.), *Cognitive development. New Directions for Child Development, No. 12.* San Francisco, CA: Jossey-Bass.

Fischer, K. W., & Bullock, D. (1984). Cognitive development in school-age children: Conclusions and new directions. In W. A. Collins (Ed.). *The years from six to twelve: Development during middle childhood.* Washington, DC: National Academy Press.

Fischer, K. W., & Elmendorf, D. (in press). Becoming a different person: Transformations in personality and social behavior. In M. Perlmutter (Ed.), *Minnesota Symposium on Child Psychology (Vol. 18)*. Hillsdale, NJ: Erlbaum.

Fischer, K. W., Hand, H. H., & Russell, S. L. (1984). The development of abstractions in adolescence and adulthood. In M. Commons, F. A. Richards, & C. Armon (Eds.), *Beyond formal operations*. New York: Praeger.

Fischer, K. W., Hand, H. H., Watson, M. W., Van Parys, M., & Tucker, J. (1984). Putting the child into socialization: The development of social categories in the preschool years. In L. Katz (Ed.), *Current topics in early childhood education (Vol. 5)*. Norwood, NJ: Ablex.

Fischer, K. W., & Pipp, S. L. (1984a). Development of the structures of unconscious thought. In K. Bowers & D. Meichenbaum (Eds.), *The unconscious reconsidered*. New York: Wiley.

Fischer, K. W., & Pipp, S. L. (1984b). Processes of cognitive development: Optimal level and skill acquisition. In R. J. Sternberg (Ed.), *Mechanisms of cognitive development*. San Francisco, CA: Freeman.

Fischer, K. W., Pipp, S. L., & Bullock, D. (1984). Detecting discontinuities in development: Method and measurement. In R. N. Emde & R. Harmon (Eds.), *Continuities and discontinuities in development*. New York: Plenum.

Fischer, K. W., & Roberts, R. J., Jr. (1984). *A developmental sequence of classification skills*. Unpublished manuscript, University of Denver, CO.

Fischer, K. W., & Silvern, L. (1985). Stages and individual differences in cognitive development. *Annual Review of Psychology, 36*, 613–648.

Flavell, J. H. (1983). Structures, stages, and sequences in cognitive development. In W. A. Collins (Ed.), *Minnesota Symposium on Child Psychology (Vol. 15)*. Hillsdale, NJ: Erlbaum.

Gelman, R. (1972). Logical capacity of very young children: Number invariance rules. *Child Development, 43*, 75–90.

Gelman, R. (1978). Cognitive development. *Annual Review of Psychology, 29*, 297–332.

Halford, G. S., & Boyle, F. M. (in press). Do young children understand conservation of number? *Child Development*.

Hand, H. H. (1981). The relation between developmental level and spontaneous behavior: The importance of sampling contexts. In K. W. Fischer (Ed.), *Cognitive development. New Directions for Child Development, No. 12*. San Francisco, CA: Jossey-Bass.

Hand, H. H. (1982). The development of concepts of social interaction: Children's understanding of nice and mean. (Unpublished doctoral dissertation, University of Denver, 1981). *Dissertation Abstracts International, 42*(11), 4578B. University Microfilms No. DA8209747).

Hanson, N. R. (1961). *Patterns of discovery*. Cambridge, England: Cambridge University Press.

Inhelder, B., & Piaget, J. (1958). *The growth of logical thinking from childhood to adolescence* (A. Parsons & S. Seagrim, Trans.). New York: Basic Books. (Originally published, 1955)

Inhelder, B., & Piaget, J. (1964). *The early growth of logic in the child* (G. A. Lunzer & D. Papert, Trans.). New York: Harper & Row. (Originally published, 1959)

Kenny, S. L. (1983). Developmental discontinuities in childhood and adolescence. In K. W. Fischer (Ed.), *Levels and transitions in children's development. New Directions for Child Development, No. 21*. San Francisco, CA: Jossey-Bass.

Kenny, S. L., & Fischer, K. W. (1984). [Optimal and functional levels in the development of abstractions for arithmetic operations.] Unpublished data, University of Denver.

Knight, C. C. (1982). Hierarchical relationships among components of reading abilities of beginning readers. (Unpublished doctoral dissertation, Arizona State University, Tempe, 1982). *Dissertation Abstracts International, 43*(2), 403A. (University Microfilms No. DA8216442)

Kohlberg, L. (1969). Stage and sequence: The cognitive-developmental approach to socialization. In D. A. Goslin (Ed.), *Handbook of socialization theory and research*. Chicago: Rand McNally.

Krus, D. J. (1977). Order analysis: An inferential model of dimensional analysis and scaling. *Educational and Psychological Measurement, 37*, 587–601.

Kuhn, D. (1976). Short-term longitudinal evidence for the sequentiality of Kohlberg's early stages of moral judgment. *Developmental Psychology, 12*, 162–166.

Lee, L. C. (1971). The concomitant development of cognitive and moral modes of thought: A test of selected deductions from Piaget's theory. *Genetic Psychology Monographs, 83,* 93–146.

McCall, R. B., Eichorn, D. H., & Hogarty, P. S. (1977). Transitions in early mental development. *Monographs of the Society for Research in Child Development. 42*(3, Serial No. 171).

McCall, R. B. (1983). Exploring developmental transitions in mental performance. In K. W. Fischer (Ed.), *Levels and transitions in children's development. New Directions for Child Development, No. 21.* San Francisco, CA: Jossey-Bass.

Peirce, C. S. (1931–1958). *The collected papers of Charles Sanders Peirce* (Vols. 1–8, C. Hartshorne, P. Weiss, & A. Burks, Eds.). Cambridge, MA: Harvard University Press.

Peters, A. M. (1977). Language learning strategies: Does the whole equal the sum of the parts? *Language, 53,* 560–573.

Piaget, J. (1970). Piaget's theory. In P. H. Mussen (Ed.), *Carmichael's manual of child psychology (Vol. 1).* New York: Wiley.

Piaget, J., & Szeminska, A. (1941/1952). *The child's conception of number* (C. Gattegno & F. M. Hodgson, Trans.). New York: Humanities Press. (Originally published 1941)

Purcell, K. (1983). *Children of divorce: The relationship between cognitive level and children's socio-emotional reactions to divorce.* Unpublished master's thesis, University of Denver, CO.

Putnam, F. W., Jr., Post, R., Guroff, J., & Silberman, E. (1983, April). *One hundred cases of multiple personality disorder.* Paper presented at the American Psychiatric Association Convention, New York.

Reese, H. W., & Overton, W. F. (1970). Models of development and theories of development. In L. R. Goulet & P. B. Baltes (Eds.), *Life-span developmental psychology: Research and theory.* New York: Academic Press.

Reichardt, C. S. (1983, October). *Assessing cause.* Paper presented at the Joint Meeting of the Evaluation Network and the Evaluation Research Society, Chicago.

Roberts, R. J., Jr. (1981). Errors and the assessment of cognitive development. In K. W. Fischer (Ed.), *Cognitive development. New Directions for Child Development, No. 12.* San Francisco, CA: Jossey-Bass.

Rubin, K. H., Fein, G. G., & Vandenberg, B. (1983). Play. In E. M. Hetherington (Ed.), *Socialization, personality, and social development.* New York: Wiley.

Sameroff, A. J., & Chandler, M. J. (1975). Reproductive risk and the continuum of caretaking casualty. In F. D. Horowitz (Ed.), *Review of child development research (Vol. 4).* Chicago, IL: University of Chicago Press.

Siegler, R. S. (1981). Developmental sequences within and between concepts. *Monographs of the Society for Research in Child Development, 46* (2, Serial No. 189).

Silverman, I. W., & Briga, J. (1981). By what process do young children solve small number conservation problems? *Journal of Experimental Child Psychology, 32,* 115–126.

Silvern, L. (1984). Emotional-behavioral disorders: A failure of system functions. In E. Gollin (Ed.), *Malformations of development.* New York: Academic Press.

Sroufe, L. A. (1979). Socioemotional development. In J. D. Osofsky (Ed.), *Handbook of infant development.* New York: Wiley.

Starkey, P., & Cooper, R. (1977). *The role of estimation skills in the development of number conservation.* Paper presented at the meetings of the Jean Piaget Society, Philadelphia.

Tabor, L. E., & Kendler, T. S. (1981). Testing for developmental continuity or discontinuity: Class inclusion and reversal shifts. *Developmental Review, 1,* 330–343.

Vaillant, G. E. (1977). *Adaptation to life.* Boston, MA: Little, Brown.

Van Parys, M. M. (in press). The relation of use and understanding of sex and age categories in preschool children. (Unpublished doctoral dissertation, University of Denver, CO, 1983). *Dissertation Abstracts International.* in press.

Vygotsky, L. (1978). *Mind in society* (M. Cole, V. John-Steiner, S. Scribner, & E. Souberman, Eds.). Cambridge, MA: Harvard University Press.

Watson, M. W., & Fischer, K. W. (1980). Development of social roles in elicited and spontaneous behavior during the preschool years. *Developmental Psychology, 16,* 483–494.

Westerman, M. A. (1980). *Nonreductionism in mainstream psychology: Suggestions for a positive hermeneutics*. Paper presented at the convention of the American Psychological Association, Montreal.

Wittgenstein, L. (1953). *Philosophical investigations*. New York: Macmillan.

Wohlwill, J. F. (1973). *The study of behavioral development*. New York: Academic Press.

11

*When Do Structural Changes Underlie Behavioral Changes?**

Tamar Globerson
School of Education
Tel Aviv University

Developmental theorists have long talked about structural changes which occur in cognitive development. In general, the structuralist approach is that cognitive development is an outcome of underlying structural changes in the cognitive system. The foremost representative of this approach is Piaget (1970). In comparison with the structuralist approach to cognitive development, nonstructuralists (e.g., Fodor, 1972, 1975; Gelman, 1983) have claimed that existing structures do not change with development but rather the same structures become increasingly accessible and applicable in a wider range of contexts.

Many other cognitive and developmental theorists have talked about changes in the cognitive system in the form of changes in control structures, metacognitive structures, operators, procedures, strategies, rules of operation, concepts, theories, and so on (e.g., Carey, 1983, in press; Case, 1978, in press; Case, Marini, Keogh, Dennis, & Goldberg, this volume; Chi, 1978; Fischer, 1980; Halford, 1970; Keil, 1979; Pascual-Leone, Goodman, Ammon, & Subelman, 1978; Siegler, 1976). Often it is not clear from these latter theories if their account of change is structural in nature. Part of the problem arises from not having a clear definition as to what a structure is and how we can know that structural change has taken place.

The present chapter is devoted to shedding some light on this issue. I will do so by asking two questions. First, what are structural changes? Second, how can we tell that structural changes have occurred? Taken together, these issues will, I hope, lead to the development of a coherent theory of structural change.

What Constitutes a Structural Change?

It is not possible to define a structural change before defining a structure. A *structure* is a distinct set of relationships among a set of elements (and their proper-

* I would like to express my gratitude to all the people who helped me shape my ideas through my conversations with them and/or their comments on earlier versions of this chapter: Susan Carey, Robbie Case, Andy diSessa, Herman Epstein, Veronica Fabian, Tom Goblick, Thomas Kuhn, Iris Levin, Moti Rimor, Yaacov Shapira, Sidney Strauss, Elliot Turiel, Sylvia Weir, and Tamar Zelniker. Some ideas expressed here reflect my own assimilation of things I learned from them; therefore the responsibility is totally mine. The article was written when I was a Bantrell Postdoctoral Fellow at the Laboratory for Computer Science, MIT. This is Working Paper No. 32 of the Tel Aviv University Study Group in Human Development and Education. Author's permanent address: School of Education, Tel Aviv University, Ramat Aviv 69978, Israel.

ties). For example, a *square* is a structure of geometric relationship (i.e., one-to-one proportions; right angles) among four elements (i.e., lines) which have certain properties (i.e., straightness, continuity, connectedness). *Face* is a structure of spatial relationship (i.e., right–left, up–down) among certain facial elements (i.e., two eyes, nose, mouth, etc.).

According to this definition of a structure, *any* change in a structure's description—be it its elements or their relationship—may be considered a *structural change*. Therefore, a change in *elements* such as drawing the square in broken lines (thus changing the properties of continuity and connectedness) or in wavy lines (thus changing the property of straightness) changes the definition of a square. So does a change in the *relationship* among elements, such as slanting the square (thus changing the angles) or elongating two opposite sides (thus changing the proportions).

Although all of these constitute structural changes by virtue of their contradicting the core definition of a *square,* most of us probably feel that the former two changes (i.e., broken and wavy lines) are different from the latter two (i.e., changes in angles and proportions). We accept quite easily the former two shapes as squares, albeit atypical ones (cf. Rosch, 1973), whereas the latter two are perceived as different shapes—parallelogram and rectangle. Why is this so?

The described examples already demonstrate the first point of departure from the claim that any change in a structure's description is a structural change: it is true that this is a structural change by our definition, but it is not necessarily *perceived* as such by us. This demonstrates a basic characteristic of the cognitive system, which applies to us both as perceivers of change in others and as subjects of cognitive change ourselves, and that is—the cognitive system is flexible in that it is "tolerant" to certain *kinds* and/or *degrees* of deviation from the core description of a structure, thus accepting these deviations as particular instantiations of the structure. Other deviations, however, are perceived differently and may be subjected to a structural change. All of these mean that there are *psychological constraints* (rules, principles, operators, etc.) which govern structural change. These set the boundaries as to the kind and amount of deviations from a given structure they allow for, within which limits no structural change is required, but when those limits are violated, then structural change is warranted. Such psychological constraints have not been specified. Very few constraints have been suggested and, when they have been, it is mostly as general rules, such as the tendency to resolve conflicts or contradictions in order to regain equilibrium (Pascual-Leone, et al., 1978; Piaget, 1970); or the tendency to be efficient, such that at minimum cost (e.g., effort, number of schemes) one could achieve maximum benefit (e.g., scope of understanding, Pascual-Leone et al., 1978); or the limitation of structural change according to certain logical constraints (Keil, 1981, this volume). However, for the most part, these constraints have not been specified sufficiently to account for specific structural changes.

The second point of departure from the definition of structural change is an explication of the previous point which has to do with the *kinds of change* in a structure's description that are most likely to be perceived or qualified as structural change. To specify these, I provide several examples. The first demonstrates a general point, and that is that not every additional information (knowledge) causes a structural change. Suppose that during a visit to a new country one learns about opossum, a new marsupial mammal similar to a familiar animal (e.g., kangaroo). This new mammal may not violate anything one knows about mammals, marsupial mammals, or animals in general. In this case, no change is introduced to the core definition of the relevant structure, in spite of the fact that the new animal has been added to one's animal list.

The next two examples present situations in which a structure's description goes through some change, but in one case the change may not be perceived as a structural change, whereas in the second one, the change may be perceived as such.

Suppose that you learn about a new fish found in South America—Caruso, the singing fish. This additional property—singing—does introduce a change in the *fish* description, but nevertheless may not be perceived by us as a significant structural change or a change at all, because it only adds another property to a list of other fish properties. We may know that although fish seem mute, they "talk" with each other, and via various techniques, one can actually hear them talk. Therefore, it may not be a violation of what we already know about fish were we to find out that they can also sing. The reason for this is that the properties presently known to us about fish define a scope of *possible* other properties which may be unknown at the moment but are nevertheless within the limits of the existing structure. This, however, will not hold true if the new breed makes room for new relationship; e.g., one merges fishes and birds together on a new basis (singing). This change will be quite readily regarded by most as a structural change.

To illustrate this last point more realistically, let's take a familiar example—the whale. Suppose that one has only three animal categories based on their means of locomotion: *fishes* for all swimming-in-water creatures, *landeers* for all moving-on-land creatures, and *birds* for all flying-in-air creatures. Suppose then that one learns that a whale is fed on mammary-gland secretion, just like many *landeers*. Here a new element has been added to the whale description. This new addition may not cause any structural change; however, it makes room for drawing new relations among animals across the existing categories. One may want to include all creatures fed on mammary-gland secretion in one category—mammals. This change in one's categorization criterion (i.e., from means of locomotion to means of feeding) may cause quite a broad change, because it makes one cut across existing structures, restructure animal members in a different way, and create new categories by ways of both merging separate (whole or part) categories into one and separating unitary categories into parts. Such structural changes have occurred in the history of many scientific concepts (cf. Kuhn, 1977; Weiser & Carey, 1983).

These examples demonstrate the second point of departure from the general definition of a structural change: By and large, it is changes in the *relationship* among a structure's elements rather than the elements themselves that are likely to be perceived by us as (significant) structural change. This has also been shown in the context of drawing analogies (Gentner, 1983). It should be stressed that the changes may take place at any level of generalization or abstraction which is relevant to consider: They may occur in the internal relationship between the elements of a particular structure or between a structure and other structures (in which case these structures are the elements of a superordinate structure). This second point explains the earlier examples about the square. The first two changes (i.e., broken or wavy lines) are changes in the properties of elements, leaving the relationship among elements intact. These do not seem to disrupt our *square* structure. In comparison, the latter two changes (i.e., slanting or elongating two sides) change the relations among elements (i.e., angles between lines or proportions of lines); this seems to violently disrupt our *square* structure and transform it into a different structure (i.e., *parallelogram* or *rectangle,* respectively).

The role of maintaining relations intact when we organize the world around us is important in everyday phenomena. Take a modern drawing of a face: as long as the basic relations are hinted at, even with a minimal set of elements (e.g., two eyes and a mouth in roughly the correct locations), it may be easy for us to apply our *face* structure to it; however, scrambled faces are hard to recognize even if many elements are present. The same is the case with tunes: take the minimal set of consecutive notes, and the tune may be recognized; but take many scrambled notes, and the tune may be beyond recognition.

This discussion may also shed light on the aforementioned structuralist/nonstructuralist debate, as in the following question: Does a change in a structure's applicability involve a structural or nonstructural change? The answer seems to depend on whether or not the specific change has bearing on a structure's relations (internal or external). As specified before, not every change in a structure's description may be perceived by us as a structural change. Let's take an example of one change—increased applicability of the conservation principle, from quantities to volumes to energy. One may claim that this increased applicability is a mere widening of the same principle, that nothing is changed in the generic structure. Maybe so, but if this increased applicability leads to drawing new connections among elements such as between mass and energy, then this increased applicability may very well reflect structural change! Truly, the change may be in the products of application rather than the conservation principle itself.

At this point, the task of assessing a structural change may seem quite clear: If a structural change is mainly characterized by changes that occur in the relationship among a structure's elements, then all we have to do is assess the structure's description in time 1, then in time 2, and finally assess the change in relationship among elements between the two times. Although on the surface this may appear to be straightforward, it is not an easy task to accomplish for the following reasons, all of which should be accounted for by a theory of structural change.

Structure's Description

Before any attempt is made to assess a change, one has to give a valid description of a structure (i.e., assessing its elements and their internal and external relations). This can be very problematic, especially in cases where it is hard to define a structure in terms of its necessary and sufficient components, or when the distribution of a structure's components is overlapping with another structure, thus making it difficult to identify the boundaries of the relevant structure (Smith & Medin, 1981). Many natural-kind and scientific concepts are of this sort. In addition, it has been shown that in order to assess a particular structure, one has to understand a great deal about the conceptual framework (e.g., theory) in which it is embedded; otherwise, it may become a futile task (Kuhn, 1977; Weiser & Carey, 1983). As a consequence, it seems that it is necessary to know a great deal about the "ecology" of a target structure before assessing it.

Extent of Change

When a structure changes, the old structure, to a certain degree, shares features (i.e., elements and relations) with the new structure. There is some agreement among developmentalists that the changes that occur in cognitive development must be *large, significant,* and *meaningful* in the sense that they must have much bearing on one's cognitive performance (cf. Fischer, Pipp, & Bullock, in press). Hence, when researchers attempt to assess structural changes, they must have explicit criteria for assessing only those changes of this sort.

Universality Versus Relativity of Criteria

To what extent can one claim that the criteria for structural change are universal or, in contrast, are relative across age, type of individual, and domain-specific task demands? It is possible that there is no one answer to the question of structural change; it is possible that these criteria change under different circumstances. Alternatively, it is possible that some criteria are universal and others are relative. All this should be well-specified before one can assess structural changes.

To sum up here, the foregoing discussion has indicated that a theory of structural change, which is what we are after here, requires more than a clear definition of what a structural change is. I have shown that not every addition of information introduces change to a structure's description, and that not every change in a structure's description is perceived by us as a structural change. In addition, a theory of structural change should include criteria for assessing a structure and the *kind* and *extent* of changes that are *significant* for cognitive development; also, it needs to specify the psychological *constraints* which govern structural changes in the cognitive system. I have already begun to specify what constitutes relevant changes in cognitive development, and those are changes in a structure's relations rather than elements. These are by no means the only relevant kind of changes in the cognitive system which occur with development, although I believe they are the most important ones. Later on, I touch on the issue of the extent of changes which are relevant to cognitive development. Note that I have not engaged in a discussion of the type of structures which are involved in a theory of structural change (e.g., procedures, rules, concepts, etc.). I have not done so because I believe that structural changes occur in different parts of the cognitive system, and that this issue should be the concern of the researcher who chooses particular structures as targets for the assessment of structural changes; this choice is made based on one's theoretical inclinations.

The Detectibility of Structural Changes

An important aspect of any theory of structural change is how we can tell that structural changes have occurred. The rest of this chapter focuses on this issue, and I consider it important for the following reasons. First and foremost, there is a danger of circularity in the theoretical definition of structural change and its assessment. We define structural change as changes in a structure's description, then we assess changes in a structure's description and claim that we proved structural changes. But surely, this is true by definition! It follows from this that to assess structural changes properly, we need *objective* indicators which are not part of a definition of structural change.

The second reason for looking for objective indicators is that a central problem in the detection of structural change is that the latter cannot be directly observed and can only be inferred from one's behavior; however, not every structural change is apparent in the person's behavior, and not every apparent change in behavior has to be a reflection of an underlying structural change. As for the first part, a person may function similarly at two different times, but the underlying cognitive structures may be different (Mounoud, this volume; Strauss, 1982). Similarly, although the same behavior may be exhibited in two different contexts, they may reflect two different intentions (Karmiloff-Smith, this volume). In such cases, it is hard to detect underlying structural changes by merely observing surface-level behavior (i.e., performance). As for the second part, it is possible that under certain task situations, one may not be able to apply a certain structure because of inaccessibility of the structure at that moment (e.g., due to fatigue or certain task conditions which prevent the application). In other words, the person may apply different structures in different contexts, but this by no means reflects structural change.

This situation complicates our task of detecting underlying structural changes, and we seem to be in a double bind here. Given the premises that: (a) structural changes can occur without behavioral manifestation; (b) there is no other way to detect structural changes except for observing behavior; and (c) behavioral changes can occur without being related to underlying structural change, then how will we be able to tell whether these observed behavioral changes are those which relate to underlying structural changes? This task seems impossible, unless we find those conditions under which structural changes and behavioral changes are in full correspondence. This is the only way out of the complicated situation. The following sections discuss this issue.

Conditions in Which Structural Changes Underlie Behavioral Changes

I chose to discuss this issue within the physical domain before I do so within the psychological domain for two reasons. First, the psychological literature regarding structural changes has escaped neither the previously described two problems. That is, many psychological arguments for structural change suffer from the circularity problem, and they do not bring any objective measures to confirm that indeed the observed behavioral changes reflect underlying structural changes. Second, in the physical domain, structures and their changes have been well defined, detected, and measured objectively. Hence, the work in the physical domain can serve as a guideline for our work in psychology and can do so without needing the claim that the two domains are analogical. Later on, I also discuss the issue within the psychological domain.

The physical phenomenon to be analyzed here is that of state transitions between solid, liquid, and gas. These are obviously three different states of the same matter. The appearance and behavior of matter are different in the different states. But does the structure of matter change when the state changes? The answer is not uniform across situations and matter. Within this domain, a structure is defined as the arrangement of atoms and is described by conditional probability functions, i.e., the probability given an atom at a certain position (the ''target'' atom) to find another atom of the same sort in another position at the same time. In a solid (e.g., crystal), the atoms are arranged in fixed positions, and the conditional probability is close to one; this means that the order of atoms does not depend on time and distance from the target atom. In comparison, in fluid states—liquid and gas—the atoms do not occupy fixed positions because they are in continual motion, and the crystalline structure is absent. The conditional probabilities are less than one and they depend on time and distance from the target atom. The conditional probabilities of liquid and gas are different from one another (Dekker, 1965; Egelstaff, 1967).

The transitions among the three states are caused by heat energy which is being supplied to matter. This heat energy increases the matter's temperature up to certain points—melting point in a solid, boiling point in a liquid—at which the temperature stays constant for a while. During these two periods of plateau in temperature, the heat energy goes into changing the matter's atomic structure, i.e., in overcoming the forces which attract and bond the particles to one another. After these changes in a matter's structure have occurred, the temperature starts to rise again. The energy supplied during the two phases of plateau is called *latent heat*.

Not all matters go through the described process of change. If we take the example of glass, although it looks and behaves like a solid (e.g., it does not take the shape of a container, it does not flow, etc.), it has an atomic structure of a liquid glass. In its solid state it is called a supercooled liquid. The transition between the solid and liquid states in glass is *not* accompanied by a plateau period of latent heat, and the temperature rises continuously. What actually happens in glass is that the molecular arrangement of the liquid state is frozen in; because of rapid cooling and a high viscosity of the liquid, crystals do not have time to grow (Dekker, 1965).

What is the difference between crystals and glass? The main difference is that the transitions between solid, liquid, and gaseous states in a crystal are marked by abrupt structural changes, whereas the transition in glass is not. An *abrupt change* is defined here when in a function $y = f(x)$, for a certain x value there can be given two

y values (and any possible y value between those two). Ideal abrupt changes or "jumps" of this sort never occur in reality and seldom only by a close approximation. However, they can often be observed in the first derivative of an empirical function. A *first derivative* is the rate of change in y as a function of x, and it gives information about the extent of change. When a function changes its direction at an angle, then the first derivative jumps at the point of change and becomes discontinuous. In the case of crystalline matter, the first derivative jumps at the onset and ending points of the plateau periods (latent heat), and these mark the existence of underlying structural changes which take place during these plateau periods. It is worth mentioning that these structural changes may take some time to occur; they in themselves do not occur suddenly. Thus, discontinuities in a function or its first derivative should not be confused with the underlying process of change which may take some time to accomplish. This cannot be said about glassy matter in which there are no abrupt changes, either in the temperature function or in its first derivative. Indeed, no abrupt, underlying structural changes occur in the transition between solid and liquid states, because the solid already has the atomic structure of the liquid.

The lesson from this analysis is that *whenever there is an abrupt change either in a function itself or in its first derivative, then underlying structural changes should be found.* This is not to say that structural changes cannot also occur under continuous, cumulative conditions. For example, with aging, fibers in a biological tissue may change their structure during which process they become less wavy and less flexible. This structural change is slow, gradual, and is reflected in behavioral changes of the entire tissue. We also cannot conclude from this claim that when underlying structural changes do not take place, other changes will not occur. For example, chemical, electrical, and other changes may take place, but these changes may not be part of the definition of the relevant structures. Therefore, the claim states only that abrupt changes seem to *guarantee* the existence of underlying structural changes. Thus, abrupt changes in a behavioral function may seem to be a good candidate for the assessment of the conditions under which behavioral changes and underlying structural changes are in full correspondence; these are the conditions that we have been seeking in order to overcome the problematic double-bind delineated earlier. In addition, the criterion of abruptness of change may give us the objective indicator we were looking for in order to pull us out of the danger of circularity in the definition and assessment of structural changes. In the next section, a case is made to support this notion in the domain of cognitive development.

Abrupt Behavioral Changes in Cognitive Development

The idea of abruptness of change is not foreign to the cognitive developmental domain. Abrupt changes in children's performances and/or reasoning have been sought and found by various researchers at specific age ranges (cf. Fischer, et al., in press; Flavell, 1977). The deliberate search for such abrupt changes in behavior has been specified as a major means for the discrimination between crucial and non-crucial qualitative changes in cognitive development. Fischer et al. claim that changes of the crucial sort must be large and quite rapid and must be reflected by discontinuities or "spurts" in the child's behavior. The obvious way to detect these spurts is to use methods which measure the amount and speed of qualitative changes. They offered two methods of measurement—scalogram analysis and analysis of multiple tasks. The main feature of these methods is that they are based on continuous scales which allow for detecting continuities versus discontinuities. This property of measurement is essential, if the only way to detect abrupt changes is by plotting either a continuous function or its first derivative, according to the foregoing discussion. A *continuous scale,* then, seems to be the first requirement for the detection of abrupt changes.

There is another requirement, however, which has resulted from psychological research. Fischer et al. (in press) have found that discontinuities are evident only in *optimal performance* in which the subject is well familiar with the testing situation and his/her upper limit of performance is made possible. This is where development shows a spurt as a new level emerges. This is a crucial condition for developmental trends to exhibit themselves because, under nonoptimal conditions, developmental trends are contaminated by other factors such as specific learning, perceptual misleadingness, and so on. The importance of these optimal conditions for assessing developmental trends have been recognized by other researchers (Goodman, 1979; Pascual-Leone, 1970), and they have been shown to be crucial for developmental research (Globerson, 1983). The converse is also the case: Testing developmental trends under impoverished (suboptimal) performance conditions (cf. Chi, 1978) is inappropriate in the assessment of developmental trends.

The main shortcoming of previous and current research on developmental discontinuities is that there has not been even one attempt, to the best of my knowledge, to investigate in a systematic way and under one theoretical framework the different discontinuities found in development along the entire age range of cognitive development. Most studies have investigated "occasional" spurts in development, in the sense that different spurts at different ages in different domains have been given different, local interpretations. Fischer et al. (in press) claim that the scarcity of spurt findings stems from the fact that most developmental psychologists who have looked for spurts have neither investigated behavior under optimal conditions nor used methods designed to detect spurts. Here I would like to add that even under optimal conditions and with appropriate methods, our understanding of cognitive development will be poor if we will not use these methods in a systematic way; i.e., across a wide age range and under one theoretical framework. In the next section I offer one such research plan.

The Investigation of Underlying Structural Changes in a Systematic Way

In this section, I suggest taking advantage of the systematic, abrupt behavioral changes that have been observed across a wide age range in the measurement of mental capacity. *Mental capacity* is a theoretical construct regarding the limited attentional capacity that people have (Kahneman, 1973) when they process information. This construct has played an important role in some recent cognitive developmental theories (Case, 1978, in press; Case et al., this volume; Pascual-Leone, 1970) in that it has been found both to grow with age and to be a powerful predictor for children's performance on various developmental tasks (cf. Case, 1974, in press; Globerson, Weinstein, & Sharabany, 1985; Pascual-Leone et al., 1978). A strong maturational component has been attributed to the mental-capacity growth because it has been found almost invariably to develop to the same extent in same-age children regardless of social class background or other within-age variations (Case, 1975; Globerson, 1983, 1985). The measures that have been developed for mental-capacity assessment have been based on continuous scales and have been administered under optimal performance conditions (Globerson, 1983; Goodman, 1979; Pascual-Leone, 1970). These rather "pure" developmental measures have revealed interesting growth functions of alternating periods of plateau and incline (Fabian, 1982; Goodman, 1979; Johnson, 1982). Although this phenomenon merits much more detailed research, these studies demonstrate that the function of mental-capacity growth may have abrupt changes of the sort discussed earlier. As a consequence, such a function seems to be an appropriate candidate for seeking underlying structural changes.

There are two relevant issues here: structural changes that (a) underlie the abrupt behavioral changes observed on the mental-capacity measures themselves; and (b) occur in the cognitive system in other domains (tasks) for which the mental-capacity measures have been shown to be relevant (i.e., predictive). We only have preliminary data regarding the structural changes that occur during the different phases of the mental-capacity growth. But, based on these limited data, it is possible to speculate on the sort of structural changes that are related to the mental-capacity measurement themselves.

Let's examine one of the mental-capacity measures—the Backward Digit Span Test, which, although not the "purest" measure of mental capacity (Globerson, 1983), still illustrates our point. In this test, the subject hears strings of digits in various lengths and then has to repeat them in a reverse order. When the mental capacity reaches a higher level (i.e., it can activate a greater number of informational units), the construction of a new strategy is allowed for, and this, in turn, structures or organizes the data in different ways than before. Suppose, for example, that a child's capacity allows the activation of only two separate units. Then when given a 3-digit string, the child would apply a 2-digit strategy and organize the three digits into two units. For example, the most common 2-digit strategy is to store the first digit in memory, repeat the last digit and then the stored digit. The most common error on a 3-digit string is to store the first pair as one unit, repeat the last digit and then the stored pair; e.g., the string 6-1-5 is repeated as 5-(6-1).

Now, suppose that the mental capacity has developed and reached the next level: three informational units. The subject now is able to construct a new strategy that includes a relational scheme, i.e., the middle digit is repeated after the last digit and before the first one. Using this strategy, the first two digits are stored separately, then the subject repeats the last digit, then the next to last, and only then the first digit. Indeed, the most common errors on a 4-digit string by a person whose capacity can only accommodate three are shown in the following example: the string 2-5-8-3 is repeated as 3-8-(2-5), 3-(5-8)-2, or 3-8- -2(Globerson, 1977).

The suggestion, then, is that the growth of mental capacity allows for the construction of a new procedural knowledge which, in turn, organizes information differently than in previous stages. It should be noted that mental-capacity measures are all measures of short-term or working memory; therefore, the strategies they involve and the kind of organized structures they allow are very limited in scope. One problem that arises here is that it is difficult to tell whether the structural changes observed on these tasks reflect more general structural changes in the cognitive system, of the sort suggested by Case (1978, in press). It is conceivable, for example, that the 3-digit strategy represents a more general structure which allows for the understanding of relational order of the kind involved in serial ordering tasks. In order to explore the generality-of-structures question, it is necessary to study the kind of changes that occur in other tasks for which mental-capacity measures have been found to be good predictors. This could be done by observing children's performance and reasoning over a wide age range and in small age differences (e.g., every two months) by cross-sectional and/or longitudinal means. Especially close notice should be made of the changes that occur during the plateau periods and around the points of abrupt changes at the beginning and end of these periods. My hypothesis is that structural changes occur after new levels of mental capacity are reached, that is, during the *plateau phases,* following the same process already described. Changes occur both in the *procedural* knowledge and the *declarative* knowledge, the latter resulting from the application of the new procedures (Anderson, 1982). This view is compatible with the classical structuralist view, especially Piaget's.

The Merits of the Proposed Research Plan. The main advantage of the proposed research plan is the possibility of conducting a systematic study of underlying structural changes that accompany systematic, abrupt behavioral changes in a relevant function of cognitive behavior (i.e., the growth of mental capacity as a function of age) across a wide range of ages (i.e., 3 to 16 years). Other studies which attempted to investigate behavioral discontinuities and underlying structural changes have not taken these two factors into account. I have claimed that abrupt changes in behavior are the only "sure" conditions under which structural changes fully correspond to behavioral changes. I further argued that the measures of mental capacity have conformed with the appropriate conditions for the detection of behavioral discontinuities—optimal testing conditions for the subject's performance and continuous scale measurements. Taken together, the function of mental-capacity growth seems to be an ideal candidate for detecting structural changes that underlie abrupt behavioral changes. It is by no means the only or best candidate. Any measure which has the same properties and which exhibits abrupt behavioral changes (not necessarily alternating plateau and incline phases) will qualify. Nevertheless, the mental-capacity measures are the only existing candidates to date which enable us to study a wide age span.

The kind and number of developmental tasks for which the measures of mental capacity have been found to be powerful predictors are so varied by now (cf. Case, in press; Case et al., this volume) that it is possible to assess not just those structural changes that underlie behavioral changes on the very measures of mental capacity but also structural changes that are associated with other cognitive tasks. This enables us to investigate structural changes in the cognitive system in a more general way. Such an investigation may supply valuable information about the *kinds* of structural change that are both relevant and significant for cognitive development.

The general claim here is that if we have an objective indicator for structural change such as abrupt behavioral changes in mental capacity growth, then we are able to assess structural changes associated with these behavioral changes. If we do so, we are then ready to develop *objective criteria* for the *kind* and *extent* of structural changes that are relevant and significant for cognitive development. From this, we can also accrue benefit in the opposite direction—from the objective empirical detection of structural changes to the theory of structural change. The development of such criteria is crucial, if one wants to assess those structural changes that occur under gradual, cumulative conditions where one does not have objective indicators for detecting structural changes such as abrupt changes in behavior. Using such criteria, which are well formulated in advance of one's empirical investigation, is essential if we want to avoid the danger of circularity mentioned earlier. My basic claim here is as follows: Let's start our investigation of structural changes under conditions where we can be certain that significant structural changes occur, i.e., under conditions of abrupt behavioral changes. And let's investigate these changes in a systematic manner and develop criteria which will describe the nature and extent of these structural changes. Only then will we be able to use these criteria as a yardstick in the assessment of structural changes that occur under conditions which are more difficult to detect objectively.

The proposed investigation also enables us to answer the question of universality versus relativity of structural changes asked earlier in that we can study the effects of age and individual differences on structural changes. For example, we can investigate whether children with different cognitive styles differ in the kind and extent of their structural changes and whether their abrupt behavioral changes are different to begin with.

Although not tested directly, we can also infer from the kind and extent of structural changes, those psychological rules that regulate structural changes and their interaction with specific organismic (e.g., cognitive style) and environmental (e.g., task characteristics, experiential background) factors. In addition, this sort of investigation enables us to assess the timing of structural change in children of different types (e.g., cognitive styles, socioeconomic backgrounds, etc.). This could give us valuable information about the maturational versus nonmaturational components involved in cognitive development. For example, suppose that we find out that all children, regardless of cognitive style, socioeconomic background, etc., go through some of the structural changes in the same manner and at roughly the same ages, whereas other structural changes are more variable. Data of this kind would tell us about the sort of changes that seem to be more maturationally related versus those which are more affected by other (environmental) factors (cf. Globerson, 1983).

The proposed research is not directly addressed to the issue of the difficulty of assessing a structure's description. However, it is my belief that this difficulty can be overcome, if we use methods already developed for this purpose (McKeithen, Reitman, Reuter & Hirtle, 1981; Nagy, 1983; Nussbaum & Novak, 1976; Reitman & Reuter, 1980; Smith & Medin, 1981; Stewart, 1979), and if we stay away from task situations that are too ambiguous.

There are many issues and questions which will be open for further investigation following the proposed research. One such issue is the relation between the systematic structural changes that can be revealed by the proposed research and those structural changes which have already been documented in the psychological literature. Questions about this issue abound. Is there a synchrony among them? Are they related to more general (e.g., Piagetian) stages of development, or are they domain-specific? Are some structural changes more permanent and/or significant than others? These and other related questions strike me as important to anybody who is interested in cognitive development, but they may take years to be answered.

Summary and Conclusions

At the beginning of this chapter, some of the debates between developmental theorists were presented. The general theme was about the sort of changes the cognitive system goes through as it develops. In the present chapter, I attempted to shed some light on these debates by asking two basic questions: What are structural changes, and how can we tell that structural changes occur? It was claimed that it is not sufficient to have a clear definition of structural change because, by definition, any change in a structure's description is a structural change. Many debates stem from the fact that not all such changes are perceived by the theorist as involving (significant) structural change. In response to this analysis, I suggested that a theory of structural change has to include criteria for deciding about the kind and extent of structural changes that are significant for cognitive developmental theory. In addition, such a theory should also include, it was argued, the psychological constraints which govern structural change.

Two major problems with the assessment of such criteria have been mentioned. The first one is the circularity that exists when the definition and assessment of structural change are confounded. The second problem is that behavioral changes—from which one has to infer underlying structural changes—and the underlying structural changes themselves are not always in full correspondence. It was suggested, therefore, to use objective indicators for structural changes. It was further suggested that these indicators should not be part of the definition of a structural change, and they should stand in full correspondence with underlying structural changes, thus guaranteeing their detection and assessment. It was claimed that one such objective indicator of change exists when abrupt behavioral changes occur, either in a relevant behavioral function itself or in its first derivative. To detect such abrupt changes, one should use a continuous measurement scale and test children's behavior under optimal conditions. We proposed such a measure—mental capacity—which exhibits the sort of abrupt behavioral changes we are seeking and conforms to the described requirements of such a measure. In addition, this measure has advantages over other measures in cognitive development: It demonstrates a developmental phenomenon of systematic abrupt changes across a wide age range, and it has a strong relevance (i.e., predictive power) for many other developmental tasks.

We proposed a research plan in which children's performance and reasoning behavior can be examined in a close relationship with changes that occur in mental capacity. In particular, it was suggested to observe children in very small age steps (e.g., every two months of age—cross-sectionally and/or longitudinally) and assess those structural changes that are associated with the abrupt changes in mental capacity, i.e., around the points of change between the plateau and incline phases and during the plateau phases where, it was proposed, most structural changes occur. It was argued that such a research program may answer many important developmental questions and may contribute to the development of a better theory of structural change by enabling us to construct objective criteria for assessing significant structural changes in situations where objective indicators do not exist.

A Word on the Uniqueness of Cognitive Development

Structural changes may occur in many contexts other than cognitive development. For example, they can occur as a result of emotional crises, acts of persuasion, and so on. It seems of utter importance to compare these changes with the systematic changes that underlie age-related cognitive development, such as those that can be revealed by our suggested research. The importance of this comparison lies in its allowing a better understanding of the uniqueness of cognitive development.

There are a few possibilities here which may distinguish the phenomenon of structural change in cognitive development from those changes in other contexts. The first possibility is that the *kind* of age-related changes that are typical of cognitive development is different from the kind of changes in other contexts. The second possibility is that the *extent* of change is different. I would like to rule out these two possibilities, based on plausibility arguments rather than on data. The kind-of-change argument does not seem to me plausible, because we know that people may go through major changes following a profound experience (e.g., emotional crisis), and often describe their new situation as a new level of understanding, which is not unlike a higher developmental stage observed in cognitive development. The same experience may be following an enlightening learning experience (e.g., insight). If this reasoning is accepted, then it does not seem necessary to assume that the kind of changes that occur in contexts other than cognitive development are different. The same argument stands for the extent of change. This sort of change may sometimes be very great and may affect the total personality; and at some other times, these changes may be quite small and insignificant. Therefore, the extent of change also does not seem to be either a plausible or an insightful criterion of distinction.

I would like to believe that all structural changes in the psychological system obey the same psychological constraints, regardless of whether they occur in the context of age-related cognitive development or other aforementioned contexts. A third possibility that reflects this belief seems to me the best criterion for distinguishing between cognitive development and other changes, and that is the possible *systematicity, regularity and predictability* of the structural changes that underlie cognitive development. This should be seen in comparison to the occasionality that signifies other sorts of structural changes. The possibility of systematic structural changes in cognitive development suggests that these changes are regulated, to some degree, in a systematic way by maturational and psychological factors whose constraints should be discovered. Systematicity, regulation, and constraints on changes in the cognitive system seem to me some of the most fascinating aspects of human phenomena and some of the most interesting domains of research.

References

Anderson, J. R. (1982). Acquisition of cognitive skill. *Psychological Review, 89,* 369–406.

Carey, S. (1983). *Acquisition of scientific knowledge: the problem of reorganization.* Unpublished manuscript, MIT, Department of Psychology.

Carey, S. (in press). Are children fundamentally different kinds of thinkers and learners than adults? In S. Chipman, J. Segal, & R. Glaser (Eds.), *Thinking and learning skills, 2.* Hillsdale, NJ: Erlbaum.

Case, R. (1974). Structures and strictures: some functional limitations on the course of cognitive growth. *Cognitive Psychology, 6,* 544–573.

Case, R. (1975). Social class differences in intellectual development: a neo-Piagetian investigation. *Canadian Journal of Behavioural Sciences, 7,* 244–261.

Case, R. (1978). Intellectual development from birth to adulthood: a neo-Piagetian interpretation. In R. S. Siegler (Ed.), *Children's thinking: what develops?* Hillsdale, NJ: Erlbaum.

Case, R. (in press). *Intellectual development: A systematic reinterpretation.* New York: Academic Press.

Chi, T. H. M. (1978). Knowledge structures and memory development. In R. S. Siegler (Ed.), *Children's thinking: What develops?,* Hillsdale, NJ: Erlbaum.

Dekker, A. J. (1965). *Solid state physics.* Englewood Cliffs, NJ: Prentice-Hall.

Egelstaff, P. A. (1967). *An introduction to the liquid state.* London: Academic Press.

Fabian, V. (1982). *Language development after five: A neo-Piagetian investigation of subordinate conjunctions.* Unpublished doctoral dissertation, University of California, Berkeley.

Fischer, K. W. (1980). A theory of cognitive development: the control and construction of hierarchies of skills. *Psychological Review, 87,* 477–531.

Fischer, K. W., Pipp, S. L., & Bullock, D. (in press). Detecting developmental discontinuities: method and measurement. In R. Harmon & R. N. Emde (Eds.), *Continuities and discontinuities in development.* New York: Plenum.

Flavell, J. H. (1977). *Cognitive development.* Englewood Cliffs, NJ: Prentice-Hall.

Fodor, J. (1972). Some reflections on L. S. Vygotsky's thought and language. *Cognition, 1,* 83–95.

Fodor, J. (1975). *Language and thought.* New York: Crowell.

Gelman, R. (1983). Recent trends in cognitive development. In J. Schierer & A. Rogers (Eds.), *The G. Stanley Hall lecture series, 3.* Washington, DC: American Psychological Association.

Gentner, D. (1983). Structure-mapping: A theoretical framework for analogy. *Cognitive Science, 7,* 155–170.

Globerson, T. (1977). Field-independence, information processing capacity, and mental effort. *Dissertation Abstracts International, 37,* (9-A), 5698.

Globerson, T. (1983). Mental capacity and cognitive functioning: Developmental and social-class differences. *Developmental Psychology, 19,* 225–230.

Globerson, T. (1985). Field dependence/independence and mental capacity: A developmental approach. *Developmental Review, 5* (3).

Globerson, T., Weinstein, A., & Sharabany, R. (1985). Teasing out cognitive development from cognitive style: A training study. *Developmental Psychology, 21* (4), 682–691.

Goodman, D. (1979). *Stage transitions and the developmental traces of constructive operators: a neo-Piagetian investigation of cognitive growth.* Unpublished doctoral dissertation, York University.

Halford, G. S. (1970). A theory of the acquisitions of conservation. *Psychological Review, 77,* 302–317.

Johnson, J. (1982). *The development of metaphor comprehension: its mental-demand measurement and its process-analytical models.* Unpublished doctoral dissertation, York University, Downsview, Ontario, Canada.

Kahneman, D. (1973). *Attention and effort.* Englewood Cliffs, NJ: Prentice-Hall.

Keil, F. C. (1979). *Semantic and conceptual development.* Cambridge: Harvard University Press.

Keil, F. C. (1981). Constraints on knowledge and cognitive development. *Psychological Review, 88,* 197–227.

Kuhn, T. S. (1977). *The essential tension.* Chicago: University of Chicago Press.

McKeithen, K. B., Reitman, J. S., Reuter, H. H., & Hirtle, S. C. (1981). Knowledge organization and skill differences in computer programmers. *Cognitive Psychology, 13,* 307–325.

Nagy, P. (1983). Assessing cognitive structure: a response to Stewart. *Science Education, 67,* 25–36.

Nussbaum, J., & Novak, J. D. (1976). An assessment of children's concepts of the earth utilizing structured interviews. *Science Education, 60,* 535–550.

Pascual-Leone, J. (1970). A math model for the transition rule in Piaget's developmental stages. *Acta Psychologica, 32,* 301–345.

Pascual-Leone, J., Goodman, D., Ammon, P., & Subelman, I. (1978). Piagetian theory and neo-Piagetian analysis as psychological guides in education. In J. M. Gallagher (Ed.) *Knowledge and development (Vol. 2).* New York: Plenum.

Piaget, J. (1970). Piaget's theory. In P. H. Mussen (Ed.) *Carmichael's manual of child development (Vol. 1.)* New York: Wiley.

Reitman, J. S., & Reuter, H. H. (1980). Organization revealed by recall and confirmed by pauses. *Cognitive Psychology, 12,* 554–581.

Rosch, E. (1973). On the internal structure of perceptual and semantic categories. In T. E. Moore (Ed.), *Cognitive development and the acquisition of language.* New York: Academic Press.

Siegler, R. S. (1976). Three aspects of cognitive development. *Cognitive Psychology, 8,* 481–520.

Smith, E. E., & Medin, D. L. (1981). *Categories and concepts.* Cambridge, MA: Harvard University Press.

Stewart, J. (1979). Content and cognitive structure: Critique of assessment and representation techniques used by science education researchers. *Science Education, 63,* 395–405.

Strauss, S. (Ed.) (1982). *U-Shaped behavioral growth.* New York: Academic Press.

Weiser, M., & Carey, S. (1983). When heat and temperature were one. In D. Gentner & A. Stevens (Eds.), *Mental models.* New York: Academic Press.

12

*How Development Works**

David Henry Feldman
Eliot-Pearson Child Study Center
Tufts University

First, a word about my title. Aside from the fact that it is outrageous on its face, what can be accomplished by choosing a title such as this one? My original reason for choosing the title was, to be honest, that I had little idea what I was going to write about at the time I was invited to participate in the Tel Aviv conference. I figured that with a title like "How Development Works" I would be unconstrained in my eventual choice, selecting from a nearly infinite set of possible topics. It is, of course, true that I am interested in the question of how development works; it is even fair to say that it is *the* question guiding my work. In this respect, then, it was not a totally whimsical choice and not without a certain appropriateness.

Once having committed myself to writing a paper with the title "How Development Works," I found myself trying to think about how I would answer such a question, given that I had had the temerity to pose it. The tension of having to try to make good on my implied promise to explain how development works was not altogether unwelcome. In a sense, I may have posed this impossible task to see just how far I could move toward carrying it off. Even if the distance I covered was small, it might be larger than I would have otherwise managed if I had not put myself in such an embarrassing position.

This technique of claiming that one will do something that one has not even come close to managing is not unprecedented. It is said of the great inventor Thomas Edison that he would call a news conference, gather together the reporters and chroniclers of the day, and announce that, say, two months later he would present to them a great invention—for example, a device that would illuminate a room without kerosene. Apparently, Edison did this a number of times in order to get his inventive juices flowing. The idea of facing an expectant world without producing what he had promised was enough to get Mr. Edison working feverishly against his own self-imposed deadline. Who knows how long it would have taken for Edison to make his contributions if he had not contrived this motivational technique for himself? While I make no claim to be an inventor of the stature of Thomas Edison, I

* This paper was prepared for the First Annual Tel Aviv University Conference on Human Development, October 1–7, 1983. Thanks to Iris Levin, Sidney Strauss, Lynn Goldsmith, and the participants in the conference for helpful comments and suggestions. I was supported by grants from the Spencer Foundation and the Andrew M. Mellon Foundation.

must confess something of a feeling of kinship with him in this technique of making a promise to heighten motivation to make good on it.

You, my readers, are the equivalent of Edison's reporters, and I have promised to tell you something about how development works. (Notice how I have already begun to reduce my monumental promise of a full explanation). Will I be able to make good on my promise? Somewhat to my surprise, I find that, to a greater degree than I would have believed when I agreed to write this chapter, I think I do have some interesting things to say about this central question. I have come to the point of view, somewhat to my own surprise, that we actually know more about how development works than I had thought when I began the process of preparing this chapter.

I will organize the topic into three major sections with an introductory interlude designed to set the stage by briefly discussing developmental domains. The first major section then focuses on things we are relatively sure of when we try to consider the question of how development works. The second includes things which we have reason to believe are true about how development works but about which we can be less confident. And finally, there are those aspects of how development works that we can speculate about, imagine, or ponder, but about which there are relatively few statements that can be supported with confidence—whether by empirical data, theoretical refinement, or conceptual analysis.

Developmental Science

Before we can turn to the matter of how development works, we must first grapple with the question of what development is. The claim made here is that development is a distinct, special, *nonreducible* phenomenon of human life. Development is, as others have claimed as well, as much a choice about a level of analysis as it is a statement about fact or truth. The argument is that there are phenomena that should be studied at their own level of complexity, neither trying to reduce them to subunits such as chunks of information (Siegler, 1981) or memory functioning (Brainerd, 1977), nor to lose them in more holistic interpretations such as person or organism. Therefore, before turning to how development works, we should try to be a bit clearer about what it is, at least in the context of this chapter.

First, developmental phenomena are not limited to human behavior changes. Indeed, I claim that there should be a field of inquiry, somewhat analogous to but very different from cognitive science, which I have provisionally called "developmental science" (Feldman, 1982), where the emphasis would be on exploring common aspects of transformation. Although different disciplines concentrate on different manifestations of change and examine phenomena at different levels of specificity, I believe that there are common, deep principles that govern all developmental change, although this is mostly an intuition at present and I think it premature to try to develop a taxonomy of such common principles. Certainly there are also properties unique to each individual discipline which attends to the issue of developmental change. A major problem for future theory and research will be to sort out those qualities which are unique to each developmental domain and those which are common to any kind of development. The purpose of a field of developmental science is, I believe, to begin to understand development both as a general phenomenon of change and transformation and as it pertains specifically to the particular situation under study. I believe that the attempt to explain how development works in any of the disciplines should be done with an eye toward considering how development might work in other realms. This, in a way, is not so different from the approach that Piaget took when he focused on a problem, reading nothing directly about the problem at hand in the discipline where it had originated; instead, he read everything available in disciplines close to or possibly relating to that target discipline, and then used this neighboring information to gain leverage over the problem at hand.

In contrast to the mechanistic metaphor of cognitive science, the notion of a science of development springs from an organic metaphor. The emphasis for developmental science would be transformation or change, whereas cognitive science has tended to focus on state. Indeed, change or transformation has been one of cognitive science's most problematic aspects (A. di Sessa, personal communication, March, 1982). A third claim I have made about developmental science, one which is admittedly more idiosyncratic, is that the study of transformation and change should, as a corollary, concern itself with novelty, newness, uniqueness, and other manifestations of something we might call creativity—more about this later.

The psychological discipline within which developmental issues have been most frequently addressed include, of course, the field of developmental psychology, although not usually with an eye toward unique or distinctive change. Clinical psychology and psychiatry also deal with change but not as the phenomenon to be studied; change is a byproduct of the efforts of clinicians. In addition, the following all study change: physical anthropology, evolutionary biology, developmental biology, archeology, paleontology (particularly brain and central nervous-system development), organizational development, developmental sociology, epistemology (especially and obviously Piaget's genetic epistemology), psychoanalysis, the history of science, philosophy of science, and organizational change. Quite possibly other disciplines as well concern themselves with issues of change. The argument is not that the identical processes operating in individual psychological development also operate in the other disciplines, but rather that there are some common problems of analysis that present themselves in a variety of disciplines focused on a variety of problems of change and transformation.

To try to characterize within a common framework all phenomena where transition and change are pertinent, it would be necessary to use something like general systems theory (von Bertalanffy, 1955) or a framework like that used by the neurophysiologist Maturana (1978). While these broad frameworks are interesting and, to some extent, give force to the argument that development is a matter of concern in numerous fields, I do not wish to claim that development works in exactly the same way in all its manifestations. While I believe that some of the principles of individual cognitive development will generalize to other realms where development is found, I do not want to put great emphasis on this point. The extent to which there is generalizability without distortion into noncognitive realms of individual human development such as emotional change or social development remains to be seen, let alone whether it may extend to such distant topics as the development of cultures or nervous systems.

Nonuniversals

As we attempt to characterize how human cognitive development occurs, it is necessary to make a further perliminary distinction, one that is familiar to those who have read *Beyond Universals in Cognitive Development* (Feldman, 1980a). Let me briefly summarize the distinction between universal and nonuniversal developmental phenomena because it is central to my purposes.

It has been a major effort of cognitive-developmental psychology to try to describe and explain those cognitive restructurings that are common to all human beings in all cultures across all time. I have called these restructurings *universals,* for they are achieved by everyone who does not suffer from major organic or environmental deficit. My use of the term differs from other cognitive developmentalists in particular Kohlberg (1971), who uses *universal* to characterize invariant sequences of restructurings, whether or not all individuals eventually achieve all levels of the domain. I use the term *nonuniversal* to categorize those developmental domains which share certain critical qualities with universal domains but which are neither universally achieved nor spontaneously acquired. This second distinction, similar to one made by Vygotsky almost fifty years ago (Vygotsky, 1934), draws attention to the conditions under which individuals move from stage to stage or level to level in a domain. (I use the term *levels* for nonuniversal domains, *stages* for universal ones.) For the achievement of universal sequences, the conditions that are necessary and sufficient for progress must be so pervasive that nothing special need be done to ensure that their achievement will occur. For nonuniversal achievements, on the other hand, the availability of various specialized conditions is critical to the acquisition of more advanced levels within the sequence.

I have argued that any domain of human intellectual development is developmental, if it can be characterized by a distinctive sequence of stages or levels and is governed by a set of qualitative change processes, be it one that is universally acquired or not. It is my claim that, within the realm of cognitive development, many of the same principles that account for transformation and reorganization in universal domains also account for transformation and reorganization within nonuniversal domains. The argument is that if one can explain how development works in nonuniversal developmental domains, one will have made a very good start in explaining how development works in universal domains as well.

In fact, the claim is still stronger: We need to understand how development works in nonuniversal domains in order to know how to ask the right questions about how development works in universal ones. I contend that Piaget remained largely baffled by the problem of novelty (how something new comes about in a cognitive structure), at least in part because he categorically ruled out the study of nonuniversal domains (Piaget, 1971, 1972, 1975/1977; Piaget & Voyat, 1979). In his theorizing and in his empirical work, Piaget focused on bodies of knowledge that he believed were universal. From time to time he wistfully regretted that he had not attended more to things like "individual differences" or "individuality," but he never focused his enormous gifts on anything that did not seem to meet the criteria for universality.

> Generally speaking—and I'm ashamed to say it—I'm not really interested in individuals, in the individual. I'm interested in what is general in the development of intelligence and knowledge, whereas psychoanalysis is essentially an analysis of individual situations, individual problems, and so forth. (Piaget, in Bringuier, 1980, p. 86)

The problem, in a nutshell, is that it is very difficult, perhaps impossible, to study how development really works in universal domains, because the process is so internalized and inevitable. Nonuniversal domains, on the other hand, require a much greater reciprocity between individual and context for achievement to occur. This, in turn, lends itself to discovering how and under what conditions new developmental levels are achieved.

Things We Know About How Development Works

While there are probably several things we know about development, I want to focus on two which seem to me to be of special importance. One of these is that development is stagelike, occurring in *qualitatively* distinct sequences of reorganized structures. The second is that development is not a constant process; it has variations in rate, intensity, organization, and balance between internal and external states. In particular, I would like to highlight the fact that development works by moving backward in order to go forward.

Stages

The first point, that development is stagelike, is in a way so obvious that it is often ignored. However, it is not a point that all would be willing to accept. Indeed, there is currently a strong antistage movement within developmental psychology. The argument against stage notions basically seems to be that they embrace development at too broad a level. Stage theories are considered to be proto-theories or early attempts at theory that should now be abandoned for the more sophisticated, precise, and powerful predictive models that come in various wrappers, but all of which tend to be more mechanistic in character, derived from computational models of the thought process (Brainerd, 1978, 1981; Siegler, 1981). Even Piaget's interpreters place less emphasis on stages than they used to (cf. Gruber & Vonèche, 1977).

This is, however, not simply a level of theory problem; it is a level of analysis problem as well. By arguing that stage theories attack the problem of development at too general a level, the focus of the field is redirected to a more microlevel. I am claiming that phenomena exist that have stagelike characteristics and that these characteristics are critical to defining the levels of theory and analysis of the field. I do not mean to suggest that phenomena that are not stagelike are uninteresting or irrelevant to our knowledge of development nor that research on more specific changes does not influence broader, qualitatively oriented stage frameworks. I do want to argue that the attempt to reduce development to smaller and smaller processes of change is one that should not be accepted without qualification. As *part* of a science of developmental psychology, the more molecular levels are useful and appropriate. As *all* of that science, these approaches are not only not welcome, they may undermine the distinctiveness of the study of development altogether.

It certainly has been the case that up to now the most influential theories of development have either been stage theories or have implicitly accepted the importance of some notion of stage as central to their framework. If we were to rate the major theories of development in terms of their impact and importance in North American circles, those of Freud, Erikson, and Piaget would certainly stand in the first rank. Going at least as far back as Rousseau's *Emile,* the concept of stages has been an explicit part of major accounts of developmental transformation and change.

Other important theories that do not themselves employ the notion of stages but seem compatible with them include those of Baldwin, Werner, and Vygotsky. These theorists tend to direct their energies toward explanations of the transformation processes between stages rather than on producing good descriptions of stage sequences. I will have more to say about these theories later on. The point is that the most important contributions to the understanding of the broad sweep of development to date have come from stage theorists or theorists who implicity accept the concept of qualitative change and transformation.

Charles Brainerd (1977, 1978, 1981) has probably been the most vocal critic of a stage theoretic approach to development, arguing that stage formulations are primitive—necessary but preliminary steps toward the more powerful, more detailed theories that will succeed them. It is possible that Brainerd's view will prove to be correct, although I'm betting against it (Feldman, 1981b). His view, in fact, offers (unwittingly, I think) a kind of stage theory of the development of the field: We will move from the stage of primitive theories, invoking the broad notion of stages, to more sophisticated, "scientific" theories which will abandon that concept in favor of a more empirical orientation. My own view, as should be more than clear by now, is that developmental psychology is, to a great extent, a discipline which needs some concept of stage as one of its "core assumptions" (Lakatos, 1970). There is more than enough reality to stage concepts of one sort or another to keep investigators and theorists busy for the foreseeable future.

There is little argument, even from critics of stage theories, that the concept of stages *has been* crucial to developmental psychology up to this point. Based on cumulative knowledge of the field, it is fair to say that at least at one level of analysis we know that development is a stagelike phenonemon. Whether or not we choose to change our minds about it in the future (I hope we will not), stages have pervaded our view of development. To say otherwise requires a massive distortion of the distilled body of knowledge about development that has been accumulated over many decades.

Asking just what kind of concept of stage makes sense for developmental psychology is a more productive question than asking whether or not there ought to be any concept of stage at all. It certainly seems reasonable to expect that refinements in our understanding of development will lead to refinements in our conceptions of how stages are to be best described. I myself have argued that there is greater heuristic power and value to a notion of stage when it is thought of as existing within domains of understanding rather than as existing within children or brains or central nervous systems (Feldman, 1980a, 1980b). Locating stages within domains allows us to define them in ways which can be directly investigated for adequacy, internal consistency, and structural integrity without demanding that such tidy organizations ever actually exist in people's minds (cf. Feldman, 1980a, 1980b). In this sense, I see stages as templates against which to look at the actual behavior of individuals.

This leads to the question of the kind of stage concept we will be using in future research: Should there be a larger number of more discrete specific stages, and should these stages reflect overall developmental shifts or more specific ones? In partial agreement with my more mechanistic, information-processing colleagues, I believe that a limitation of stage theories to date has been that they have focused on the broadest, most general levels of reorganization, and that this tendency has left the field with rather Olympian preoccupations. I do not see the opposite extreme as the most promising solution to the problem, however. My own line of thought has been to redefine and reconceptualize the concept of stage at a more "middle" level, as Strauss and others have suggested (Strauss, 1982, 1984; Sternberg, 1983). This does not mean that every day in every way the child is getting better, but it does mean that the child is continually gaining access to ways of thinking better suited to the demands of various bodies of knowledge. As a number of investigators have shown, there are ways in which the less mature organism is actually cognitively stronger than the more mature. In certain memory functions, for example in eidetic imagery and in some problem-solving tasks, the younger child actually may do better than the older. Thus, it is with respect to a specific set of challenges that we can say that a child becomes more adequate, not necessarily in all cognitive functions (Gardner, 1983).

The notion of stage also tends to now bring with it the idea of inevitable progress, springing as it did in its modern form from the nineteenth century misapplication of both Darwin's theory of evolution and Newtonian physics. In this cosmology western European culture was seen as the pinnacle of evolution and Caucasian males the end product of natural selection. Wherever western European civilization went was *eo ipso* the next stage of evolution. There has been this tendency to see stages in idealistic, romantic ways. It is important to specify that a stage theory need not partake of the social Darwinistic march toward progress in order to be a viable concept. A stage theory requires a fixed sequence or several known patterns of sequences of stages, and these must move from a less- to more-adequate representation of a domain of activity (Gruber, 1982). In this sense, stages do imply progress. But the relatively specific concept of stages I have been advocating certainly does not suggest that there is an overall, coordinated, hierarchical arrangement of cultures and human beings, let alone that any one set of these is the most evolved or adapted.

Much more accurate as an image of how development works is a vast array of developmental domains each with stages or levels of various sorts. How development works, then, is substantially more complex and, to my mind, more interesting than a unidimensional sequence of stages would suggest (Gruber, 1982).

Constant Rate

The second thing we know about how development works is that, while progressive on the whole, development does not occur at a constant rate. This, too, may seem so obvious as to hardly need mentioning, but my impression is that many developmentalists have forgotten about it. In some ways the traditional notion of stage tends to make the rate of change seem more constant than it actually is. If one focuses on a single sequence of very broad stages of development, then it is not difficult to imagine that there is inexorable, steady movement through these stages. This, in turn, leads to an assumption that the rate of change is more or less constant and that it is only by looking back that one can see that major change has occurred.

Even within broader stage frameworks like Piaget's, it is nonetheless true that when one looks more carefully at them, it is possible to see periods of relatively more or less rapid transformation. As is the case in Freud's theory, these more rapid periods of change may be brought on by physiological maturation, although this is a somewhat controversial point (cf. Beilin, 1971).

Although it is not directly about transitions between stages per se, there is a compelling body of evidence to suggest that sensitivity to learning or skill development is not constant. The idea of the "critical period," for example, has received substantial support in studies of humans and nonhumans as well. Gardner (1983) summarizes the idea of the critical period as follows:

> A second, related principle underscores the importance of so-called critical periods during the process of development. For example, in the cat, there is a critical period in visual development from the third to the fifth week postnatally. If, during this time, one eye is deprived of form or light, then the central connections of the eye change and the ill-seeing eye will be suppressed from functioning. Such interference seems to be permanent. As a general point, it seems that the most vulnerable time for an organism occurs during these sensitive periods. Irreversible damage to the central nervous system seems particularly likely to occur in the wake of even mild restrictions during such a critical period. Conversely, rapid development will occur if the proper conditions obtain during the critical period. (p. 40)

Thus it seems to be fairly well established that the child is not equally capable of reorganization or enhancement of his or her capabilities at any time. At least this seems to be true at the broader, more general levels.

I would also like to suggest, although with less empirical support, that within more specific realms of achievement and developmental change, there are variations in the rate of progress an individual makes through a domain. Prodigies, for example, evidence extremely rapid movement through the levels of a domain (Feldman, 1980a, 1980b, 1982, in press). There is also the fact that some individuals make extraordinary or creative breakthroughs in a domain, whereas most do not. This simply could not happen, unless there are differences in the resistance of structures to transformation from individual to individual, a matter we will come back to later.

The most interesting question to me is whether there are orderly, regular, and clear patterns to these periods of less and greater developmental change, either between or within individuals. Here, I am afraid, we move from things which we can assert with reasonable conviction to things which seem to be true but which do not have sufficient weight of evidence to be considered established processes of development change.

Things We Probably Know About How Development Works

We probably know two things about how development works: that systematic, orderly *transition* between levels or stages is the rule rather than the exception, and that systems for representing, communicating, and expressing information are *constructed.* As suggested in the previous section, we know that there are variations in the rate of movement through developmental levels and stages within and among individuals, and probably that critical periods for the acquisition of new information exist as fundamental constraints on what can be learned and when (Keil, 1981, this volume). But what we have only begun to determine empirically is the extent to which there are predictable rhythms of movement from less to more rapid rates of transformation and back again as individuals progress through stages or levels.

Transformation

At the level of theory, we of course have Piaget's very suggestive equilibration model of transition between stages of thought (Piaget, 1975/1977). This model certainly implies that children are not equally capable of restructuring their mental apparatus at all times. The model includes the notion of greater relative internal stability at each successive stage as one of its central premises and also includes periods of relative instability between the more stable plateaus in development. This notion of alternating phases of disequilibrium and equilibrium is illustrated in Fig. 1. Figure 1 shows straightforward, progressive movement from one plateau to another through relatively brief periods of disorganization and disequilibrium (see also Flavell, 1971). Although Piaget's model is much more complex and more subtle than this—especially his last refinements of it (Piaget, 1975, 1981)—this figure gives an overall sense of the kind of rhythm of change that I want to attend to. Things become much more complicated, of course, when we introduce the idea, as Fischer (1980) and others have done, that development proceeds at different rates in

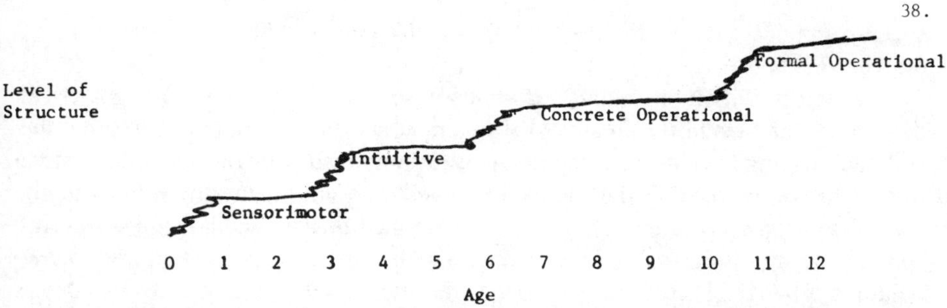

Figure 1. A representation of the equilibration model of Piaget with alternating periods of instability and stability of structure.

different domains. Piaget's model deals with overall general cognitive restructuring. When one begins to think about alternating periods of greater or lesser disequilibrium across several domains, the situation becomes much more complex. For present purposes, however, let us keep things artificially simple in order to draw attention to possible regularities in transition phases or rhythms, even though we will be able to say very little about how these patterns might interact or connect with each other (Fischer, 1981).

Our own research on the phases of transitions has focused on developmental levels within the domain of map drawing, a task that has been used by the Genevans for exploring the construction of systems that integrate logical/mathematical and spatial/temporal concepts (Snyder & Feldman, 1977, 1980, 1984). This research borrows the basic sequence of levels from Piaget and Inhelder's earlier work (Piaget & Inhelder, 1948/1956) and elaborates upon it so that a more detailed look could be taken at movement between levels. I will not describe the work in detail here, because descriptions of the techniques we have used are available elsewhere in the literature (cf. Snyder & Feldman, 1977, 1980a, 1980b, 1984). Briefly, we did a conceptual task analysis of the map drawing problem, arriving at a set of spatial concepts that must be applied to groups of features of a physical landscape in order to draw a map (Snyder, Feldman, & LaRossa, 1976).

Rather than attempt to score a subject as "being at" a particular stage or level of map drawing as is usually done in work of this sort, we focused instead on the problem of detailing the transition process from level to level. By using twenty specific indicators of developmental levels in the child's map-drawing system (four spatial concepts applied to five groups of features), we were able to chart changes in the distribution of "partial structures" (Turiel, 1977, 1983) in each subject's drawings. Five drawings of the same landscape were collected from a sample of fifth-grade children over a 3-year period. Using changes in frequencies of map-drawing skills at the six levels of development of map drawing in the child's system, five "phases" of transition were identified, representing five characteristic steps toward the consolidation of a new developmental level. A subject must, we believe, pass through each of these phases as s/he moves from one developmental level (or stage) to another. We have found that each of the five phases of a transition cycle has specific implications for the succeeding set of changes in the child's system for drawing maps.

The five transition phases are illustrated in Fig. 2, which shows a hypothetical distribution of capabilities in a child's system at six succeeding points in time, the sixth duplicating the pattern of the first phase but with all components of the system moved to the succeeding developmental level. The data themselves are actually drawn from group results, but we have determined that individual subjects correspond quite well (with about 75% accuracy) to this overall group pattern (Snyder & Feldman, 1984). The five phases in the transition cycle correspond to the steep, jagged line shown in Fig. 1; in other words, the period of transition between stages is differentiated and elaborated to reveal a sequence of distinct phases.

The phases we have identified, though more discrete than other descriptions of transitions, are not altogether separate, one from the other. There seems to be a more or less continuous process of transformation and change which leads to certain distinct periods of reorganization of the child's system. These findings are con-

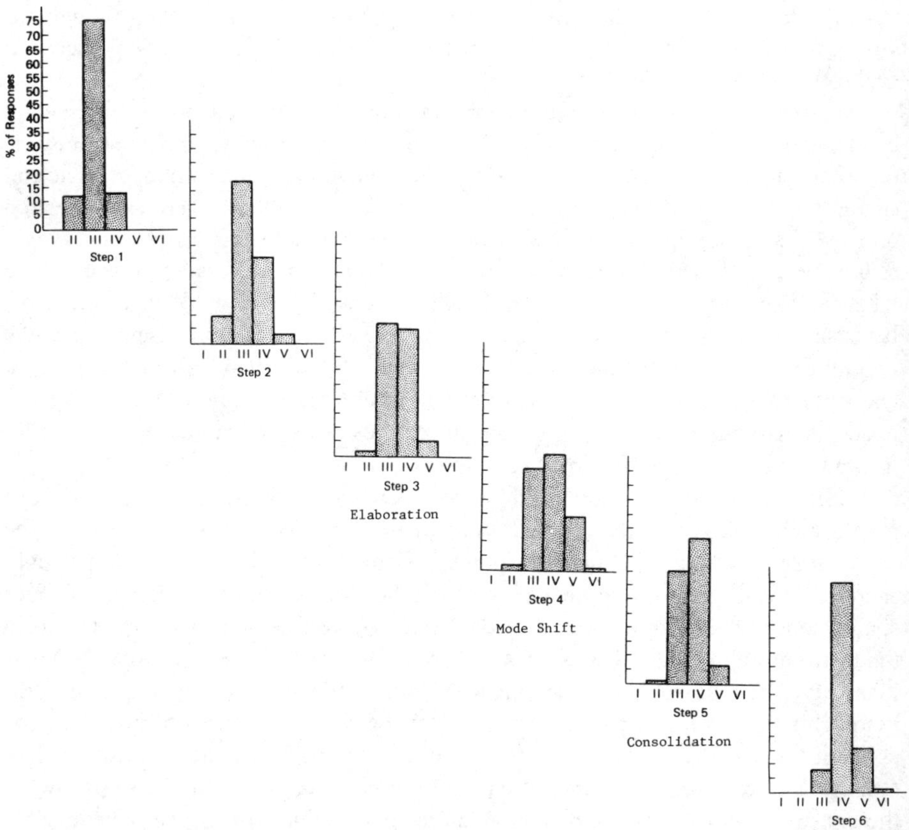

Figure 2. An illustration of the six phases in the transition process based on group data. (Note that 1 and 6 are essentially the same, indicating a complete transition cycle.)

sistent with Flavell's recent musings that perhaps stage has become a *less* central notion, at least in the sense of children ever really being in one (cf. Feldman, 1980b; Flavell, 1971, 1982). The effect of these phasic reorganizations, however, is to render the child's system quite distinctive with respect to how she or he deals with various map-drawing problems during each of the six phases.

We have described phases 2 and 3 of the transition cycle as an *elaboration* of a strong modal stage or level (in this example, Modal Level III in the drawing sequence). Incremental forward movement is particularly frequent at this point in the cycle. This is in contrast with the kinds of changes that occur between phases 3 and 4, when a new mode emerges. While there continue to be some incremental, single shifts forward, there is, in addition, *more backward activity,* i.e., movement of specific partial structures or elements of the system backward to lower levels. There is also, in phase 4, an enhanced likelihood of a breakthrough to a new level or *novelty* appearing in the child's system. Examine step 4 in the phase sequence (Fig. 2) and you will see that there is one element scored at Level 5. This represents the appearance of a novelty in the child's developmental progression toward a new level. Novelties were observed to occur relatively infrequently and were most often associated with the *fourth* phase of transition. We are able to say, therefore, that while a transition cycle may include five or more distinct phases, at only one point in the cycle are there actually likely to be qualitatively new (in the sense of unprecedented) advances or restructurings. For the rest, it is more a reorganizing and redeploying of specific capabilities among the different levels of capability already present in the child's map-drawing repertoire.

These are the kinds of results that have led us to assert, although not without qualification, that development takes place in distinctive transition phases. We have even gone so far as to speculate that the likelihood of a child's operating with a consistent stage system in any domain is minimal, at least until the very late (or very early) stages of a particular domain. In other words, we have reached the conclusion that transition is the rule in development, even within specific domains, whereas equilibrium or stable state conditions occur only rarely. These conclusions are not dissimilar from those reached more than 10 years ago by John Flavell (1971) in his paper about stage and sequence. We have simply added some empirical weight as well as additional conceptual distinctions to support Flavell's claim that structures as a whole probably do not operate as a governing system.

Construction

The second thing that we probably know about how development works is that complex systems of thought are *constructed* and that construction of such systems is the central purpose of cognitive development. Needless to say, there are those who would dispute this claim. Documenting the constructivist argument was a major theme of Piaget's work. To my mind, any reasonable assessment of Piaget's attempt to establish a constructivist approach would have to grant that it has been largely successful. Not so much because the approach is "true" in any absolute sense, but rather because it is, first, a reasonable possibility, and second, because it has had enormous heuristic value (cf. Lakatos, 1970; Piatelli-Palmarini, 1980). This is not to say that empirical support for the constructivist position is nonexistent—only that the argument rests less on an empirical than on a conceptual/theoretical base (but see Murray & Ames, 1982, for some compelling empirical evidence).

The constructivist stance asserts that the child confronts his or her experiences and attempts to make sense of them by using a set of regularized internal techniques for taking in and dealing with events. The child also sets about making changes in those characteristic responses to various kinds of stimulating conditions. Depending on the state of one's internal structures, the catalyzing events can come from the child's own mind, from events and conditions in the environment, or, almost certainly in most cases, from both sources. Out of this continuous process of filiation of internal structurings with external conditions comes an increasingly sophisticated, differentiated, articulated, and veridical set of structures—not, of course, without reversals and difficulties, but nonetheless primarily in a progressive fashion.

It has been stylish in some circles, particularly the north side of the Charles River in Cambridge, Massachusetts, to claim in Kantian fashion that there can be no construction worth paying attention to that is not already part of the natural basic equipment that the organism is given upon entering the human community (Chomsky, 1980; Fodor, 1983). Chomsky is the most prominent spokesman for this radical nativist view, although Jerry Fodor has, of late, been the most intransigent. The arguments often center around children's acquisition of spoken language. Chomsky claims that all of the necessary structures for language must be preorganized and pretuned at birth, if not earlier. Experience or stimulation or attempts at construction play no more important role than to activate or stimulate already present capabilities. For this reason, Chomsky argues, they are of relatively little interest or significance. For Chomsky, the major question to answer is: What is the nature of the original equipment? All other questions are derivative, if not trivial (Chomsky, 1980).

In some sense, of course, the potential for all that follows *must* be present at the moment of conception. It may even be present in both the sperm and ovum prior to conception. Or it may even be earlier than that. But what Chomsky's *reductio ad absurdum* argument does is to eliminate virtually all psychological questions from "basic" or "significant" science. This would be like saying that all potential for life on earth was present at the moment of the big bang or that only studying the most basic, microscopic particles should be of interest to the physicist or the biologist, because everything "derives" from these principles or processes.

Without arguing the obvious merits of looking for fundamental principles, or particles, or events, it is to my mind absurd to claim that there are no interesting questions, no issues of major significance to be investigated that are not patently of this sort. What is critical is to be guided by the search for deep issues and questions, to be alert to the implications of whatever one studies for basic principles. Carlton Gajdusek won a Nobel Prize because his curiosity about certain rare tropical diseases led him to discover some deep principles about how viruses work. Had he considered such oddities not "basic science," we would be in much greater ignorance (and illness) than we are today. Essentially, the Chomsky/Fodor position eliminates psychology altogether as a topic of inquiry. Needless to say, I do not accept this criticism of our field, nor, to his immense credit, did Piaget.

The question of whether constructivist explanations or more nativist ones best account for changes in mental capabilities is probably not one that will ultimately lend itself to conclusive proof. Yet, as soon as one considers domains of activity that are nonuniversal, it is difficult (I would even say impossible) to hold that nothing new can possibly occur: nonconstructivist views of cognitive development simply cannot account in a satisfying way for the discovery of new ideas. Piaget was very fond of using such examples to illustrate his thoughts about the necessity of a constructivist interpretation of development from the fields of mathematics and logic. For example, as I mentioned earlier, Piaget (e.g., 1971) frequently mentioned Cantor's set theory and Boole's algebra. These, he claimed, represent genuine novel constructions. And indeed they do. When one thinks about it, qualitative transformation and change of a constructivist sort are not unique to such esoteric examples; they abound when one looks at virtually any product of knowledge and skill. Every object of inquiry, every concept, every technique, many objects, even many "facts" have been transformed in significant ways over the course of time. To suggest that these changes are somehow uninteresting strikes me as a remarkable exercise in the denigration of human capability. That in some sense the "potential" for all of these things inheres in the common physical stuff of all human beings in no way reduces the significance of what happens with that stuff at every moment during its functioning.

How a domain or field changes its sets of assumptions, concepts, techniques, approaches, theories and even facts is not simply a matter of history or philosophy. It is a quintessentially psychological matter each time an individual seeks to master a domain. The psychological reality of this process lies in my claim that the mastery of domains occurs in orderly and patterned phases through stages or levels. Piaget's brilliant intuition about process was that each individual must become an "epistemic subject," a seeker of knowledge (Brief, 1983). This deceptively simple notion is at the very heart of the question of how development works. We cannot say with utter certainty that knowledge is constructed, but to say anything else at this time seems likely to promote a more confining and distorting approach to explanations of change and transformation in thought (cf. Murray & Ames, 1982). Twentieth-century science seems to me to point in the constructivist direction, and Piaget was extremely finely tuned to the landmarks of contemporary scientific thought (Gruber & Vonèche, 1977).

To summarize this section, then, I have discussed two aspects of development about which we can be reasonably certain. The first of these is that development occurs through orderly phases of transition between stages or levels. Although we only have systematic empirical data on one nonuniversal domain (map drawing), the most parsimonious argument is that substantially similar phases will be found for other domains as well, including those domains where development is universal, as described in Piaget's theory. In fact, one conclusion that seems reasonable from this work is that transition is the rule in development and that plateaus or stable states are relatively rare, except perhaps during the later stages of mastery of a domain (Flavell, 1971).

We do not yet know much about the impact of an individual's level of functioning in one domain on progress through other domains (Fischer, 1981). Presumably, at some later point we will be able to find clusters or families of domains which have predictable impact on each other. Following Howard Gardner's (1983) recent work, we may find that some domains are more distant and independent from each other and others less, with the picture varying among individuals or clustering into groups. In any case, it seems that development probably works through transition phases. To say that a person is in transition does not mean that there is nothing more to be learned about his or her mental state. It simply says that a more refined pattern must be detected to locate where that subject is in the transition process itself and what the impact of his or her mental system will be on subsequent changes in the system (Snyder & Feldman, 1980, 1984).

The second thing we can say with conviction if not certainty is that development is *constructed* much in the manner that Piaget has claimed it is. In spite of arguments that construction must, as Aristotle said, always be out of that which exists, it is nevertheless true that some things are indeed new. In particular, those constructions which transform a body of knowledge are new. Constructivism is also a view which has had and, it seems to me, is likely to continue to have, considerable heuristic power. It gives the epistemic subject significant control of the structures that are formed in the mind in contrast to both the radical nativist and radical positivist perspectives which it was designed to replace. If for no other reason, this feature of individual control seems to make constructivism a desirable assumption to place at the center of any late twentieth-century theory of the developmental process. While we cannot prove that constructivism is a fact in the sense that it will withstand every alternative interpretation, it is nonetheless a critical quality of any theory that attributes to individuals an important role in their own development, on the one hand, and the need to confront the world of experience on the other (Cowan, 1978).

This does not mean that individuals necessarily have total control—or even that they have the majority of control—over the course of their own development. It simply says that individuals can be self-conscious agents in the creation of their own cognitive structures. To say anything else requires one to ignore the accumulated works of humanity, great and small. The arts, scientific discoveries, refinements in method and technique, the wisdom of the crafts and trades, professions such as law and medicine, and virtually anything else that makes any real difference in the fabric of human culture turn on the issue of construction. We may not be certain of it in a scientific sense, but we can be certain that we cannot do without it as a guide.

Things We Do Not Know About How Development Works But Would Like To

Now we turn to some questions that we would like to have answers to but which we can do little more than speculate about at this time. In fact, and perhaps unfortunately, this category contains the largest number of questions—there are still many more things about which we know very little than about which we know a great deal. I am particularly interested in two of these questions: What are the conditions that *catalyze* developmental change, and how does one account for the emergence of *novel ideas*? These are also problems that preoccupied Piaget (Bringuier, 1980; Piaget, 1971).

Catalysts of Change

To begin, we again raise the distinction between developmental progress that is spontaneous versus that which is not: this distinction was originally made by Vygotsky long ago (Vygotsky, 1934/1962). For Vygotsky the idea of nonspontaneous development was central to his criticism of Piaget and other universalists. Vygotsky argued that some developmental changes occur spontaneously (that is, in the absence of systematically applied interventions from others), but that these developments are neither the only ones nor necessarily the ones which "drove" spontaneous change. He claimed that development of both spontaneous and nonspontaneous varieties occurred in continuous reciprocal influence, making it impossible to say that one or the other is primary, prior, or prerequisite.

As long as interest in developmental change focused on universal aspects, as it has for the past several decades, the likelihood of being able to study conditions that catalyze qualitative change was low. When development is believed to be universal and spontaneous there is little point in trying to detail the conditions bringing about change, since it follows that the necessary and sufficient conditions must be available for all individuals in all cultures and at all times. Although there has been some general speculation about what these conditions are (Flavell, 1971, 1982), the fact of their universality virtually precluded the possibility of directing systematic inquiry toward the specific experiences that bring about development (Feldman, 1980a, 1981a).

By once again drawing attention to the existence of *nonuniversal* developmental domains (the equivalent of Vygotsky's nonspontaneous development), it is feasible to examine the conditions which catalyze cognitive change (Beilin, 1971). Since it is simply not true that all individuals will advance through all the various stages of nonuniversal domains, it is possible to vary the external conditions that individuals encounter and to examine their consequences for progress from stage to stage. These interventions may, in turn, be studied in relation to characteristics of the subject's existing system of understanding of a given domain more or less along the lines suggested in previous sections of this paper and elsewhere (Feldman, 1980a, 1980b, 1981a, 1982, in press).

We are not totally in the dark about how to bring about developmental change.

The notion of "optimal discrepancy" as a means of inducing change has been in the literature for a number of years (Hunt, 1961; Snyder & Feldman, 1977; Turiel, 1974, 1983). The idea here is that the conditions provided in the environment should be developmentally just enough in advance of the subject's own organization to be stimulating but not confusing. A number of intervention studies have been done using optimal discrepancy with promising but mixed results (Damon, 1980; Kuhn, 1972, 1974, 1976; Inhelder, Sinclair, & Bovet, 1974; Karmiloff-Smith & Inhelder, 1975; Levin, 1978; Snyder & Feldman, 1977; Strauss, 1972; Turiel, 1966, 1974, 1983).

These studies tend to define discrepancy as a gap between overall stages in the child's current system (say concrete operations) and the external environmental conditions confronting the system (say formal operations). No doubt this is an important part of the equation, but it is far from sufficient. For example, given that we now know that subjects rarely use consistent structures to represent any particular problem (Cowan, 1978; Damon, 1980; Feldman, 1980a; Turiel, 1966, 1974, 1977, 1983); which of the structures available to the child should be used to establish discrepant environmental conditions? Should they be from the most common level of analysis the child uses, from the child's most advanced though less robust level of thinking, or perhaps from some estimation of the shape of the child's overall configuration of responses? It is simply not enough to say that there should be a discrepancy between a child's thinking and the environment. Now that we have a more articulated notion of what "stagelike" thinking is like for a child (Damon, 1980, Feldman, 1980a; Turiel, 1974, 1977, 1983), we must specify interventions as discrepant from a specific part of the child's system of thought (Snyder & Feldman, 1984). Although changes in our conceptions of cognitive structures would presently seem to make it impossible to specify the conditions sufficient for developmental change, they do suggest several experiments that would help give precision to the "optimal discrepancy" idea.

For anyone interested in education (as I have been), or more broadly, interested in enhancing possibilities for the realization of human potential, the issue of identifying the conditions necessary and sufficient for developmental progress is absolutely central (Gardner, 1982, 1983). In many ways, the reason for my own interest in developmental change revolves around this aspect of how development works. I can think of no more important task for theory and research during the next few decades than to specify more completely the kinds of external conditions that catalyze developmental progress. Relatedly, the exploration of internal transition states that interact with environments will, eventually, lead to an educational developmental psychology that makes current theories based on classical learning paradigms as obsolete as steam-powered automobiles (Case, 1974; Feldman, 1981a; Strauss, 1972, 1984). At present, we have woefully little information about what stimulates major cognitive changes—all the more reason, from my point of view, why the best efforts of developmental psychology and related fields should be directed toward them.

Novelty

Finally, I turn to another question about how development works for which there is currently no adequate answer. I refer to the problem of creativity, or how it is that a genuinely new idea is constructed. As I mentioned earlier, this was a major preoccupation of Piaget, although he never made much headway with it, and its solution remains one of the great unknowns in the field of psychology (Piaget, 1971; see also Piaget & Voyat, 1979). Howard Gardner (1980), has expressed the almost singular absence of significant work on creativity in these words:

> To my mind, the greatest challenge confronting any science of cognition is the task of explaining superior and revolutionary intellectual achievements—the plays of Shakespeare, the symphonies of Beethoven, the scientific system devised by Einstein, and, yes, the achievements, almost equivalent, of investigators like Chomsky and Piaget. The height of creativity, as exemplified by the scientist or the artist, remains one of the most enigmatic issues facing any student of human psychology. Until now, neither academic psychology (with its embarrassing tests of creativity), nor psychoanalysis nor behaviorism have had much of importance to say about the substance of individual contributions; as Freud was candid enough to admit, "The nature of artistic attainment is psychoanalytically inaccessible to us." Indeed those who seek some explanation for the cognitive processes involved in the creation of a major work would still be better advised to read Thomas Mann's *Doktor Faustus* or to examine Beethoven's sketchbooks than to sequester themselves in a psychology library. (p. 35)

I think it not at all an exaggeration to say that this is *the* critical question for understanding how development works. Among the major figures in psychology, Piaget has been the most persistent in pointing out just how critical it is to try to comprehend how a new idea is made (Piaget, 1971, 1975/1977; Piaget & Voyat, 1979). I believe, however, that his emphasis on universals was an impediment toward understanding this problem. Piaget claimed that the structures he was studying were independent of specific subject matter or content, and yet, as I mentioned earlier, time and again he used examples from bodies of knowledge which are anything but universal: Cantor's invention of set theory, Boole's algebra, great works of literature and art (Piaget & Voyat, 1979). He also resisted dealing with all matters pertaining to individual differences, although from time to time he acknowledged that he was uncomfortable at having made this decision (Bringuier, 1980; Piaget, 1971). Had he attended to these two components more closely, Piaget might have made substantial progress on the problem, for I think that he actually had some of the key elements in place to deal with the question of novelties: they can be seen in his rare but illuminating writing about what I call nonuniversals.

I have attempted in my own research to approach the issue of how new ideas are constructed, although I think these efforts have been only minimally successful to date. Using the same map-drawing technique described earlier in this chapter, for example, my co-workers and I have found a way to operationalize some aspects of novelties in the child's system for cartographic representation of three dimensional space. The novelties that emerge in this system are, to be sure, of the mundane variety, inasmuch as they are already well-known solutions to map-drawing problems. But for a given child on a given occasion, discovering how to show elevation using contour lines, for example, may be as dramatic a thought to that child as the solution to a major scientific question is to a practicing scientist. Our technique is one which detects instances of novelties and then examines retrospectively the conditions that gave rise to their appearance, combining to some degree the two issues discussed in this section. This technique allows us to examine the consequences of having achieved a novel construction on the rest of the child's map drawing system (Feldman, 1980a; Feldman, Snyder, & Goldsmith, 1984).

Ours is not the first set of studies to attempt to deal with new ideas, but it is the first, so far as I know, to try to do so within a theory positing an overall evolving system of developmental transformations known to occur with regularity. While there are technical problems to overcome in this work (primarily statistical ones), it does attempt to attack the problem of novelties directly. It offers the possibility of getting answers to questions that bear quite directly on the issue of how novelties are formed. For example, we have some intriguing results regarding the state of a child's map-drawing system just prior to the emergence of a novelty which suggest that the novelty is part of a broad conceptual transformation related to a particular portion of the system (a spatial concept). Other aspects of the concept also seem to be catalyzed to advance by the novelty.

It also appears that novelties have characteristic patterns of movement within the child's overall map-drawing system. Novelties seem to "leap" from the middle of

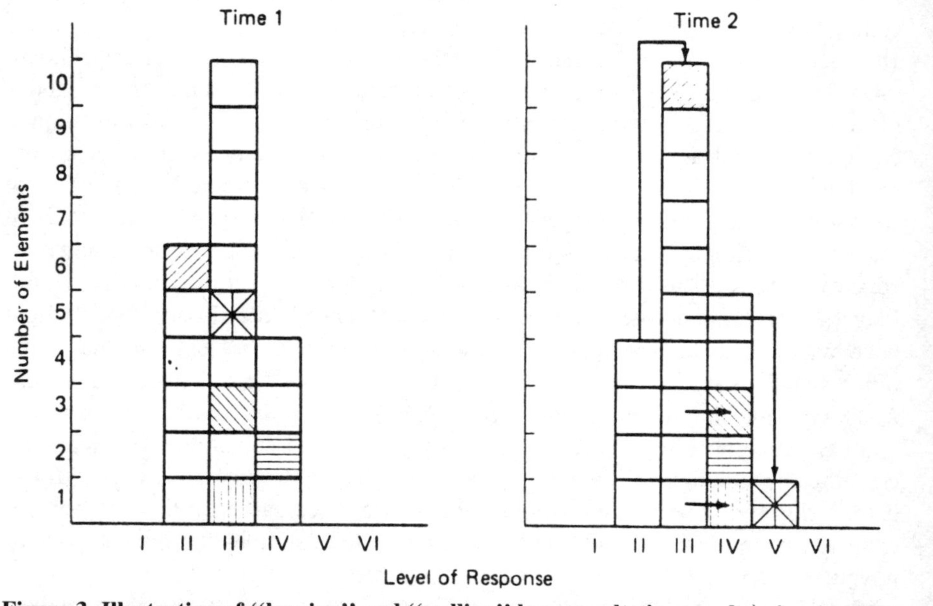

Figure 3. Illustration of "leaping" and "pulling" by a novelty in map drawing (novelty is the cross-hatched element).

the system more frequently than would be expected by chance, take their place at the leading edge of the child's system only briefly, and then "revert" back into the center (see Fig. 3). This pattern seems extremely unlikely to occur by chance and is distinctly characteristic of novelties, almost as if the novelty leaves a unique trail through the child's map-drawing system. These generalizations should be taken with caution, because statistical tests are not yet available to actually determine the robustness of the findings. While the effects seem to us very clear, they must await further refinement in both technique and analysis for us to assess their durability. Still, there seem to be promising possibilities for the study of novel ideas through the investigation of transitions between developmental levels in nonuniversal developmental domains. These domains, after all, are the places where truly creative work—the great breakthroughs—are going to occur.

Let me just briefly summarize the themes that seem to emerge from the novelties issue. The basic point is that novelties are produced through something like an equilibration process or something very much like it. Just how this process works is, of course, one of the great unanswered questions of our time. Piaget sketched a plausible beginning scenario, and during his final years tried to formalize that account (Piaget, 1975/1977; Piaget & Voyat, 1979). But essentially our understanding of equilibration is still very limited. I believe that some of the work my colleagues and I have done on novelties, transition phases, elements moving within systems, etc., will help refine Piaget's preliminary description of equilibration. And surely the emphasis in Geneva on watching children closely and making fine-grained descriptions of their activities will help a great deal (cf. Karmiloff-Smith & Inhelder, 1975). The common core of all construction processes is an equilibrationlike process of transformation and change. What varies is the unprecedented nature of the transformation or lack of it and the sets of conditions that are necessary and sufficient to bring such varying transformations about (Feldman, 1980a, 1981a).

Mention should be made in closing that the work of Howard Gruber and his associates on the psychological processes involved in the construction of a new theory such as Darwin's represents one important line of inquiry into how novelty occurs. Some of our own work on child prodigies represents another, while the novelties research in map drawing yet another (Feldman, 1980a; Gruber, 1981). What needs to be done is to study transitions in numerous nonuniversal domains across all levels of those domains. To do so represents a great challenge for future research, but one which seems to me as important as any that might be proposed. Although there is a great deal more knowledge about how development works than I originally thought, that knowledge will remain inadequate until a better account of creativity or novelty is available to us.

References

Beilin, H. (1971). Developmental stages and developmental processes. In D. Green, M. Ford, & G. Flamer (Eds.), *Measurement and Piaget*. New York: McGraw-Hill.

Brainerd, C. J. (1977). Cognitive development and concept learning: An interpretive review. *Psychological Bulletin, 84,* 919–939.

Brainerd, C. J. (1978). The stage question in cognitive-developmental theory. *The Behavioral and Brain Sciences, 1,* 173–182.

Brainerd, C. J. (1981). Stages II: A review of *Beyond universals in cognitive development. Developmental Review, 1,* 63–81.

Brief, J. C. (1983). *Beyond Piaget: A philosophical psychology.* New York: Teachers College Press.

Bringuier, J. C. (1980). *Conversations with Jean Piaget.* Chicago, IL: University of Chicago Press.

Case, R. (1974). Structures and strictures: Some functional limitations on the course of cognitive growth. *Cognitive Psychology, 6,* 544–573.

Cowan, P. (1978). *Piaget with feeling.* New York: Holt, Rinehart & Winston.

Chomsky, N. (1980). *Rules and representations.* New York: Columbia University Press.

Damon, W. (1980). Patterns of change in children's social reasoning: A two-year longitudinal study. *Child Development, 51,* 1010–1017.

Feldman, D. H. (1980a). *Beyond universals in cognitive development.* Norwood, NJ: Ablex.

Feldman, D. H. (1980b). Stage and transition in cognitive-developmental research: Getting to the next level. *The Genetic Epistemologist, 9,* 1–6.

Feldman, D. H. (1981a). Beyond universals: Toward a developmental psychology of education. *Educational Researcher, 10,* 21–31.

Feldman, D. H. (1981b). The role of theory in cognitive developmental research: A reply to Brainerd. *Developmental Review, 1,* 82–89.

Feldman, D. H. (1982, March). *Developmental science: An alternative to cognitive science.* Paper presented at a colloquium. Cambridge, MA: MIT.

Feldman, D. H. (in press). *Nature's gambit: Child prodigies and the development of human potential.* New York: Basic Books.

Feldman, D. H., Snyder, S. S., & Goldsmith, L. T. (1984). *On the transformation of psychological structure: An exploratory study of novelties in development.* Unpublished manuscript.

Fischer, K. W. (1980). A theory of cognitive development: The control and construction of hierarchies of skills. *Psychological Review, 87,* 477–531.

Fischer, K. W. (1981). The last straw for Piagetian stages? A review of *Beyond universals in cognitive development. Contemporary Psychology, 26,* 338–339.

Flavell, J. H. (1971). Stage-related properties of cognitive development. *Cognitive Psychology, 2,* 421–453.

Flavell, J. H. (1982). On cognitive development. *Child Development, 53,* 1–10.

Fodor, J. (1983). *The modularity of mind.* Cambridge, MA: MIT Press.

Gardner, H. (1980). Foreword. In M. Piatelli-Palmarini (Ed.), *Language and learning: The debate between Jean Piaget and Noam Chomsky.* Cambridge, MA: Harvard University Press.

Gardner, H. (1982). Giftedness: Speculations from a biological perspective. In D. H. Feldman (Ed.), *Developmental approaches to giftedness and creativity.* San Francisco, CA: Jossey-Bass.

Gardner, H. (1983). *Frames of mind.* New York: Basic Books.

Gruber, H. (1980). Afterword. In D. H. Feldman, *Beyond universals in cognitive development.* Norwood, NJ: Ablex.

Gruber, H. (1981). *Darwin on man: A psychological study of scientific creativity. Second edition.* Chicago: University of Chicago Press.

Gruber, H. (1982). On the hypothesized relation between giftedness and creativity. In D. H. Feldman (Ed.), *Developmental approaches to giftedness and creativity.* San Francisco, CA: Jossey-Bass.

Gruber, H., & Vonèche, J. (Eds.). (1977). *The essential Piaget.* New York: Basic Books.

Hunt, J. McV. (1961). *Intelligence and experience.* New York: Ronald Press.

Inhelder, B., Sinclair, H., & Bovet, M. (1974). *Learning and the development of cognition.* Cambridge, MA: Harvard University Press.

Karmiloff-Smith, A., & Inhelder, B. (1975). If you want to get ahead, get a theory. *Cognition, 3,* 195–212.

Keil, F. (1981). *Semantic and conceptual development: An ontological perspective.* Cambridge, MA: Harvard University Press.

Kohlberg, L. (1971). From is to ought: How to commit the naturalistic fallacy and get away with it in the study of moral development. In T. Mischel (Ed.), *Cognitive development and epistemology*. New York: Academic Press.

Kuhn, D. (1972). Mechanisms of change in the development of cognitive structures. *Child Development, 43,* 833–844.

Kuhn, D. (1974). Inducing development experimentally: Comments on a research paradigm. *Developmental Psychology, 10,* 590–600.

Kuhn, D. (1976). Short-term longitudinal evidence for the sequentiality of Kohlberg's early stages of moral judgment. *Developmental Psychology, 12,* 162–166.

Lakatos, I. (1970). Falsification and methodology of scientific research programmes. In I. Lakatos & Musgrave (Eds.), *Criticism and the growth of knowledge*. Cambridge, England: Cambridge University Press.

Levin, D. (1978). *Peer interaction as a source of cognitive developmental change in spatial representation.* Unpublished doctoral dissertation, Tufts University, Medford, MA.

Maturana, H. (1978). Biology of language: The epistemology of language. In E. Lenneberg (Ed.), *Psychology and biology of language and thought.* New York: Academic Press.

Murray, F. B., & Ames, G. J. (1982). When two wrongs make a right: Promoting cognitive change by social conflict. *Developmental Psychology, 18,* 894–897.

Piaget, J. (1971). The theory of stages in cognitive development. In D. R. Green, M. P. Ford, & G. B. Flamer (Eds.), *Measurement and Piaget.* New York: McGraw-Hill.

Piaget, J. (1972). Intellectual evolution from adolescence to adulthood. *Human Development, 15,* 1–12.

Piaget, J. (1973). *The child and reality.* New York: Grossman.

Piaget, J. (1975/1977) *The development of thought: Equilibration of cognitive structures.* New York: Viking. (Original work published in 1975)

Piaget, J. (1981). *Intelligence and affectivity.* Palo Alto, CA: Annual Reviews Monographs.

Piaget, J., & Inhelder, B. (1956). *The child's conception of space.* New York: Norton. (Original work published in 1948)

Piaget, J., & Voyat, G. (1979). The possible, the impossible, and the necessary. In F. Murray (Ed.), *The impact of Piagetian theory on education, philosophy, psychiatry, and psychology.* Baltimore, MD: University Park Press.

Piatelli-Palmarini, M. (Ed.). (1980). *Language and learning: The debate between Jean Piaget and Noam Chomsky.* Cambridge, MA: Harvard University Press.

Siegler, R. S. (1981). Developmental sequences within and between concepts. *Monographs of the Society for Research in Child Development, 46* (Serial No. 189).

Snyder, S. S., & Feldman, D. H. (1977). Internal and external influences on cognitive developmental change. *Child Development, 48,* 937–943.

Snyder, S. S., & Feldman, D. H. (1980). Individual developmental transitions: A film metaphor. In D. H. Feldman, *Beyond universals in cognitive development.* Norwood, NJ: Ablex.

Snyder, S. S., & Feldman, D. H. (1984). Phases of transition in cognitive development: Evidence from the domain of spatial representation. *Child Development, 55,* 981–989.

Snyder, S. S., Feldman, D. H., & La Rossa, C. (1976). A manual for the administration and scoring of a Piaget-based map drawing exercise. In O. Johnson (Ed.), *Tests and measurements in child development: A handbook (II).* San Francisco, CA: Jossey-Bass. (abstract)

Sternberg, R. (1983). Components of human intelligence. *Cognition, 15,* 1–48.

Strauss, S. (1972). Inducing cognitive development and learning: A review of short-term training experiments. I: The organismic developmental approach. *Cognition, 4,* 329–357.

Strauss, S. (Ed). (1982). *U-shaped behavioral growth.* New York: Academic Press.

Strauss, S. (1984, June). *Developmental and educational psychology.* Paper presented at the Jean Piaget Society Annual Symposium, Philadelphia, PA.

Turiel, E. (1966). An experimental test of the sequentiality of developmental stages in the child's moral judgments. *Journal of Personality and Social Psychology. 3,* 611–618.

Turiel, E. (1974). Conflict and transition in adolescent moral development. *Child Development, 45,* 14–29.

Turiel, E. (1977). Distinct conceptual and developmental domains: Social convention and morality. In

C. B. Keasey (Ed.), *Nebraska Symposium on motivation.* Lincoln, NE: University of Nebraska Press.

Turiel, E. (1979, June). *Social/cognitive development.* Paper presented at the Jean Piaget Society Annual Symposium, Philadelphia.

Turiel, E. (1983). *The development of social knowledge.* New York: Cambridge University Press.

von Bertalanffy, L. (1955). An essay on the relativity of categories. *Philosophy of Science, 22,* 243–263.

Vygotsky, L. (1962). *Thought and language.* Cambridge, MA: MIT Press. (Original work published in 1934).

Author Index

Italics indicate bibliographic citations.

Subject Index